Frameworks for ICT Policy:

Government, Social and Legal Issues

Esharenana E. Adomi
Delta State University, Nigeria

A volume in the Advances in IT Standards
and Standardization Research (AITSSR)
Book Series

Director of Editorial Content:	Kristin Klinger
Director of Book Publications:	Julia Mosemann
Acquisitions Editor:	Lindsay Johnston
Development Editor:	Joel Gamon
Publishing Assistant:	Natalie Pronio
Typesetter:	Michael Brehm / Travis Gundrum
Production Editor:	Jamie Snavely
Cover Design:	Lisa Tosheff

Published in the United States of America by
Information Science Reference (an imprint of IGI Global)
701 E. Chocolate Avenue
Hershey PA 17033
Tel: 717-533-8845
Fax: 717-533-8661
E-mail: cust@igi-global.com
Web site: http://www.igi-global.com

Library of Congress Cataloging-in-Publication Data

Frameworks for ICT policy : government, social and legal issues / Esharenana E. Adomi, editor.
      p. cm.
  Includes bibliographical references and index.
  Summary: "This book provides policy-relevant insights and information to aid those engaged in systematic research and teaching about ICT policy formation and implementation"--Provided by publisher.
   ISBN 978-1-61692-012-8 (hardcover) -- ISBN 978-1-61692-013-5 (ebook) 1. Information technology--Government policy--Cross-cultural studies. 2. Information technology--Security measures--Cross-cultural studies. 3. Computer crimes--Prevention--International cooperation. I. Adomi, Esharenana E.
  HC59.72.I55F73 2010
  303.48'33--dc22
                                        2010007131

This book is published in the IGI Global book series Advances in IT Standards and Standardization Research (AITSSR) Book Series (ISSN: 1935-3391; eISSN: 1935-3405)

British Cataloguing in Publication Data
A Cataloguing in Publication record for this book is available from the British Library.

All work contributed to this book is new, previously-unpublished material. The views expressed in this book are those of the authors, but not necessarily of the publisher.

Advances in IT Standards and Standardization Research (AITSSR) Book Series

Kai Jakobs (Aachen University, Germany)

ISSN: 1935-3391
EISSN: 1935-3405

MISSION

IT standards and standardization are a necessary part of effectively delivering IT and IT services to organizations and individuals as well as streamlining IT processes and minimizing organizational cost. In implementing IT standards, it is necessary to take into account not only the technical aspects, but also the characteristics of the specific environment where these standards will have to function.

The **Advances in IT Standards and Standardization Research (AITSSR)** book series seeks to advance the available literature on the use and value of IT standards and standardization. This research provides insight into the use of standards for the improvement of organizational processes and development in both private and public sectors.

COVERAGE

- Analyses of standards-setting processes, products, and organization
- Descriptive theory of standardization
- Emerging roles of formal standards organizations and consortia
- Intellectual property rights
- Management of standards
- National, regional, international and corporate standards strategies
- Open Source and standardization
- Risks of standardization
- Technological innovation and standardization
- User-related issues

IGI Global is currently accepting manuscripts for publication within this series. To submit a proposal for a volume in this series, please contact our Acquisition Editors at Acquisitions@igi-global.com or visit: http://www.igi-global.com/publish/.

Titles in this Series

For a list of additional titles in this series, please visit: www.igi-global.com

IGI GLOBAL
DISSEMINATOR OF KNOWLEDGE

www.igi-global.com

701 E. Chocolate Ave., Hershey, PA 17033
Order online at www.igi-global.com or call 717-533-8845 x100
To place a standing order for titles released in this series, contact: cust@igi-global.com
Mon-Fri 8:00 am - 5:00 pm (est) or fax 24 hours a day 717-533-8661

Editorial Advisory Board

Table of Contents

Detailed Table of Contents

Chapter 1

Much research has focused on formulating frameworks for ICT management in general and there is a paucity of guidelines in literature for ICT security policy management, in particular. This chapter explores ICT security management issues faced in different environments and proposes an integrated framework for managing ICT security policies in an iterative manner. The framework provides the flexibility and adaptability for different organisations to follow the guidelines effectively as it emphasises on policy alignment with business objectives. Since the framework underpins the continuous improvement philosophy, it caters to ICT security policy reform and implementations for the future as well.

Chapter 2

This chapter proposes a model for building trust of citizens in e-government. The proposed model is premised on five trust pillars: ethical/human; information/content; technical; policy/legal; and political/ governance. Each of these pillars has several dimensions adapted from various existing user satisfaction tools or frameworks such as Service Quality (SERVQUAL), trust formation framework, technology acceptance model, information systems success model, and many more. The chapter also covers history of e-government; models of e-government interactions; e-government trust related models; theoretical basis of trust; causes of distrust in e-government; challenges of applications of ICT in e-government; framework for building trust in e-government; and future trends of trust in e-government

This chapter addresses issues related to the mobile phone evolution in Indonesia, pedagogical effectiveness of using it in educational context with empirical study on Indonesian teachers' perception of the aforementioned in the classroom for science and mathematics learning. The useful research evidence of the value of ICT in education thereby serves as empirical basis for the formulation of ICT policy in educational institutions, particularly ICT integration in instruction.

This chapter reports a research, conducted at a computer centre in the Indian state of Tamil Nadu, which centred on the perception of participants with respect to whether the centre had played a role in any improvements in the community and whether they could see a role for it in changes they would like to see, or aspirations they may have for their communities. A key finding of the field research was that participants valued the centre mainly for its contribution to education of their children. Education was appreciated beyond its instrumental utility and included intrinsic value, i.e. value that exceeds its potential as a path to higher incomes. Participants frequently referred to how a higher level of literacy would empower them to deal with government officials without intermediaries. This finding is consistent with the CA's emphasis on development as a process facilitating capabilities that enable people to lead lives they have reason to value.

Chapter 5

This chapter focuses on the issues evolved out of the Indian Information Technology Act of 2000; the key subject related to authentication of digital signatures with special reference to India based on case studies; the benefits of strong information technology infrastructure in India for advancement of future technologies and expansion of domestic market worldwide as well as the vital suggestions on advantages of electronic and digital signatures in enriching and ensuring swiftness in business desires and security.

Chapter 6

Udo Richard Averweg, Information Services, eThekwini Municipality
and University of KwaZulu, South Africa
Geoff Joseph Erwin, The Information Society Institute (TISI), South Africa

This chapter discusses that information and communication technologies (ICTs) can (and should) be used to disseminate information and participation to disadvantaged communities in order to foster socio-economic development in South Africa. The objective of this chapter is twofold: (1) how should ICT policies and frameworks in South Africa be implemented (e.g. by a "top-down", "bottom-up" or "mixed approach" paradigm) in order for the South African government to achieve its socio-economic goals?; and (2) can socio-economic development in South Africa be effectively assisted by the use of ICT? A discussion of these points may assist in the formulation of national ICT policies in South Africa and thereby spawn the setting up of social appropriation of ICT advancement programs. Such programs are particularly relevant to the digital divide, for fostering socio-economic development and in promoting an inclusive information society in South Africa.

Chapter 7

Alex Ozoemelem Obuh, Delta State University, Nigeria
Ihuoma Sandra Babatope, Delta State University, Nigeria

This chapter discusses cybercrime and cybercrime regulation in the Nigeria. It gives the meaning to cybercrime, types of cybercrimes (of which advance fee fraud is the most prevalent in Nigeria), means of perpetrating cybercrimes, the current situation and efforts towards combating cybercrime in Nigeria.

Chapter 8

Saul F.C. Zulu, University of Botswana, Botswana

This chapter reviews emerging ICTs and their potential for application in leveraging Africa's efforts towards meeting development efforts. The digital divide barriers that may inhibit emergent ICTs in Africa are highlighted. The current ICT policies of selected African countries are reviewed and the review indicates that the policies are geared towards application of ICTs other than their production. The review also reveals a lack of appreciation of emerging ICTs in Africa both at the national as well as sub-regional economic bloc levels. The chapter proposes policy framework for emerging ICTs for Africa that are necessary for creating an enabling environment for harnessing the emerging ICTs that will propel the continent into the 21st Century and beyond. And that such an emerging ICT environment must be anchored on a number of strategic policy frameworks including the legal, regulatory/administrative institutional framework, infrastructure, technology advocacy, human resources, education and research frameworks. It concludes that Africa can prepare for its future by creating an appropriate environment for fostering the adoption and application of emerging technologies

This chapter describes these ethical issues and how to deal with them as an individual or an organization. It provides information on the concept of ethics and the technological advancements responsible for the ethical concern. It discusses privacy, information rights, and intellectual property rights and ethics policy. The Nigerian national intellectual property right laws were examined in line with World Trade Organization/Trade Related Aspects of Intellectual Property Rights (WTO/TRIP) compliance.

Chapter 10

The chapter explores variety of ICT infrastructure/services that are available in university libraries, sources of ICT funding, ICT policy priority areas, key ICT policy issues and strategies for formulating ICT policy. Questionnaire survey was used for the study. The findings of the study indicate that, there is a widespread use of Internet in the 14 surveyed university libraries; and only 5 of these libraries have computerized library services. NUC/ETF, library development fund (LDF) and university management are major ICT funding sources. ICT funding/budgeting, ICT infrastructure procurement/maintenance, ICT literacy/capacity building and ICT use are the highest ranked ICT policy priority areas in the surveyed libraries. And annual budgetary allocation to ICT in university libraries, training/capacity building for librarians, as well as, organization of ICT literacy programme for patrons are the highest ranked key ICT policy issues in university libraries. The chapter recommends that, each university should formulate relevant ICT policy for its library, besides the national ICT policy.

This chapter is devoted to discussion of ICT and gender policy. Itt explores the need for gender consideration in ICT policy, gender issues in ICT policy, adoption of gender perspective in ICT policies, challenges, for adopting gender perspective information, implementation of ICT policies, case studies of gender and ICT policies in Asia, Africa, Europe, Latin America and Caribbean and Australia, gender approaches to ICT policies and programmes, guidelines for policy-making and regulatory agencies ,and future

The purpose of this chapter is to give an international perspective and overview of the theory, standardization processes and following practices in the field of authority control, with particular view on the name authority control since the 1960s to the present. In the focus of interest of this chapter is paradigm shift in the field, and the possibilities of semantic web technologies in meeting library users' needs, as well as librarians' tasks to produce tools convenient to the user in the network environment.

This chapter examines and analyses the Indian ICT laws and policies in the backdrop of cyber-crime prevention and regulation, with the aim of offering a comprehensive model of ICT policy. It will discuss the extent of legal framework in the light of classification and criminalization of various cyber-crimes. Also, while examining the policy instruments, it will bring out the public and private initiatives on protection of information infrastructures, incident and emergency response and the innovative institutions and schemes involved.

This chapter discusses national information and communication technology policy process in developing countries. It describes the need for information and communication technology policy, ICT policy development process, national ICT policy in developing countries, the role of an ICT policy in the developing country, factors affecting the formulation of national ICT policies and the future of national ICT policy was also discussed.

This chapter focuses on regulation of Internet content. It presents the arguments for and against Internet content regulation, approaches to content regulation on the Net, how Internet content is regulated in different parts of the world, issues inherent in content regulation, choice of content regulation mechanism as well as future trends.

This chapter describes the history, development and operation of the New Zealand censorship system, as it applies to Internet content. It is likely to be of interest to policy-makers, law enforcement officers and media regulators in other countries.

Chapter 17

Organizations should recognize that its records are a vital business resource and are key to the effective functioning and accountability of the organization. Efficient management of records is essential in order to support organization's core business activities, to comply with legal and regulatory obligations, and to provide a high quality service to citizens. Electronic records management programme ensures that the organizational business activities are well documented, organized and managed, given access, protected from unauthorized access and disposed off (either destroyed or archived). Credible and dependable information systems are desired to achieve this. Also, adequate skills sets are required by personnel working with and managing electronic records.

Chapter 18

This chapter investigates the Ugandan ICT policy approach to promoting access to and the empowerment of the poor majority, remote and "under-accessed" communities in Uganda. The chapter highlights the strengths of the policy framework while at the same time draws attention to its weaknesses. For instance, while the chapter acknowledges the fact that the ICT policy framework recognises and has pursued strategic approaches to expanding access to remote areas, a closer scrutiny of the policy framework indicates disparities that may delimit its pragmatism. These disparities, it is argued, mainly emanate from the fact that the policy framework is not entirely holistic in its outlook, not only because the processes (of policy making) left out the rural users, it also fails to address the gender dynamics. In addition to divorcing itself from political and democratic aspects necessary for development, the policy framework seems shorthanded on sustainability fundamentals that are conjectured to delimit its pragmatism at the grassroots.

Foreword

There is little doubt we are in the midst of a communication and information revolution. But what type of revolution will this be? Fifty years from now will we speak of the current era as an edifying moment in human history when the ICTs of the day were harnessed to effectively address society's social, political and economic ills? Or will we speak forlornly of lost opportunities and of an era that exacerbated inequalities and wrought new forms of insecurity across the globe? Or will our cogitations lie somewhere in between?

This book accepts the following premise as foundational: Where we end up, and how the revolution unfolds will be determined to a very significant extent by the policy frameworks devised to guide the development, deployment and use of ICTs. As with all new technology systems, recent and emerging ICTs portend both significant potential benefits and serious potential problems for organizations and societies. Wherever there are "winners", there are likely to be "losers" unless steps are taken to prevent such developments. Appropriately designed ICT policies can increase the possibilities for benefit, reduce the risk of loss and harm, and ensure that the implications for various sectors of society are considered. The book also takes as axiomatic the premise that the formation of these policies should be informed by scholarly work, whether this takes the form of relevant empirical findings or of theoretical and analytic frameworks that inform the questions and issues policy makers prioritize.

Taken together, the various contributions in the Handbook provide an engaging and timely analysis of a number of pressing questions that come in to plain sight when these eminently reasonable premises are accepted. What values, normative commitments and priorities should inform the design of ICT policy frameworks? Whose interests should be incorporated into the frameworks and whose needs should receive priority? What processes shape how these frameworks are developed, implemented and reformed? Whose voices should be present in the policy formation process and how can their voices be institutionalized in the process? Given that ICT developments cut across many if not all sectors of society, how should we ensure ICT policies take into consideration and harmonize with priorities in other policy domains?

Twenty-eight experts from eight countries and four continents and an array of disciplines address these and other important questions in a variety national contexts and in relation to numerous topical issues including: ICT security and cybercrime policy; citizen trust and e-government; ICTs and socioeconomic development; privacy and intellectual property issues associated with ICTs; gender considerations and ICT policy; internet content regulation. While some contributions are concerned with policy issues at the international or national level others train their sights on the regional or organizational context.

The book provides a highly engaging and instructive cross-national comparative perspective on ICT policy issues. In doing so, it fills an important lacunae in the extant literature on ICT policy. It contains a wealth of policy-relevant insights and information that will be of great use to those who engage in or systematically research and teach about ICT policy formation and implementation. More specifically, the book includes important observations on the philosophical, ethical and moral touchstones that should guide sound ICT policy, it provides useful research evidence that can provide the basis for good policy decision making, and it provides the reader with specific, granular-level information on designing particular kinds of ICT policy initiatives. In short, the book provides a body of knowledge and information that will prove to be a valuable resource for those seeking to maximize the benefits of ICT while minimizing potential negative consequences.

Peter Shields
Eastern Washington University, USA

Peter Shields *is Professor of Communication and Director of the Masters of Science in Communications program at Eastern Washington University in Cheney, Washington. His areas of research include electronic surveillance and national security, communication technology policy, and the political economy of telecommunication networks. He is the author of forty journal articles and book chapters on topics such as communication privacy, telecommunication and development, money laundering regulation, and media reform in Eastern Europe. He is also co-editor of* International Broadcasting in Asia: Economic, Political and Cultural Implications *(University Press of America, 1998).*

Preface

ICT policy framework is a set of principles and goals intended to govern the development, implementation, adoption, monitoring, evaluation and application of ICTs in organizations, institutions, societies or nations. It provides the rationale and philosophy to guide the planning and development and utilization of ICTs in a particular setting.

ICTs are always evolving and have been contributing immensely to economic, political, social, scientific and educational development in every society where they are deployed. It is the existence and utilization of appropriate policy framework that would enable individuals, institutions, organizations, nations, or regions to benefit from the developments propelled by the application of ICTs. ICT policy framework is capable of bridging or reducing the gap between those who do and do not have access to ICT. An ICT policy framework is an essential step towards creating an enabling environment for the deployment of ICT for the development of the society. Absence of ICT policy can impede the development of information infrastructure in affected organizations and societies. The availability of ICT policy has the potential of building capability and capacities in the community, thereby enabling individuals to participate in an economy and society that increasingly relies on ICT. Though ICT policies are very important, not all institutions, countries, regions, etc., have been able to formulate, adopt and implement policies. Some that have put ICT policies in place may even have defective ones which would not enable them to derive desirable developmental benefits. Review of literature has not been able to reveal any book that has addressed different ICT policies. This book will therefore bridge this literature gap.

ICT as a field is interdisciplinary in nature. Every field of human endeavour depends on as well as derives benefits from it. It is therefore not surprising that contributors to this book are from various subject disciplines – computer science, education, library and information science, mass communication, law, etc. These scholars have addressed ICT policy from their various backgrounds. This has therefore made the book to be relevant to different fields of study.

The book also has contributors from different world regions – both developed and developing countries. This shows that ICT and its policies have been embraced and adopted in different parts of the world. The contents of the book would therefore be useful to scholars, researchers and practitioners in different parts of the world.

This book aims to provide the most complete and reliable source of information on current developments in the field. Specifically, the book will be a source book on ICT policy framework; be a guidebook to those who are involved in ICT policy formulation, implementation, adoption, monitoring, evaluation and application; provide background information to scholars and researcher who are interested in carrying out research on ICT policies; furnish teachers of information technology with necessary knowledge which they can impart to their students/trainees; provide ICT users with information that can enable them to understand the policies which guide technology and how they can make use of ICT components for their enhancement..

The book will be essential reading for professionals, governmental and non-governmental officials involved in ICT matters; teachers/academics in the field of information science, technology and management; students, scholars and researchers in the field of information science, technology and management; ICT users; library and information service users, etc. The publication will provide the audience access to information that will advance research in ICT policies. It will enable individuals to become acquainted with ICT policy process, which will then assist those concerned to formulate and implement appropriate policies. This book will provide teachers, students, scholars and researchers in the field of information science, education, technology and management with useful material on curricular offering. It will enable different ICT users to apply ICTs for the advancement of different areas of their lives.

Esharenana E. Adomi
Delta State University, Nigeria

Acknowledgment

The editor would like to acknowledge the help received from all those involved in the collation and review process of the book, without whose support the project could have not been completed satisfactorily.

Due appreciation and gratitude is due to the authors of the included chapters, most of whom served as reviewers for chapters written by others. Thanks go to members of the editorial board and all others who provided constructive and comprehensive reviews.

Special thanks go to Professor Peter Shields of the Eastern Washington University, USA for writing the foreword.

Special thanks also go to the management and staff of IGI Global, whose contributions throughout the whole project from inception of initial ideas to final publication has been very invaluable. In particular, the editor is grateful to Beth Ardner for the invitation to edit this book, Christine Bufton, Joel Gamon and Jan Travers for their support and encouragement.

In closing, I wish to thank all the authors for their insights and excellent contribution to this handbook. I also want to thank all of the people who assisted me in the reviewing process. Finally, I must thank my wife and children for their support, patience and encouragement throughout this project.

Esharenana E. Adomi
Delta State University, Nigeria

Chapter 1
A Framework for ICT Security Policy Management

Sitalakshmi Venkatraman
University of Ballarat, Australia

ABSTRACT

Organisations around the world are increasingly relying on the potential of information and communication technologies (ICTs) for their business operations as well as competitiveness. Huge amounts of money and time are invested on ICT infrastructure as there exists a high level of business dependency on ICT. Hence, protecting the ICT resources using effective security policies is of utmost importance for the sustenance of organisations. With the recent exponential rise in ICT security threats witnessed worldwide, governments and businesses are trying to successfully develop ICT security policies for their internal and external operations. While ICT security best practices are quite similar globally, ICT security policy management is very much localised and specific to different business scenarios and applications. Moreover, ICT security policies in an organization keep evolving from time to time and more recently changes take place at a much faster pace. This situation warrants a pragmatic framework for the development and management of ICT security policies in an organisation. Much research has focused on formulating frameworks for ICT management in general and there is a paucity of guidelines in literature for ICT security policy management, in particular. This chapter explores ICT security management issues faced in different environments and proposes an integrated framework for managing ICT security policies in an iterative manner. The framework provides the flexibility and adaptability for different organisations to follow the guidelines effectively as it emphasises on policy alignment with business objectives. Since the framework underpins the continuous improvement philosophy, it caters to ICT security policy reform and implementations for the future as well.

DOI: 10.4018/978-1-61692-012-8.ch001

INTRODUCTION

With information and communication technology (ICT) becoming part and parcel of business environments worldwide, the role of ICT security and its policies are of paramount importance in this globally inter-connected world (OECD, 2002; IT Governance Institute, 2005; Bojanc & Jerman-Blaic, 2008). Organisations, both private and public, have the critical objective of protecting their ICT infrastructure as well as business information assets from intrusions and risks (Conklin, 2007). Hence, it is mandatory for them to have in place suitable ICT security policies that could facilitate in achieving this objective.

Traditionally, before the onset of the Internet, ICT security policies were not given high priority among the various business strategies. Organisations were able to sustain with or without such policies. However, with the Internet explosion opening up global market opportunities, more and more businesses are harnessing the benefits of the inter-connectivity of ICT. Hence, formulating ICT security policies have become mandatory for the sustenance of every organisation (Caruso, 2003; Drevin, Kruger & Steyn, 2006). In addition, as shown in Figure 1, ICT security has become highly complex with three main dynamic changes taking place in the following areas that create new issues to focus on:

1. **Rapid ICT innovations:** new devices (mobile, wireless, etc) are getting inter-connected to traditional computing systems with different hardware and software platforms / protocols for businesses to operate on (Sathish Babu & Venkataram, 2009);
2. **Growing security threats:** recent security breaches are increasing exponentially creating a race between the hackers and the anti-virus solutions architects (Jahankani, Antonijevic & Walcott, 2008);
3. **Varying social and legal focus:** Globalisation requires business interaction between traditional and modern societies having different social and ethical beliefs / laws (Small, 2007).

Hence, ICT security policies have to be modified frequently to deal with the above said dynamic changes that impact both technological and non-technological areas. It is, therefore, important to incorporate security monitoring and planning steps that include protection measures, security standards, risk analysis and contingency plans so as to ensure information security in an organisation for the present as well as the future. All these steps require considerable effort, time and money and hence developing and managing an effective ICT security policy has become one of the main concerns for businesses worldwide.

This chapter aims to discuss the prevailing issues that face ICT security management and to propose an integrated framework for managing ICT security policies effectively. The main objectives of this chapter are:

- To explore the major issues and challenges that are surrounding ICT security policies,
- To understand the key global trends in developing and managing ICT security policies,
- To appreciate the need for a guideline towards ICT security policy management,
- To propose an effective framework for managing ICT security policies and for continuously reforming them,
- To understand the implementation details of the framework through the governance of ICT security policy at the strategic, tactical and operational levels of management, and
- To provide an overview of the future trends in the challenges, concerns and issues that organisations would face while managing their ICT security policies.

Figure 1. Influences on ICT security policies

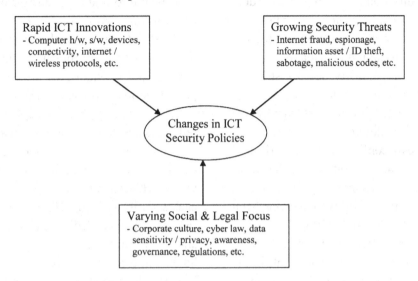

BACKGROUND

Developing an ICT security policy is the first step towards enhancing an organisation's information security and hence it should clearly identify goals, directions, procedures, and stakeholder responsibilities (Fulford & Doherty, 2003). A good ICT security policy should try to balance rigid information control that creates constraints and limitations to the ICT services of the organisation against an open information control that creates scope for security compromises. It should provide the necessary procedures and controls to ensure that the ICT systems in an organisation are protected against information loss or ICT service disruptions that could be caused by inadvertent or malicious actions. Essentially, the ICT policy is a comprehensive document that should provide at least the following baseline information for protecting the ICT resources of an organisation:

- The ICT security policy statement and coverage;
- The objectives and the strategies behind the ICT security policy as a business enabler of the organisation;

- The security risks and the roles / responsibilities of staff;
- The standards and procedures to be adopted in protecting the information assets of the organisation and in reporting incidents.

For the past two decades, much research and practice have been focussing on the technology aspects of information security (Von Solms, 2006). Research studies were concentrating on tools and techniques in achieving the three main objectives of ICT security, namely confidentiality, integrity and availability of information that are required to be achieved during transaction processing or in transit over the network or within storage devices. Hence, ICT security policies were also developed around the technological aspects outlining the general use of computer resources, access controls, staff responsibilities in handling sensitive information and incident management procedures. However, such ICT security policies have served their purpose only in the short-run as they cater to ICT resource management alone and lacked risk management, performance measurement, strategic alignment and other non-technical aspects associated with the ever changing business needs of an organisation (Dhillon & Torkzadeh, 2006). ICT security policies are also required to comply with a number of government regulations that span multiple industries and could be uniquely specific to industry sectors such as, finance, healthcare and small businesses (Greene, 2006). Hence, information security standards have also been raised by adding on to the traditional security objectives of confidentiality, integrity and availability of information, some more objectives such as authenticity, accountability, compliance, non-repudiation and reliability that have been introduced recently (ISO/IEC 27000, 2007).

In the past, organisations perceived ICT security as a technical problem and did not relate information security problem as a risk to its business objectives (Siponen & Oinas-Kukkonen, 2007). There has been lack of management involvement and commitment in developing ICT security policies and procedures. According to studies conducted by Dong, Neufeld & Higgins (2009) on various enterprises, the level of outcomes in such ICT initiatives varies very much with the level of top management support rendered. Hence, management support is essential for a successful implementation of ICT policies.

Today, we are in a networked world that provides openness and connectedness through a variety of gateways that could lead to escalated security breaches. Recent evidences indicate that there is enormous potential for intrusions through the network, which may result in business service disruption and loss of strategic / asset information by organisations (Jahankani, Antonijevic & Walcott, 2008; Whitman & Mattord, 2009). Security breaches have negative impacts on achieving business objectives and result in huge financial implications (Hansche, 2001; IT Governance Institute, 2005; Bojanc & Jerman-Blaic, 2008). Hence, organisations need to change from the traditional relaxed approach shown towards the management of ICT security policies to a more involved and integrated approach that connects the technical and non-technical (socio-economic)

aspects of information security with the business operations.

The key to success with a security policy is not the technology alone, but the ability to effectively manage and safeguard the information assets as part of an organisational culture (Caruso, 2003). In order to develop and manage an ICT security policy successfully, it is important for the top management of an organisation to understand the cost and measures involved in information security: where, when and how much money to allocate for each security-related task, what are the levels of risks associated with different business operations and how to strategically align policies with stakeholders interests (IT Governance Institute, 2005; Bojanc & Jerman-Blaic, 2008). Security alignment with stakeholders' interests is also important in the success of a security policy, as humans are the ones connecting the technology with the business activities. In other words, information security should be considered as a business driver and an ICT security policy should become a vehicle in establishing the bridge between the human and the technological paradigms of an organisation.

In broad terms, an ICT security policy is a plan that involves both the technology component and the management component. It should outline not only the techniques that are to be adopted to protect the ICT resources and information assets of an organisation, but should also explain the delegation of the associated duties and responsibilities of participants at all levels, strategic, tactical and operational levels of an organisation in a co-operative manner in order to achieve the desired security culture.

ICT SECURITY POLICY ISSUES

Over a period of time, the ICT era has posed many challenges that require much attention. One of the major concerns that has surfaced recently is the insufficiency of the current ICT security policies to tackle the multi-dimensional issues posed by the interconnected world of today with much more diversity. While formulating ICT security policies, organisations need to be aware of the recent trends and issues that arise from internal as well as external business operations. Some of the major ICT security policy issues surrounding a globally connected organisation are given below:

1. Though information security standards are available in literature, there are no standardised procedures for developing ICT security policies that meet all the security requirements of an organisation (Hone & Eloff, 2002; Saint-Germain, 2005; Pinder, 2006). Well-established guidelines such as Information Technology Infrastructure Library (ITIL), Control Objectives for Information and Related Technology (COBIT), Capability Maturity Model (CMM), and International Organisation for Standardisation (ISO) 1779 provide best practices for managing ICT infrastructures, in general, and they have complementary purposes (IT Governance Institute, 2005). While ICT security best practices are quite similar worldwide, the ICT configurations, business needs and the corporate culture are much different from one organisation to another. Hence, none of the studies indicate how these standards and guidelines fit together in providing a management framework for ICT security policies, in particular.

2. Information security needs for an organisation are generally seen from merely the technical context while the much-required non-technical (socio-economic) aspects are ignored (Dhillon & Torkzadeh, 2006). The social concerns, corporate cultural issues and economic impacts of a security problem are not considered by many policy makers (Chang & Chin-Shien, 2007). The recent trends in computer attacks through the Internet have resulted in changing the security policies to be closely tied to the legal issues of privacy (Whitman & Mattord, 2009). Hence, in general, ICT security policies are framed around tackling technical issues and do not address the socio-economic risks and privacy barriers that could impact the organisation's business objectives.

3. There is a lack of shared responsibility with regard to ICT security issues. Normally, ICT security policies are merely conveyed to the employees of an organisation in the form of an information communication session. Many organisations do not have in place any in-depth security awareness program that could help management and employees get involved and participate in the ICT security policy implementation and that could facilitate in taking ownership of their security roles and responsibilities. The lack of focus on user awareness could result in deliberate and accidental introduction of malicious activities by the employees (Abu-Musa, 2007).

4. Even a single incidence of security breach or virus attack on a company's ICT could lead to denial of service or asset information leakage resulting in a serious damage to the integrity of the organisation and denial of service leading to a huge financial loss (Von Solms, 2006; Conklin, 2007; Jahankani, Antonijevic & Walcott, 2008). However, until the recent security awareness wave, ICT security policies have not been given a risk-managed outlook and therefore have not been given high priority among the business objectives of an organisation.

5. Effective governance of ICT security is possible with planned interventions such as security acts / government laws or even internal / external audits, which require changes in corporate culture. Employees need to believe that such external interventions or internal audits would in fact strengthen corporate accountability (Pinder, 2006).

6. There is no international law that can be adopted to prosecute cybercrime and information asset thefts of an organisation (Aljifri, Pons & Collins, 2003; Shalhoub, 2006). Electronic commerce poses legal risks that cannot be ignored by organisations as they involve the understanding of the different cyber laws practised in other countries.

With so many ever-changing technological as well as non-technological issues impending on information security, organisations are in need of a management framework that caters to these issues and acts as a guideline for creating, implementing and continuously reforming ICT security policies. The next section proposes a framework to achieve this.

PROPOSED FRAMEWORK

Developing a good ICT security policy for an organisation requires an iterative approach as it is an ongoing journey towards a risk-reduced environment. It cannot be a one-time endeavour as it requires continuous improvements from time to time in an iterative manner. This is due to increased business dependencies with ICT services, increased security threats and diverse socio-economic issues. Such a dynamic situation could warrant major policy upgrades from one iteration of development to the next. The framework proposed below captures this requirement of continuous improvement with an iterative approach for the development of ICT security policy:

1. **Requirements Analysis:** This step involves identifying various issues related to the organisation's ICT systems, possible security threats and risks, their impact on the business operations and more importantly the drivers for an ICT security policy. Such an analysis of the organisation's environment would then lead to defining the vision and goals of an

ICT security policy that is geared towards meeting all the organisation's information security requirements that were identified.

2. **Policy Recommendations:** This step would involve researching the best practices to deal with the organisation's security issues identified in the requirements analysis step. This would help the policy makers in identifying alternate security solutions / strategies that are adopted in other similar environments. All the solution options are then considered with a cost-benefit analysis leading to certain viable recommendations for the policy. In other words, the recommendations are drawn based on the pros and cons that are weighed through the cost-benefit analysis of each solution option.

3. **Policy Design:** This step involves outlining how the recommended security solution would be put into practice in the organisational context. The making of the policy involves identifying the scope of the policy, the risks to be addressed, the detailed security measures to be adopted and the support / guidance required. It outlines the procedures, checklists, standards, rules and regulations, user roles, training and support that are essential for implementing the ICT security policy within the organisation.

4. **Policy Communication:** This step focuses on promoting discussions about the information provided in the policy as part of the security awareness program of the organisation. It enables individuals and stakeholder groups to identify policy issues, inconsistencies across various departments of the organisation and possible implementation hitches. This step also enables staff participation in policy making and develops a sense of policy ownership and security responsibility among individuals of the organisation. Through such participation techniques one could achieve employee empowerment (Whyte & Macintosh, 2002; Hart-Teeter, 2003).

5. **Policy Endorsement:** The policy gets refined based on the feedback obtained through the policy communication step. The refined ICT security policy then gets endorsed by a security governance committee, which could consist of internal and external security auditors. The committee would ensure that the policy complies with the international standards and practice. The International Organisation for Standardization (ISO) and the International Electrotechnical Commission (IEC) have published standards for information security termed as ISO/IEC 27000 series, which provides high-level advice on a broad range of controls, risk assessment processes and certification standards for information security management. The security governance committee of an organisation could adopt policy compliance and governance by verifying that the policy guidelines conform to the best practices recommended by the ISO/IEC 27000 series of information security standards as well as the government / local security organisations such as Information Systems Audit and Control Association (ISACA). It is also important for the committee to check that the policy captures the hardware and software guidelines suggested by the organisation's IT industry partners such as Microsoft, Sun, IBM, Cisco, etc., who have provided the ICT infrastructure.

6. **Policy Implementation:** The ICT security policy that has been formed from the previous steps would go through the implementation process. The most important aspect in this step is to provide education and training for staff to always think and act in a secure manner. The security awareness program would also involve changing their roles and duties that are required for protecting the information assets of the organisation that are related to their everyday business operations. The protocols to be followed, contingency steps to be undertaken, procedures for reporting and handling incidents, etc. are laid out clearly and communicated to the staff. The deployment of the policy also involves different ways of establishing the communication lines between the stakeholders and the Information Security Team of the organisation for continuous updates and support.

7. **Policy Feedback and Evaluation:** It is important to monitor and collect feedback periodically in order to evaluate the success of the policy. Monitoring event logs plays an integral part in the evaluation of ICT policy compliance and governance (Greene, 2006). Evaluation is an important step in an information security management initiative as it helps in collecting information on its efficiency and effectiveness, to define follow-up actions and to justify the investments made (Schlienger & Teufel, 2005). Such a feedback system is highly essential because, all organisations undergo changes in this globally competitive world and that could have an impact in the implementation of the policy. There could be organisational cultural barriers that form roadblocks in the implementation of certain aspects of the policy. The effectiveness of the policy could be determined based on certain key metrics such as the number of security incidents identified and resolved, system audit compliance achievements and the business service continuity performance. Finally, the review process in this step would result in major recommendations for policy improvements that would lead back to step 2 for the next iteration of the ICT security policy reform.

ICT security is not a mere technical problem but the entire organisation's problem. Many studies indicate that the ICT security issues are not given due importance at all levels of management, namely strategic, tactical and operational levels.

The lack of budget allocation for ICT security (Wiander, Savola, Karppinen & Rapeli, 2006) at the strategic level indicates that the organisational management, in general, has failed to understand the complexity of information security risks and their implications on business operations. Instead, the information security problem has been delegated to the ICT departments, which try to develop ICT security policies for addressing the multi-dimensional issues of protecting the entire organisations information assets. This has resulted in misalignment of policy objectives as against business objectives. However, the proposed framework outlining the seven-step iterative process of ICT security policy development clearly indicates the participation of all levels of management. Figure 2 depicts the iterative continuous improvement process cycle of the proposed framework. It shows that the feedback loop paves way for flexibility and adaptability of the policy from time to time as changes take place in the business environment. In addition, the next section provides the implementation details and governance of the ICT security policy at the strategic, tactical and operational levels of management.

ICT SECURITY POLICY GOVERNANCE

A good ICT security policy has three levels of policy formulation and governance that are aligned with the management objectives, namely, strategic level, tactical level and operational level (Figure 3). At the strategic level, the policy states the vision, mission and overall objectives of the security policy by drawing close relationship with the organisations business objectives. These high-level statements provide the overall management expectations, scope and directions of the policy. They indicate a comprehensive set of areas covered by the policy such as risk management, legal advisory, security standards and reference objects that relate to security awareness programs and the

Figure 2. Iterative process cycle of ICT security policy development

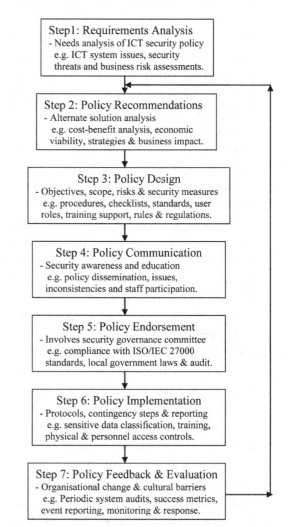

procedures for preventive management, control and recovery of ICT assets. At the strategic level, the policy would cover the technical aspects of security such as defining the information assets, the ICT infrastructure and the user groups that could have an impact on the business. Similarly, examples of non-technical aspects would include defining the business risks, legal matters, audits and standards that would be associated with the information security of the organisation.

At the tactical level, the policy determines non-technical aspects such as risk assessment

Figure 3. ICT policy alignment with management objectives

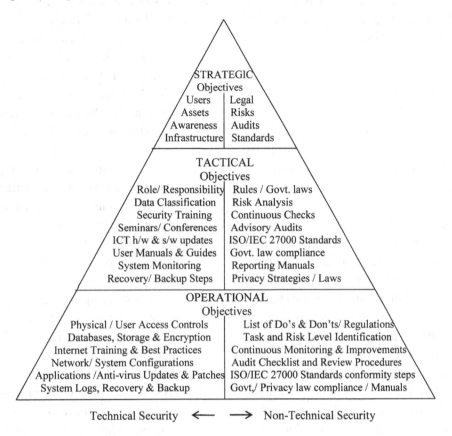

Technical Security ⟵ ⟶ Non-Technical Security

guidelines, security standards, audit procedures, and incident reporting manuals along with governing rules. It also covers technical aspects of security such as identification of user groups, roles and responsibilities, classification of sensitive data, training programs, system monitoring and recovery mechanisms. While the policy at the strategic level provides the motivation (why) of the security initiatives and objectives, at the tactical level it provides the procedures (what) that would be adopted to realise these objectives.

Finally, at the operational level, the policy elaborates the details (how, when and where) of the implementation mechanisms to be adopted by the organisation. It provides the mechanisms and guidelines for determining the risk levels (low, medium, high) of each business functional task and allocating roles (user login, password,

privileges, etc.) for each user. It clearly prepares the procedures, checklists and templates that are essential in the day-to-day business operations for user groups to adopt. The policy documents the training details and support available to staff when they perform their functions of protecting, monitoring and reporting. Further, the policy determines the metrics for evaluation and continuous improvements. In other words, at the operational level, the ICT security policy provides detailed steps, both preventive and proactive, for dealing with technical and non-technical security problems. Some examples on the technical arena that would be included in the security policy are procedures for latest anti-virus software updates, user account rights created in databases, and incident monitoring and reporting. Some examples on the non-technical arena that would be included in the security policy are procedures for communicating new security initiatives, protecting stakeholders' privacy and documentation.

FUTURE TRENDS

This section gives an overview of some of the future trends in the challenges, concerns and issues that organisations would face while managing their ICT security policies.

1. **Management Challenges:** The previous sections have clearly established that security cannot be delegated to a specific department within an organisation and it requires an active participation of representatives from all departments to interact towards the iterative development of an ICT security policy that covers technological, economic, and social issues pertaining to its business information security. This requires a drastic change in the mindset of the management and staff to buy-in the security culture. The policy framework should promote a security compliance culture within an organisation so as to satisfy government regulations that are specific to the industry sector (Greene, 2006). The security attacks seen so far are only the beginning steps to a more advanced maliciousness of the future that end-users are unaware of (Whitman & Mattord, 2009). Hence, top management should understand that with security problem worsening as years go by, such a participatory management of ICT security policy would definitely enhance the end-user security awareness and the success of the policy implementation. In addition, it provides the necessary inputs to the feedback loop of the policy development framework (Figure 2) that would result in a well-structured policy improvement from time to time.

2. **Overhead and Training Concerns:** Developing an ICT security policy should not be considered as an additional overhead. As emphasised in the proposed framework, the security policy objectives should be aligned with the business objectives, and security mechanisms should be integrated with the everyday business operations. In such a case, all the system logs and generated data could automatically serve as inputs into the final evaluation process. This would pave way to an automatic approach to measuring the success metrics and effectiveness of the policy and hence implementing such a security policy would not become an additional overhead. It would also greatly assist in studying and analysing the security behaviour of ICT users so as to fine tune the training program and reform the policy for incorporating tighter security measures for the future.

3. **Cyber Compliance Challenges:** In this globally interconnected business environment, almost all organisations handle customer information in electronic form and protecting such data during transaction processing or in transit over the network or within storage devices should be given high priority as any unintended disclosure could lead to privacy and legal issues that have immense business implications. While the policy would state the necessary security controls that deal with compliance standards and local government regulations, any violations could lead to not only business liabilities but could also result in local court litigation claims based on the type of security breach. In addition, cyber security laws vary from one country to the other and with business transactions taking place over the internet, the ICT security policy of an organisation should be developed with proper understanding of the cyber requirements and regulations of different countries. This

could pose a major challenge in developing security policies for the future.

4. **Accountability and Audit Issues:** Another concern for the future ICT security policy management would be to conduct ICT security audits and to create an accountability framework that could work hand-in-hand with both internal and external audits. Periodic system audits are essential for continuous improvements in the ICT security policies. However, there is a general resentment among staff to cooperate with regard to any kind of audit and accountability, which calls for a co-ordinated effort among all levels of participants for a sound security review (OECD, 2002; Pinder, 2006). Hence, staff must be educated that internal and external audits would rather help to strengthen corporate accountability and to minimise liability exposures / court litigation claims. In future, external reviews and international accreditation of ICT security policy could even become the law for many companies.

CONCLUSION

Information security objectives have been modified from time to time to meet up with the socio-technological changes taking place in this globally interconnected environment. This affects the ICT security policy of an organisation. The key factors that warrant frequent changes in the ICT security policy of an organisation are, i) rapid ICT innovations, ii) growing security threats and iii) varying social and legal focus. Such a situation indicates that protecting the information assets of an organisation is not merely a technical problem, but involves privacy, social, economic and management issues that are unique to each business setting. Today, organisations require a good management framework for developing their ICT security policy. Hence, this chapter has

proposed an integrated management framework which contains a seven-step practical guideline for developing, implementing and reforming ICT security policies.

The salient features of the proposed framework are:

- **Flexibility:** The policy development goes through a continuous improvement feedback loop that allows the policy to get reformed through iterations.

- **Consultative:** The framework uses participative techniques that facilitate the involvement of stakeholders in the policy-making process.

- **Business Alignment:** By mapping the security tasks with the business functions at the strategic, tactical and operational levels, the framework promotes the alignment of the security risks with the business objectives.

- **Conformity:** Policy endorsement through internal and external auditing processes facilitates in achieving conformity and compliance with security standards and regulations.

- **Integrated:** Through policy communication initiatives and policy evaluation, the framework is able to integrate the different requirements (technical, social, economic) of various user groups from different departments of the organisation.

- **Adaptability:** The generic nature of the framework supports adaptability of the framework to any organisation.

This chapter is by no means intended to be a complete reference for ICT security policy management. Instead, the main purpose is to provide an awareness of the multi-dimensional issues and global trends that impact upon ICT security and the need for an effective framework for its policy development. It has proposed an effective framework that provides practical steps for the implementation, enhancement and governance of ICT security policies and procedures in an organisation.

REFERENCES

Abu-Musa, A. A. (2007). Evaluating the security controls of CAIS in developing countries: An examination of current research. *Information Management & Computer Security*, *15*(1), 46–63. doi:10.1108/09685220710738778

Aljifri, H. A., Pons, A., & Collins, D. (2003). Global e-commerce: A framework for understanding and overcoming the trust barrier. *Information Management & Computer Security*, *11*(3), 130–138. doi:10.1108/09685220310480417

Bojanc, R., & Jerman-Blaic, B. (2008). Towards a standard approach for quantifying an ICT security investment. *Computer Standards & Interfaces*, *30*(4), 216–222. doi:10.1016/j.csi.2007.10.013

Caruso, J. B. (2003). Information technology security policy keys to success. *Center for Application Research Bulletin, 2003*(23). Retrieved January 2009 from http://www.educause.edu/ir/library/pdf/ERB0323.pdf

Chang, S. E., & Chin-Shien, L. (2007). Exploring organizational culture for information security management. *Industrial Management & Data Systems*, *107*(3), 438–458. doi:10.1108/02635570710734316

Conklin, W. A. (2007). Barriers to adoption of e-government. In *Proceedings of the 40th Annual Hawaii International Conference on System Sciences*. Washington, DC: IEEE Computer Society.

Dhillon, G., & Torkzadeh, G. (2006). Value-focused assessment of information system security in organizations. *Information Systems Journal, 16*, 293–314. doi:10.1111/j.1365-2575.2006.00219.x

Dong, L., Neufeld, D., & Higgins, C. (2009). Top management support of enterprise systems implementations. *Journal of Information Technology*, *24*(1), 55–80. doi:10.1057/jit.2008.21

Drevin, L., Kruger, H. A., & Steyn, T. (2006). Value-focused assessment of ICT security awareness in an academic environment. In S. Fischer-Hubner, K. Ranneberg, L. Yngstrom, & S. Lindskog (Eds.), *IFIP International Federation for Information Processing. Volume 201, Security and Privacy in Dynamic Environments*, (pp. 448-453). Boston: Springer.

Fulford, H., & Doherty, N. F. (2003). The application of information security policies in large UK-based organizations: An exploratory investigation. *Information Management & Computer Security*, *11*(2/3), 106–114. doi:10.1108/09685220310480381

Greene, S. (2006). *Security Policies and Procedures: Principles and Practices*. Upper Saddle River, NJ: Prentice Hall.

Hansche, S. (2001). Designing a security awareness program: Part 1. *Information System Security*, *9*(6), 14–22.

Hart-Teeter. (2003). The new e-Government equation: Ease, engagement, privacy and protection. *A Report prepared for the Council for Excellence in Government*. Retrieved February 10, 2009 from http://www.cio.gov/documents/egovpoll2003.pdf

Hone, K., & Eloff, J. H. P. (2002). Information security policy - What do international information security standards say? *Computers & Security*, *21*(5), 402–409. doi:10.1016/S0167-4048(02)00504-7

ISO/IEC 27000. (2007). *The ISO/IEC 27K Toolkit – Implementation guidance and metrics*. Retrieved January 20, 2009 from http://www.iso27001security.com/html/iso27k_toolkit.html

IT Governance Institute. (2005). Aligning CO-BIT, ITIL and ISO 17799 for business benefit. *A management briefing from ITGI and OGC*. Retrieved February 4, 2009, from http://www.nysforum.org/documents/pdf/itil-6-6-06/AligningCOBITITIL.pdf

Jahankani, H., Antonijevic, B., & Walcott, T. H. (2008). Tools protecting stakeholders against hackers and crackers: An insight review. *International Journal of Electronic Security and Digital Forensics, 1*(4), 423–442. doi:10.1504/IJESDF.2008.021459

OECD. (2002). Guidelines for the security of information systems and networks: Towards a culture of security. *OECD Information Privacy and Security*. Retrieved February 10, 2009 from http://www.oecd.org/dataoecd/16/22/15582260.pdf

Pinder, P. (2006). Preparing information security for legal and regulatory compliance (Sarbanes-Oxley and Basel II). *Information Security Technical Report, 11*(1), 32–38. doi:10.1016/j.istr.2005.12.003

Saint-Germain, R. (2005). Information security management best practice based on ISO/IEC 17799. *Information Management Journal, 3*(4), 60–66.

Sathish Babu, B., & Venkataram, P. (2009). A dynamic authentication scheme for mobile transactions. *International Journal of Network Security, 8*(1), 59–74.

Schlienger, T., & Teufel, S. (2005). Tool supported management of information security culture: Application in a private bank. In Sasaki, R., Okamoto, E. & Yoshiura, H. (Eds.) *IFIP International Federation for Information Processing. Volume 181, Security and Privacy in the Age of Ubiquitous Computing*, (pp. 65-77). Boston: Springer

Shalhoub, Z. K. (2006). Trust, privacy, and security in electronic business: The case of the GCC countries. *Information Management & Computer Security, 14*(3), 270–283. doi:10.1108/09685220610670413

Siponen, M. T., & Oinas-Kukkonen, H. (2007). A review of information security issues and respective research contributions. *ACM SIGMIS Database, 38*(1), 60–80. doi:10.1145/1216218.1216224

Small, B. (2007). Sustainable development and technology: Genetic engineering, social sustainability and empirical ethics. *International Journal of Sustainable Development, 10*(4), 402–435. doi:10.1504/IJSD.2007.017912

Von Solms, B. (2006). Information security - The fourth wave. *Computers & Security, 25*(3), 165–168. doi:10.1016/j.cose.2006.03.004

Whitman, M. E., & Mattord, H. J. (2009). *Principles of Information Security* (3rd ed.). Boston: Thomson Course Technology.

Whyte, A. & Macintosh, A. (2002). Analysis and evaluation of e-Consultations. *e-Service Journal, 2*(1), 9-34.

Wiander, T. Savola, R. Karppinen, K. & Rapeli, M. (2006). Holistic information security management in multi-organization environment. In *Proceedings of the International Symposium of Industrial Electronics*, (pp. 2942-2947). Montreal, Canada: IEEE.

KEY TERMS AND DEFINITIONS

ICT Security: The techniques to protect information, computer systems and communication technologies from unauthorised access, misuse, service disruption, destruction and malicious manipulation or disclosure.

Security Policy: A definition of what, how and why the information and communication technologies of an organisation are being secured including the constraints and the mechanisms.

Security Objective: A projected state of the ICT security that a security policy of an organisation wishes to achieve with the underlying common goals of protecting the confidentiality, integrity and availability of information and communication technology services.

Security Issue: The exposure of ICT to concerns, problems and vulnerabilities from technology, management, social and ethical perspectives.

Management Framework: A coordinated component set that would endeavour to meet the policy objectives of an organisation.

Business Alignment: The alignment of the strategic goals, objectives and plans of the business to the enterprise-wide activities and supporting systems.

Continuous Improvement: It is one of the philosophies of Total Quality Management and refers to the belief that an organisation requires to constantly measure the effectiveness of its processes and products with the objective of increasing quality and reducing waste.

Security Awareness: The knowledge and attitude possessed by an individual with regard to realising the importance of security of information assets of an organisation, one's responsibilities, the procedures to be followed and the implications of security failure.

Security Audit: A measurable assessment of a security system that checks for operational and technological problems and security vulnerabilities of an organisation.

Chapter 2
A Model for Building Trust in E-Government

Stephen M. Mutula
University of Botswana, Botswana

ABSTRACT

E-government is a new and complex field which is yet to be clearly understood because of lack of a well developed theoretical framework. E-government provides a platform for different forms of interactions such as business to business (B2B), citizens to government (C2G), government to government (G2G), business to government (B2G), etc. These interactions raise several ethical issues which have implications on citizens' trust in e-government. This chapter proposes a model for building trust of citizens in e-government. The proposed model is premised on five trust pillars: ethical/human; information/content; technical; policy/legal; and political/governance. Each of these pillars has several dimensions adapted from various existing user satisfaction tools or frameworks such as Service Quality (SERVQUAL), trust formation framework, technology acceptance model, information systems success model, and many more. The chapter also covers history of e-government; models of e-government interactions; e-government trust related models; theoretical basis of trust; causes of distrust in e-government; challenges of applications of ICT in e-government; framework for building trust in e-government; and future trends of trust in e-government.

INTRODUCTION

E-government is increasingly being perceived as panacea for a transparent and accountable administration. Yet e-government is a complex phenomenon that is not yet fully understood and grasped by most people. The concepts of e-government and e-governance can better be understood in the context of government and governance respectively. Government refers to exercising state power, the governing body of a state, or the way by which a state is governed. The term government also refers to an administration, regime, executive, rule, leadership, command or

DOI: 10.4018/978-1-61692-012-8.ch002

control (Hawker, 2002). Governance is a noun from the term government which Neumayer (2006) defines as the exercise of economic, political and administrative authority to manage a country's affairs at all levels. It comprises the mechanisms, processes, and institutions through which citizens and groups articulate their interests, exercise their legal and human rights, meet their obligations, and mitigate their differences. Governance encompasses political legitimacy and accountability, administrative accountability, financial and budgetary accountability, transparency, openness, and the rule of law. Governance therefore entails various things, among them, that those who hold public trust should be able to account for the use of the trust to citizens or their representatives. This signifies the superiority of the public interest over the private interest. Secondly, it presupposes the competence of the state to exercise administrative and political power in a fair, transparent and equitable manner, including the protection of personal rights and the rule of law. Thirdly, it implies that no manner of accountability can be realized unless there is the rule of law, transparency and the free flow of information among all stakeholders in the governance process (Mutula & Wamukoya, 2009).

Transparency being a major component of governance means that decisions taken and their enforcement are executed based on clearly stipulated rules and regulations that are known to all stakeholders. This requires that the institutions of governance in both public and private sector including civil society organizations interact and engage with citizens in a manner that contributes to good governance. Governance can be both good and bad depending on whether it brings positive benefits to the governed or the benefits merely serve to benefit the personal interests of a few individuals in government. Therefore, the government as the major actor in the governance process must create an enabling environment for citizen participation in the decision making process and service delivery systems (Mutula &

Wamukoya, 2009). E-government is expected to help governments meet these goals.

This chapter is intended to discuss history of e-government, models of e-government interactions, e-government trust related models, theoretical basis of trust, causes of distrust in e-government, challenges of applications of ICT in e-government, framework for building trust in e-government, and future trends of trust building in e-government.

BACKGROUND

E-government, a derivative term from governance, refers to the application of information and communication technology within public administration to optimise its internal and external functions, [thereby providing] government, the citizen and business with a set of tools that can potentially transform the way in which interactions take place, services are delivered, knowledge is utilized, policy is developed and implemented, the way citizens participate in governance, public administration reform and the way good governance goals are met (United Nations Department of Economic and Social Affairs, 2006). E-government may also be perceived as the application of ICTs to facilitate social governance processes or objectives, such as information for political participation, consultation and consensus-seeking among governments, public servants, politicians and citizens (Sheridan and Riley, 2006).

E-governance is an advanced level of e-government where citizen and government engagement takes place electronically. The United Nations (2008) discusses various forms of e-government. *Connected governance* (sometimes referred to as networked governance) is a deepened form of e-governance and refers to governmental collective action to advance the public good by engaging the creative efforts of all segments of society. E-governance aims at improving cooperation between government agencies, allowing for an enhanced, active and effective consultation and engagement with citizens, and a greater involvement with multi-stakeholders regionally and internationally. *Connected governance* provides better organized, aligned and often integrated information flows, new transactional capacities, as well as new mechanisms for feedback, consultation and more participative forms of democracy. Connected governance has various approaches through which it is achieved. E-*government-as-a*-whole is a form of connected governance which focuses on the provision of services at the front-end, supported by integration, consolidation and innovation in back-end processes and systems to achieve maximum cost savings and improved service delivery. Connected governance can be achieved through the *whole-of-government* approach which refers to 'public service agencies working across portfolio boundaries to achieve a shared goal and an integrated government response to particular issues. Through *whole-of-government* approach government agencies and organizations share objectives across organizational boundaries, as opposed to working solely within an organization. It encompasses the design and delivery of a wide variety of policies, programmes and services that cut across organizational boundaries.

HISTORY OF E-GOVERNMENT

The history of using information and communication technology in government operations especially in developed countries spans several decades. In Britain, the Technical Support Unit of the telecommunication services was engaged by government to evaluate and advise on the use of computers as far back as 1957 (Interoperable Delivery of European E-government Services to Public Administration, Businesses and Citizens-IDABC, 2005). However, the use of the word 'e-government' in literature is a recent phenomenon, having been given impetus in many ways by the growth of the Internet and World Wide Web in the 1990s. E-government evolved with the aim

of enhancing progress towards an information society status (Interoperable Delivery of European E-government Services to Public Administration, Businesses and Citizens -IDABC, 2005) as customers increasingly expected government to be accessible and convenient. In Canada, the Government On-Line Strategy that paved way for e-government was created in 1999 (Public Works and Government Services Canada, 2004). While in the United States, Government Paperwork Elimination Act of 1998 required federal agencies to provide the public with the option of submitting, maintaining and disclosing required information electronically rather than on paper. The e-government act of 2002 in the United States established broad measures that required using Internet-based information technology to enhance citizen access to government information and services and for other purposes (Relyea & Hogue, 2003).

In Asia, since the year 2000, governments have made significant progress in adopting information and communication technologies to improve financial management, streamline the delivery of government services, enhance communication with the citizenry, and serve as a catalyst for empowering citizens to interact with the government (Wescott, Pizarro, & Schiavo-Campo, 2001). In contrast, though African governments have been using information technology for more than 40 years, key innovations such as computer networks, intranets and the Internet started to emerge on the continent in the late 1990s. However, e-government as we know it today is only starting to slowly diffuse within the continent because high levels of e-readiness needed to realise e-government implementation is not yet in place (Heeks, 2002).

MODELS OF E-GOVERNMENT INTERACTIONS

E-government consists of several online interactions such as business to business (B2B), citizens to government (C2G), government to government (G2G), business to government (B2G), and more. Through these interactions, e-government links people to people, people to business, and people to government. The different digital government applications are referred to as e-government interactions. The applications of various e-government interactions include e-administration, e-business, e-services, and e-society (Mutula, 2009). These applications define the levels of interactions that define relationships between the government and the respective stakeholders. The e-administration applications arise from the interactions within and between government agencies, as well as between different spheres of government at local and provincial levels. These interactions are sometimes referred to as government-to-government or simply G2G. These applications seek to improve the operational efficiency and effectiveness of government. E-administration applications provide facilities to enable electronic communications and sharing of information and knowledge. They also permit simultaneous access to information, thus shortening the red tape often associated with access to, and transmission of, information. The e-business applications interaction can take place with business entities, citizens or any other legal entity with which the government has a business interest. E-business applications take most government procurement and disposal of assets to the electronic medium, thus cutting the red tape, the middlemen and the time required, and reducing operational costs, unnecessary delays, paperwork and redundant data capture (Oyomno, 2003).

The electronic services (e-services) applications enable interactions and relationships between the government and citizens, through which the latter gain access to a range of public services. These applications are referred to as government-to-citizen or G2C. However, these interactions and relationships may not be limited to citizens alone. They include non-citizens and other legal entities that the government interacts with in the process of delivery of public services. E-services enable

all branches and levels of government to function as a single coordinated entity, thus expanding government availability and accessibility, moving government in the direction of anytime, anywhere and by any means. The e-society applications on their part enable government to engage with the collective membership of communities and societies that comprise the nation. These applications are generally referred to as government-to-society or simply as G2S. E-society applications give communities a collective voice in their dealings with the government. The capacity of e-society applications to enable interactive participation and multiple consultations between the government and all its stakeholders is sometimes referred to as e-democracy (Oyomno, 2003). The highest echelon of e-government is e-participation which measures the extent to which governments are able to establish more transparency by allowing citizens to use new channels to influence policy. E-participation looks at how governments are able to create an environment that allows citizens to voice their views online and more importantly, to create a feedback mechanism (United Nations, 2008).

ISSUES, CONTROVERSIES, PROBLEMS

Despite the increasing adoption of electronic governments by most countries around the world (United Nations, 2008), little attention has been paid to the ethical dimension of e-government and in particular how trust of citizens can be nurtured in such e-government environments. Carbo (2007) concurs that little if any attention has been paid to two critical questions in e-government especially with regard to whether people trust e-government or how the cultural differences affect individuals' trust in government and their perceptions of how government affects their human dignity. The low priority accorded to the subject of trust dynamics in e-government is perhaps attributed to the fact that e-government evolved with emphasis on technological rather than on institutional, organisational and human considerations. The failure to take the earliest opportunity to differentiate between e-government and e-governance has not helped in focusing attention to citizen-government engagement as a priority and consequently on issues of trust and ethics. The limited attention paid to ethics and trust in e-government is not made any better by the fact that there is paucity of analytical tools including a theoretical framework for studying and understanding the subject.

The lack of trust in e-government has various undesirable consequences. For example, in countries that have been involved in long drawn wars such as Eritrea, Democratic Republic of Congo, Uganda, and many others, it is difficult to access data in the custody of government because of fear that such information could fall in wrong hands. If one happens to access data in the custody of government, the process is lengthy and tedious and such information if given is vigorously vetted. Moreover, because of lack of transparency by government with regard to enabling access to information in its custody, systems that would facilitate access to information are deliberately not in place. Consequently, it is difficult to find information because the collation, filing and storage systems are not of good quality (Mutula & Ocholla, 2009). Government that have restrictions to information in their custody, do not often have freedom of information laws in place. Denial of access to information is justified on account of protecting national security and sovereignty. In states where trust in government by the people is low, the public will often be cautious about intruding in matters they believe may lead the state to retaliate against them. In such circumstances, there is bound to be lack of freedom of expression as well as limited political activities. Such states would also exercise excessive monitoring (e.g. phone tapping or monitoring Internet) of what people are doing. This kind of state repression is common in Tunisia, Morocco, Syria, China,

Zimbabwe, Swaziland and Burma to mention but a few (Mutula & Ocholla, 2009). Lack of trust between the people and government, will often lead to the state providing limited information on government web portals thus denying citizens adequate access to information and also incapacitating them to question the decisions of government. Moreover, governments that do not have trust with its people will not strive to create awareness or develop skills needed for the people to use information and communication technologies effectively. Lack of trust of governments for its people often will lead to a top down approach to governance where citizens are hardly involved in decision making of governance matters. Moreover, lack of trust between government and its people will not enhance e-government as there will be limited information sharing between people and the state, and even among government departments. Governments that do not enjoy the trust of its people resort to regulating the media as in Zimbabwe, Kenya, Swaziland, and Botswana among others (Mutula & Ocholla, 2009).

E-GOVERNMENT TRUST RELATED MODELS

There is little consensus on the notion of what a model is. In the context of this chapter however, a model is perceived in the context of Microsoft Corporation's (2009) which defines it as 'an abstract representation of an item or concept'. The purpose of a model is to help visualise something that cannot be directly observed. It helps to understand the complexity of a phenomenon or a thing. A model helps to predict and explain the behaviour of a system. E-government is a concept that is complex in nature and is yet to be well understood. The concept is made even more complex when issues relating to trust are added to it. In this chapter, the concept of model will be used interchangeably with the notion of 'framework' which is a real or conceptual structure intended to serve as a support or guide for the building of something that expands the structure into something useful (Whatis.com, 2009). Genoud & Pauletto (2004) have developed a model for an e-society known as *a repository for implementing e-government strategy* to help in understanding the complexity of e-government. The repository structure has three layers each with five dimensions. The first layer is known as the *Project* which has the following dimensions: security, economic factors, information policy, and technology. The second layer is the *Organisation* and covers interoperability, transversality of processes, transversality of data, e-government processes, and knowledge management dimensions. The final layer is the *e-society* and covers legal framework, e-inclusion, society component, ethics and information concept. The dimensions in the *Project* layer collectively deal with issues related to security and new technologies.

The granularity of each layer is encapsulated in the dimensions making up the layers. For example, security dimension under the *Project layer* covers confidentiality, integrity and availability. Confidentiality refers to limiting information access and disclosure to a set of authorised users. Data integrity implies that data has not been changed inappropriately whether accidentally or deliberately. Availability represents the requirement that an asset can be accessible to authorised person, entity or device. Economic factors are perceived as return on investment or costs benefits such as changes in waiting times or fees. Information policy relates to information channels through which project leaders, end users and IT specialists can communicate. Knowledge management dimension under the *Organisation layer* creates value, increases productivity and fosters innovation (Malhotra, 2002). Transversality of processes refers to reengineering existing processes to create completely new ones going beyond internal borders of an administration. Transversality of data ensures that data is consistently shared and used, and that its quality is preserved. Interoperability refers to the ability of information and commu-

nication technology systems and that of business processes they support to exchange data and to enable sharing of information and knowledge. E-inclusion under the *eSociety layer* refers to achieving information society for all by enabling access (to government website for all including the disabled), enhancing usability (e.g. users experience with an application/website for ease of learning, efficiency of use, satisfaction, etc); training, etc. Ethics dimension under the *eSociety layer* covers privacy, security and the enhancement of confidence (European Commission, 2002).

Though the e-Society repository provides a framework for understanding the complexity of e-government, it does not consider the subject of trust in e-government. The other model that explains online government development is the *eGovernment maturity models* which focus more on the progress of maturity of online applications than on issues of trust or ethics.

THEORETICAL BASIS OF TRUST

The foundation of trust can be traced in part to key theories such as the Social-Psychological Model; the Social Cultural Model and the Institutional Performance Model. The Social-Psychological Model posits that trust and institutional confidence (or distrust and lack of confidence) is a basic aspect of personality types. The Social Cultural Model posits that the ability to trust others and sustain cooperative relations is the product of social experiences and socialization. This Model states that society has the ability to inculcate 'habits of the heart" such as trust, reciprocity, and co-operation (Bellah, Madsen, Sullivan, Swidler, & Tipton, 1985). The Institutional Performance Model opines that actual performance of government is the key to understanding citizens' confidence in it. Government institutions that perform well are likely to elicit the confidence of citizens; those that perform badly or ineffectively generate feelings of distrust and low confidence. The general public, the model assumes, recognizes whether government or political institutions are performing well or poorly and reacts accordingly. Collectively, these three models generate various meanings of trust namely: reliance on the integrity, ability, or character of a person or thing; reliance on somebody or institution; to have faith or a feeling of certainty that a person or thing will not fail; depth and assurance of feeling that is often based on inconclusive evidence (Mutula, 2009).

Trust has origin in such disciplines as business management, psychology, management, sociology and economics (Williamson, 1981). From business management perspective, the common concepts associated with or related to trust building include (White, 2008) customer satisfaction, interaction, honesty, moral values, responsiveness, and confidentiality. In the context of e-learning, trust building is associated with or leads to (Daniel & West, 2006) privacy, data protection, confidentiality, data security, accuracy, choice (opting in or opting out to the use and sharing of information), redress (means for filing a complaint), access (giving people to view information held about them), consistency (data integrity); appropriateness, authentic (accredited), affordability, efficiency, effectiveness, benefits, transparency, mobility (any time, anywhere), and ubiquitous interaction between service provider and consumer.

Service Quality Framework (SERQUAL) which has origin in management is useful in understanding user satisfaction. It identifies attributes of trust to include tangibles (appearance of facilities; reliability, responsiveness, assurance, courtesy, credibility, security); empathy (access, approachability, communication, and understanding the customer). Others include: competence (possession of requisite skill and knowledge to perform service); courtesy (politeness, respect, consideration and friendliness of contact person); credibility (trustworthiness, believability, honesty of service provider); feeling of security (freedom from danger, risk or doubt); communication (lis-

tens to customers and acknowledges their comments, keeps customers informed in a language they understand); understanding the customer (making the effort to know customers and their needs) (Zeithaml, Parasuraman, & Berry, 1999).

Trust formation framework a model from mobile communications examines 'trust' from the perspectives of quality of product and service, interface design and customisation; brand and online feedback mechanism (Shao, Ma, & Meng, 2005). Lee (2005) points out that interactivity is a source of trust. Interactivity is characterised by ubiquitous connectivity which refers to the continuance of a service activity irrespective of user's time and location. Trust formation as observed from the perspective of 3G mobile services also include responsiveness and brand image. Moreover, responsiveness refers to providing speedy feedback to service subscribers (Dholakia, Miao, Dholakia, & Fortin, 2000) and is used to measure service quality and is also a diagnostic tool for uncovering areas of service quality strengths and weaknesses. Responsiveness influences user satisfaction towards the use of the service because it can convey the trustworthiness of the service provider or system to customers.

Information systems success model (DeLone & McLean, 1992) is widely used for studying information systems success. The model intimates that IS quality characteristics (system quality), quality of IS output (information quality), consumption of IS output (usage) and user reaction to the IS (user satisfaction) are important to IS implementation success. Whyte & Bytheway (1996) in a UK study of business people on their perception of the success of information system in business identified success constructs to include: user friendliness, responsiveness of personnel, reliability of system and personnel, data accuracy, system response time, system accuracy, usefulness, and net benefits. Technology acceptance model posits that acceptance and use of a given system will be influenced by the degree to which the person believes that using such a system would

Table 1. Trust theoretical models (Adapted from: Easton; 1965; Bellah, Madsen, Sullivan, Swidler, & Tipton, 1985; Erikson, 1963; Dirks & Ferrin, 2002; Williamson, 1981)

Model Name	Source	Coverage
The Social-Psychological Model	Sociology/Psychology	trust, confidence, distrust
The Social Cultural Model	Sociology	trust, social experiences, socialization. reciprocity, co-operation.
The Institutional Performance Model	Management	confidence, distrust
SERVQUAL	Management	user satisfaction; tangibles (appearance of facilities; reliability, responsiveness), assurance (competence, courtesy, credibility, security); empathy (access, approachability, communication, and understanding the customer); courtesy (politeness, respect, consideration and friendliness of contact person); credibility (trustworthiness, believability, honesty of service provider); feeling of security (freedom from danger, risk or doubt); communication) ; understanding the customer
Technology Acceptance Model	Information Technology	technology acceptance, ease of use, usefulness, attitudes
Information Systems Success	Information Systems	system quality, information quality, usage, user reaction, user satisfaction, success, user friendliness, responsiveness, reliability of system and personnel, data accuracy, system response time, system accuracy, usefulness, net benefits
Trust Formation Framework	Mobile Communications	web quality, product quality, service quality, interface, customisation; brand, online feedback, interactivity, ubiquitous connectivity; continuance of a service, responsiveness, user satisfaction, trustworthiness
Online Learning	Education	privacy, data protection, confidentiality, data security, accuracy, choice, redress, access, consistency (data integrity); appropriateness, authentic (accredited), affordability, efficiency, effectiveness, benefits, transparency, mobility (anytime, anywhere),ubiquitous interaction
Governance	Public Administration	democracy, intellectual property rights; freedom of information; universal access; privacy/confidentiality; accountability; transparency; respect for human rights, openness, free press, freedom of speech
e-society repository for implementing e-government strategy	Inter-disciplinary (e.g. sociology, IT, education, management, etc)	security, economic factors, information policy, interoperability, transversality, legal framework, e-inclusion, ethics, acceptance, confidentiality, system availability, data integrity, costs benefits, interoperability, access, usability, ease of use, efficiency, user satisfaction, privacy, security, confidence.

enhance his or her job performance. Moreover, acceptance will be influenced by the extent to which the person believes that using a particular technology would be free from using much effort. Shih (2004) in a study found that individual attitudes towards e-shopping (an online transaction similar to e-government's C2B) are strongly and positively correlated with user acceptance. The trust theoretical models described above are summarised in Table 1.

CAUSES OF DISTRUST IN E-GOVERNMENT

There are several anecdotes why citizens distrust their governments and by extrapolation e-government. In 2006, Yahoo the widely used online content provider was compelled to provide user information to Chinese authorities who subsequently used it to imprison online activists for years. In addition, Google, Microsoft, and Yahoo operating on Chinese soil, censor their content in line with the Chinese government expectations. In Egypt, government officials beat Ahmed Maher

Ibrahim, a 27-year-old civil engineer for using Facebook to support calls for a general strike on May 4, 2008-President Mubarak's 80th birthday. In Russia, Saava Terentev - a musician was put on trial for speaking out about corruption in a blog posting. The 2007 Burma crackdown on monks by government when it shut down the country's internet connections to make sure no information got into the country and that little information got out of the country is another affront to freedom of citizens. The Syrian government also regularly restricts free flow of information on the internet and subsequently arrests individuals who post comments that the government deems too critical (Mutula, 2009).

E-government in developing countries is unlikely to score highly on citizen trust. This is because by and large, e-government is still in the confines of government enclaves. While information on the "who is who" in the ministry, its organizational structure and mission will often be abundantly available, the average website will not have public service information, for instance on how to go about applying for a particular service, who the right officer/ person to approach is, and where on the website to download and even electronically submit these application forms (Rotberg, 2007:20). During the WSIS Summit in Tunis in 2005, civil society held demonstrations against the Tunisian and Moroccan governments complaining about brutal state repression against free press and freedom of expression (World Information Society Report, 2006). In Kenya, the founding father of the nation Mzee Jomo Kenyatta was fond of stating, 'Serikali' ni 'Siri Kali' (translated literary from Swahili to English to mean 'government is top secret'). Governments that have traditionally restricted access to public information cannot be expected to embrace e-government whole-heartedly and avail information that would empower citizen to question public policy, accountability and integrity in government.

Though most governments in developing world are implementing e-government systems, little

awareness is being created among the people. This is exacerbated by lack of involvement of citizens in the planning process. Xiong (2006) in the context of e-government in China noted that if people have never heard of e-government, have no interest, have no equipment, insufficient knowledge or skills to access and use the online government information and services, e-government is meaningless. Most developing countries also face problems of poor organization of knowledge in government portals thus denying citizens the benefits that could accrue from sound information and knowledge management practices. Identifying the right processes to capture, store and share knowledge is an essential aspect of knowledge management (United Nations, 2008).

Distrust by citizens of government or e-government in many countries is caused by failure on the part of the state to involve the public in electronically-enabled decision-making processes because politicians fear that e-democracy may result in a loss of power (Mahrer and Krimmer, 2005).

CHALLENGES OF THE APPLICATION ICT IN E-GOVERNMENT

Mason (1986) identifies four main ethical and trust issues obtaining in digital environment that affect its use. These include: Privacy: what information about one's self or one's associations must a person reveal to others, under what conditions and with what safeguards? Accuracy: who is responsible for the authenticity, fidelity and accuracy of information? Who is to be held accountable for errors in information and how is the injured party to be made whole? Property: who owns information? What are the just and fair prices for its exchange? Who owns the channels, especially the airways, through which information is transmitted? How should access to this scarce resource be allocated? Accessibility: what information does a person or an organization have a right or a privilege to obtain, under what conditions and with what safeguards?

By and large, the technologies including ICTs used in developing world are imported. Such technologies are often implanted rather than transferred because of a number of constraints, including the inadequate or complete lack of indigenous mechanisms to adapt to such technologies. This leads to limited enthusiasm and the consequence underutilisation of the digital resources. Gerhan & Mutula (2005) in a study of bandwidths problems in Botswana found that shortage of computers and poor connectivity were major factors hampering effective access. Lenhart, Horrigan, Rainie, Allen, Boyce, Madden, & O'Grady (2003) note that not all "have nots" necessarily want to be "haves" and neither do they view engagement in ICTs as a positive force that would transform the quality of their life. Research work in the area of digital government in the US revealed how the lack of appropriate access amongst communities hinders the provision of social services by forcing individuals, often the poor, to travel long distances between offices (Bouguettaya, Ouzzani, Medjahead, & Cameron, 2001).

The Executive Secretary of the South African Development Community (SADC) in 2003, observed that while efforts had been channelled towards the development of ICTs in the region, many challenges still prevailed. For example, there were a low proportion of electrified households in most member states. Moreover, telecommunications facilities were generally poor and fixed line teledensity was low, with less than five percent of the population in the majority of SADC countries (SADC E-readiness Task Force, 2002). In Mauritius, Aubeelack (2002) identifies problems that have to be overcome in order to put the country on a stronger and sustainable e-government footing. These problems include: user resistance, IT skill shortages; frequent policy changes; bureaucracy leading to technological obsolescence; limited funding; lack of awareness about the potential of ICTs. In Tanzania problems relating to inadequate

infrastructure and low literacy level estimated at 60% have been reported (Sawe, 2005). Comparatively, in Lesotho, (Sehbalaka, 2004) cites problems of low internet penetration in the population, restructuring within government departments, lack of commitment from senior officials and shortage of IT skilled personnel. For Botswana, most of internet connectivity is in government. The country faces the problems of fast changing industry and low acceptance of ICT in a predominantly resources-based economy. In South Africa 45% of the population is estimated to live in rural areas with less developed infrastructure compared to the urban areas. In addition, PC penetration and human poverty index according to the 2004 estimates are 6.2% and 20% of the population respectively. The country has also diversity of languages with 11 official ones that need to be converted to the language of the Internet (Geness, 2004). Globally, Africa has the lowest internet penetration in the world at 3.6% (Internet World Statistics, 2007).

The United Nations (2008) points out that in some instances, governments have spent vast amounts of money building online systems and products only to observe that their citizens do not fully utilize them. This challenge in part is occasioned by the multiple aspects of ICT implementations that have to be dealt with or people simply not willing to use it for their own reasons. Ngulube (2007) points out that the major ingredients of e-government are infrastructure, human resources and information. However, the reality shows that in all these ingredients sub-Saharan Africa is deficient. In particular, the ICT infrastructure is not widely available to rural populations, and in most cases, both government officials and the people who may want to use government services online lack basic skills. The World Bank in a report on 'African Region Communications Infrastructure Programme' released in April 2007 (Nyasato & Kathuri, 2007) observes that east and southern African region suffers bandwidth deficiency as it accounts for less than one per cent of the world's international bandwidth capacity. Nyasato and

Kathuri further note that telecommunication users face some of the highest costs in the world. The international wholesale bandwidth prices in the east and southern Africa region are 20 to 40 times higher than in the United States, and international calls are on average 10 to 20 times more costly than in other developing countries.

SOLUTIONS AND RECOMMENDATIONS FOR BUILDING TRUST IN E-GOVERNMENT

One way to promote trust in e-government is through enhancing e-readiness environment. E-readiness refers to a community that has high-speed constant access and application of ICTs in government offices, businesses, healthcare facilities and homes; user privacy and online security; and government policies which are favourable in terms of promoting connectedness and the use of the network (Bridges.org, 2001). The e-readiness status of the nation is also measured by education levels of the people, use of appropriate technologies (e.g. radio and television), private public partnerships for resource mobilisation, research and development and creating relevant content accessible and useable by all citizens.

A policy framework that ensures universal access is a necessary precondition for enhanced digital inclusion and trust-building of citizens in e-government. As part of universal access strategies, diversity of choices of accessing content should be encouraged by allowing users to use a variety of technologies with which to gain access, such as the telephone, fax, e-mail, kiosks, face-to-face interaction, etc. Universal access should be accompanied by freedom of access to information through constitutional guarantees that enhance sharing of information (Farelo and Morris, 2006). The European Union member states (European Commission, 2005) have undertaken measures to ban the sale of inaccessible technology products

while enhancing the growth of assistive technology as one way of promoting universal access. Moreover, national strategies of member states emphasise interoperability of products, universal service policies for electronic communications, affordable pricing of network, and interactive content. The European Commission through the eEurope initiative recognizes accessibility for disabled users such as the blind, deaf people or people with learning impairments.

Trust in digital environment can be enhanced further if content providers can respond to cyber crime through effective investigation and prosecution of misuse; protecting privacy, data and offering consumer protection. A legal oriented framework is necessary to cater for cyber laws, consumer protection, and the security of transactions online (Department of IT eTechnology Group -India, 2003). Content providers must in addition, promote cultural and linguistic diversity online with regard to identity, traditions and religions. Sensitivity to cultural values can be buttressed by local content development, providing local content that is relevant to the people and in languages they understand.

Observing ethics too can help build trust among users with regard to managing and using digital content. Ethics refers to the universal or commonly held values of persons, despite different moral or cultural backgrounds (Fallis, 2007). Ethics focuses on the norms and standards of behaviour of individuals or groups within a society based on normative conduct and moral judgment, principles of wrong or right. Perceived benefits and ease of use of an information system are key determinants of trust in the actual usage of a technology. Information systems must be easy to use for diverse customers. The content must also be useful and beneficial to the customers if they have to be motivated to use it (Baeza-Yates and Ribeiro-Neto, 1999). The effective use of government services online starts with the involvement of people in the design of e-government applications. This calls for close consultation among stakeholders, creating customer awareness so that an environment is developed from the beginning where user's demands and the mechanisms of service delivery and feedback are provided for. User satisfaction analysis with government web portals need to be regularly carried out as is the periodic monitoring of administration of e-government services. The European Union member states have set good practices of e-government. For example, in 2005 the European Ministers produced an 'e-Government Declaration' setting targets for citizen inclusion, user-orientation and trust (European Commission, 2005).

Some countries have made good progress in implementing e-governance systems that are tailor-made to cater for various categories of citizen needs. For example, Singaporean government portal provides information services on culture, recreation, sports, defence and security, education, employment, family, community development, health and environment. The portal also includes user-centric hot links such as 'give us your feedback on national issues and policies' (Government of Singapore, 2004). The Canadian e-government portal on the other hand enables public participation that allows individuals to share their opinions on specified subjects, or to participate in various activities (Government of Canada, 2006). Building trust in e-government also requires branding of government web portals to make them people-centric and friendly. For example, in South Africa, the e-government web portal is branded 'Batho Pele' (translated to mean people first) (Department of Public Service and Administration, 1996).

MODEL (FRAMEWORK) FOR BUILDING TRUST IN E-GOVERNMENT

In Table 2, an integrated model/framework for building trust in e-government deriving from the various frameworks described in Table 1 is proposed. The framework consists of five pillars and their corresponding dimensions modelled on *a repository for implementing e-government strategy* (Genoud & Pauletto, 2004). The pillars include: Ethical/Human Pillar; Information/Content Pillar; Technical Pillar; Policy/Legal Pillar; and Political/Governance Pillar.

Each of the pillars in Table 2 serves a particular role in enhancing trust in e-government as follows:

- **Ethical/Human Pillar:** This pillar deals with issues that distinguish right from wrong actions. Ethics in general is concerned with the universal or commonly held values of persons, despite different moral or cultural values. Ethics focuses on the norms and standards of behavior of individuals or groups within a society based on normative conduct and moral judgment, principles of wrong or right.
- **Information/Content Pillar:** This pillar focuses on commitment by the state to the free flow of information and knowledge.
- **Technical Pillar:** The technical Pillar is concerned with what users perceive as benefits from the technology. Perceived ease of use is a key determinant of actual usage of a technology (Davis, 1989). E-government systems must be easy for diverse customers to use.
- **Policy/ Legal Pillar:** This pillar is concerned with a legal oriented framework necessary to cater for cyber laws, consumer protection, and the security of transactions. National strategies that emphasise interoperability of products, universal service policies for electronic communications, affordable pricing of network and interactive content are crucial.
- **Political/Governance Pillar:** The political and governance pillar is concerned with the exercise of economic, political and administrative authority to manage a

Table 2. Trust building model/framework in e-government

Ethical/Human Pillar	Information /Content Pillar	Technical Pillar	Policy/Legal Pillar	Political/Governance Pillar
Ethical/Human Dimensions	**Information/Content Dimensions**	**Technical Dimensions**	**Policy/Legal Dimensions**	**Political/Governance Dimensions**
Ethics	Friendly information architecture	Online support	Consumer protection	Transparency
Trust/distrust	Diverse formats	Feedback mechanism	Data protection	Accountability
Trustworthiness	Diverse retrieval tools	Interactivity	Intellectual property rights	Integrity
Technology acceptance	Diverse delivery channels	Interoperability	Freedom of information	Democracy
Education and training	Information security	Service /system availability	Standardization	Good governance
Cultural values	Knowledge management	Online security	Universal access	Respects for human rights
Special needs	Information policy	Friendly interface	Privacy	Commitment
Privacy /confidentiality	Information quality	Transversality of processes	Confidentiality	Responsible
Benefits	Data accuracy	Transversality of data	E-inclusion	Branding image
Customer satisfaction	Data integrity	Appearance	Redress	Free press
Quality of service	Choice	Reliability	E-government	Freedom of speech
Approachability of service provider	Access	Responsiveness	ICT policy	Free market economy
Courtesy	Appropriateness	Consistency of interface	Affordability	Service quality
Politeness	Local content	Efficiency /effectiveness	E-signatures	Product quality
Love	Multilingual content	Availability/mobility	Freedom of information legislation	Rule of law
Affection	Free flow of information	Customisation		Free and fair elections
Kindness	Usefulness of content	Friendliness		Protection of minorities
Gentleness	Usability of content	Ease of use		Consultations
Respect	Current content	Web quality		Competence
Honesty	Trustworthy sources	System quality		
Social experiences	Availability	Response time		
Socialisation	Ubiquity	Ubiquitous connectivity		
Reciprocity		Authentic		
Cooperation		Website ease of learning		
Communication		Help features		
Attitudes				
User reaction				
Moral values				
Human dignity				
Relationships				
Interdependence				
Faith				
Beliefs				

country's affairs at all levels. It comprises the mechanisms, processes, and institutions through which citizens and groups articulate their interests, exercise their legal and human rights, meet their obligations, and mitigate their differences (Neumayer, 2006). E-government is expected to help achieve this goal.

FUTURE TRENDS AND TRUST IN E-GOVERNMENT

E-government implementations around the world have gained impetus. However, the issues of trust in electronic governments will have to be addressed through interventions involving technology, capacity building, legal and regulatory frameworks. An e-government founded on the pillars of trust as enunciated in Table 2 would help increase the overall quality of public sector service delivery and assists in enhancing customer satisfaction from services they receive from government. Such satisfaction would enhance citizen-government engagement and make governance easier.

As governments make the transition from the traditional to e-government environment, the emphasis has largely been on improving access to information and transaction-based services for the public, clients and partners. But through e-government founded on trust there is potential to:

- Enhance confidence of people while transacting online
- Improve internal management and administrative processes such as service delivery
- Create partnerships between government and the broader public sector
- Increase efficiency
- Foster digital democracy
- Increased citizen involvement in governance
- Promote e-business by enabling the people deal with governments electronically
- Increase integrity in government
- Encourage citizens to bring pressure on their governments to demonstrate accountability and transparency.

The people's perceived trust in e-government will in part be predicated on how governments observe the law and maintain integrity in satisfying citizen needs through the delivery of relevant, value-added and high-quality services. E-government system that enhances transparency, especially with regard to the procurement of services; provision of opportunities for people to work in partnership with the private sector and citizens to influence policy decisions will likely be valued and adopted by the people. A trusted e-government system would inspire people to come forward and use services provided by government thus, reducing the cost and time of service delivery; making it easier for business and individuals to deal with government; enabling government to offer services and information

through new media like the Internet; improving communication between different parts/levels of government, making it possible for government to provide a variety of services around the clock, and enhancing democratic processes. The importance of trust in e-government need not be over emphasised. Foreign investors look at different locations and issues relating to democracy, accountability, skills, technology and infrastructure before making investment decisions. They also look at how helpful and efficient government is, and how easy it is to get things done such as applying for trade licenses and the efficiency of the legal system.

In developing countries where there are several challenges of infrastructure nature, government should consider the use of any appropriate technology to facilitate access to information such as batteries and solar panels to generate energy, using radio and television for public education and use of CDMA (Code Division Multiple Access) cordless telephone suitable for rough terrain where cabling would be difficult to lay and also provide high channel capacity than other commercial mobile technologies. Moreover, very small aperture terminals (VSAT) could be used to provide access to rural areas and others which do not have telephone connectivity. The future of e-government lies in the private public partnerships to mobilise resources needed to implement such governments. The potential of e-government and its utilisation also lies in the increased penetration of mobile phone services across the world both in developing and developed countries. This would be completed by increased establishment of cybercafés in most developing countries, the liberalisation of telecommunication sector as well as increased democratisation of hitherto undemocratic states.

CONCLUSION

This chapter has demonstrated that e-government is a growing and complex field which is yet to be well understood. E-government presents a new method where there is minimal contact between the citizens and government except through the intermediation of information and communication technologies. Such an environment is bound to have far reaching trust implications unless the system is completely transparent, efficient and interactive. E-government is predicated upon improving overall service delivery and good governance to citizens. However, the human aspect with regard to trust has received little attention despite the fact that governments collect a lot of information on citizens and private enterprises (e.g. corporate filing information, tax information, and regulatory information). The information in the custody of the state when stolen may erode the trust that citizens may have in their governments.

Hitherto, e-government as a field of study has no specific model available to study the issue of trust building. Studies that have made attempts to examine the issue of trust have applied models from such field as sociology, psychology, management, etc. E-government as we know it is a field that strides different disciplines and a theoretical framework or model that is cross-disciplinary is needed to help understand it. This chapter has attempted to develop an integrated model that can help to study the subject of building citizens' trust in e-government. The model consists of five pillars - ethical/human; information/content; technical; policy and legal; and political/governance with each pillar having several dimensions. The integrated model by virtue of its pillared structure provides the flexibility to be applied in whole or part depending on the issues that are to be studied. It can also be expanded to accommodate more pillars or dimensions as they arise. It is hoped that this model will fill up the lacuna that has to date existed in the field of e-government with respect to trust building among customers.

REFERENCES

Aubeelack, P. (2004). *Mauritius. Proceedings of workshop of the regional consultation on national e-government readiness in Gaborone, Botswana.* From 14-16 April 2004. Retrieved July 16, 2005, from http://www.comnet-it.org/news/CESPAM-Botswana.pdf

Baeza-Yates, R., & Ribeiro-Neto, B. (1999). *Human-computer interaction.* Retrieved September 15, 2006, from http://www.ischool.berkeley.edu/~hearst/irbook/10/node3.html

Bellah, R. N., Madsen, R., Sullivan, W. M., Swidler, A., & Tipton, S. M. (1985). *Habits of the heart.* Berkeley, CA: University of California Press.

Bouguettaya, A., Ouzzani, M., Medjahead, B. & Cameron, J. (2001). Helping citizens of Indiana: Ontological approach to managing state and local government databases. *IEEE Computer,* February.

Bridges.org. (2001). *Comparison of E-readiness Assessment Models: Final Draft.* Retrieved July 16, 2003, from http://www.bridges.org/eredainess/tools.html

Carbo, T. (2007). Information rights: trust and human dignity in e-government. [from http://www.i-r-i-e.net]. *International Review of Information Ethics, 7*(9), 1–7. Retrieved December 8, 2008.

Daniel, J., & West, P. (2006). *E-learning and free open source software: The key to global mass higher education?* International Seminar on Distance, Collaborative and eLearning: Providing Learning Opportunities in the New Millennium via Innovative Approaches, Universiti Teknologi Malaysia Kuala Lumpur, Malaysia, 4-5 January 2006.

Davis, F. D. (1989). Perceived usefulness, perceived ease of use, and user acceptance of information technology. *Management Information Systems Quarterly, 13*(3), 319–340. doi:10.2307/249008

Delone, W. H., & Mclean, E. R. (1992). Information system success: The quest for the dependent variable. *Information Systems Research, 3*(1), 61–95. doi:10.1287/isre.3.1.60

Department of IT eTechnology Group (India). (2003). *Assessment of central ministries and departments: e-governance readiness assessment 2003: Draft Report 48.*

Department of Public Service and Administration. (1996). *Green paper transforming public service delivery.* Pretoria, South Africa: GCIS.

Dholakia, R. R., Miao, Z., Dholakia, N., & Fortin, D. R. (2000). *Interactivity and revisits to websites: a theoretical framework.* RITIM Working paper. Retrieved December 15, 2008, from http://ritim.cba.uri.edu/wp/.

Dirks, K. T., & Ferrin, D. L. (2002). Trust in leadership: meta-analytic findings and implications for research and practice. *The Journal of Applied Psychology, 87*(4), 611–628. doi:10.1037/0021-9010.87.4.611

Easton, D. (1965). *A systems analysis of political life.* New York: John Wiley.

Erikson, E. H. (1963). *Childhood and society* (2nd ed.). New York: W. W. Norton.

European Commission. (2002). *eEurope 2005 action plan: e-Inclusion.* Retrieved December 20, 2008, from http://europa.eu.int/information_society/eeurope/2005/all_about/einclusion/index_en.htm

European Commission. (2005). *Transforming public services.* Report of the Ministerial eGovernment Conference, Manchester, UK. Retrieved December 12, 2008, from http://www.egov2005conference.gov.uk/documents/pdfs/eGovConference05_ Summary.pdf

Fallis, D. (2007). Information ethics for the twenty–first century library professionals. *Library Hi Tech, 25*(1), 23–36. doi:10.1108/07378830710735830

Farelo, M., & Morris, C. (2006). *The working group on e-government in the developing world: Roadmap for e-government in the developing world, 10 questions e-government leaders should ask themselves.* Retrieved December 24, 2008, from http://researchspace.csir.co.za/dspace/bitstream/10204/1060/1/Morris_2006_D.pdf

Geness, S. (2004). E-government, the South African experience. In *Proceedings of workshop of the regional consultation on national e-govt readiness in Gaborone, Botswana.* From 14-16 April 2004. Retrieved July 16, 2004 from http://www.comnet-it.org/news/CESPAM-Botswana.pdf

Genoud, P., & Pauletto, G. (2004). *The e-society repository: Transforming e-government strategy into action.* Paper Presented at the 4th European Conference on E-government, Dublin Castle, Ireland, 17-18 June 2004.

Gerhan, D., & Mutula, S. M. (2005). Bandwidth Bottlenecks at the University of Botswana: Complications for Library, Campus, and National Development. *Library Hi Tech, 23*(1), 102–117. doi:10.1108/07378830510586748

Government of Canada. (2006). *On-line forms and services.* Retrieved May 13, 2007, from http://canada.gc.ca/form/e-services_e.html

Government of Singapore. (2004). *E-citizen: Your gateway to all government services.* Retrieved May 14, 2007, from http://www.ecitizen.gov.sg/

Hawker, S. (Ed.). (2002). *Oxford dictionary thesaurus: wordpower guide.* New York: Oxford University Press.

Heeks, R. (2002). *E-government in Africa: Promise and practice.* Manchester, UK: Institute for Development Policy and Management University of Manchester.

Internet World statistics. (2007). *Internet Usage Statistics: World Internet Users and Population Stats*. Retrieved September 12, 2007, from http://www.internetworldstats.com/stats.htm

Interoperable Delivery of European E-government Services to Public Administration. *Businesses and Citizens-IDABC*. (2005). Retrieved July 19, 2005, from http://europa.eu.int/idabc/en/chapter/383

Lee, T. M. (2005). The impact of perceptions of interactivity on customer trust and transaction intentions in mobile commerce. *Journal of Electronic Commerce Research*, 6(3), 165–180.

Lenhart, A., Horrigan, J., Rainie, L., Allen, K., Boyce, A., Madden, M., & O'Grady, E. (2003). *The ever–shifting internet population: A new look at internet access and the digital Divide*. Retrieved May 11, 2003, from http://www.Pewinternet.org/

Mahrer, H., & Krimmer, R. (2005). Towards the enhancement of e-democracy: Identifying the notion of the middleman paradox. *Journal of Information Systems*, 15(1), 27–42. doi:10.1111/j.1365-2575.2005.00184.x

Malhotra, Y. (2002). Why knowledge management systems fail? Enablers and constraints of knowledge management in human enterprises. In Holsapple, C. W. (Ed.), *Handbook on knowledge management 1: Knowledge matters* (pp. 577–599). Heidelberg, Germany: Springer-Verlag.

Mason, R. O. (1986). Four ethical issues of the information age. *Management Information Systems Quarterly*, 10(1), 5–12. doi:10.2307/248873

Microsoft Corporation. (2009). *What is a model*. Retrieved September 19, 2009, from http://msdn.microsoft.com/en-us/library/dd129503 (VS.85, printer).aspx

Mutula, S. M & Ocholla, D. (2009). *Trust, Attitudes and Behaviours in E-Government*. Paper Presented at the African Information Ethics and e-Government Workshop, Held from 23-27 February 2009 at Mount Resort Magaliesburg, South Africa

Mutula, S. M. (2009). *Digital Economies, SMES and E-readiness*. New York: Business Science Reference.

Mutula, S. M., & Wamukoya, J. (2009). Public sector information management in east and southern Africa: implications for FOI, democracy and integrity in government. *International Journal of Information Management*, 29(5), 333–341. doi:10.1016/j.ijinfomgt.2009.04.004

Neumayer, E. (2006). Self-interest, foreign need and good governance: Are bilateral investment treaty programmes similar to aid allocation? *Foreign Policy Analysis*, 2(3), 245–268. doi:10.1111/j.1743-8594.2006.00029.x

Ngulube, P. (2007). The nature and accessibility of e-government in sub Saharan Africa. *International Review of Information Ethics, 7*(09/2007). Retrieved November 25, 2007, from http://www.i-r-i-e.net/inhalt/007/16-ngulube.pdf

Nyasato, R., & Kathuri, B. (2007). High Phone Charges Hamper Region's Growth, Says W Bank. *The Standard*. Retrieved April 10, 2007, from http://www.eastandard.net/hm_news/news.php?articleid=1143967136

Oyomno, G. Z. (2003). *Towards a framework for assessing the maturity of government capabilities for 'e-government'*. Retrieved July 20, 2005, from http://link.wits.ac.za/journal/j0401-oyomno-e-govt.pdf

Public Works and Government Services Canada. (2004). *Government online history*. Retrieved July 19, 2005, from http://www.communication.gc.ca/gol_ged/gol_history.html

Relyea, H. C., & Hogue, H. B. (2003). A brief history of the emergency of digital government in the United States. In Pavlichev & Garson (Eds.), *Digital Government,* (pp. 16-33). Hershey, PA: IGI Global.

Rotberg, R. (2007). The Ibrahim index on African governance: how we achieved our rankings. *African Business, November* (336), 20.

SADC E-readiness Task Force. (2002). *SADC e-readiness review and strategy.* Johannesburg, South Africa: SADC, 1060.

Sawe, D.J.A. (2005). *Regional e-governance programme: Progress from Tanzania.* 2nd EAC Regional Consultative Workshop held in Nairobi, Grand Regency Hotel from 28-29 June 2005.

Sehlabaka, C. (2004). The Lesotho government. In *Proceedings of workshop of the regional consultation on national e-govt readiness in Gaborone, Botswana*, 14-16 April 2004. Retrieved July 16, 2005, from http://www.comnet-it.org/news/CESPAM-Botswana.pdf.

Shao, B., Ma, G., & Meng, X. (2005). The influence to online consumer trust: an empirical research on B2C ecommerce in China. In *Proceedings of the fifth International Conference on Computer and Information Technology,* (pp. 961-965). Washington, DC: IEEE Computer Society.

Sheridan, W., & Riley, T. B. (2006). *Comparing e-government and e-governance.* Retrieved December 12, 2006, from http://www.electronicgov.net/pubs/research_papers/SheridanRileyComparEgov.d

Shih, H. P. (2004). An empirical study on predicting user acceptance of e-shopping on the web. *Information & Management, 41*(93), 351–368. doi:10.1016/S0378-7206(03)00079-X

United Nations. (2008). *UN e-government survey 2008: From e-government to connected government.* Retrieved January 28, 2008, from http://unpan1.un.org/intradoc/groups/public/documents/UN/UNPAN028607.pdf

United Nations Department pf Economic and Social Affairs (2006). *E-government readiness assessment methodology.* Retrieved December 12, 2006, from http://www.unpan.org/dpepa-kmb-eg-egovranda-ready.asp

Wescott, C. G., Pizarro, M., & Schiavo-Campo, S. (2001). *The role of information and communication technology, in improving public administration.* Retrieved July 21, 2005, from http://www.adb.org/documents/manuals/serve_and_preserve/default.asp

Whatis.com. (2009). *Framework.* Retrieved September 18, 2009, from http://whatis.techtarget.com/definition/0,sid9_gci1103696,00.html

White, B. (2008). *Business management skills-trust building tips for managers.* Retrieved December 24, 2008, from http://ezinearticles.com/?Business-Management-Skills--Trust-Building-Tips-for-Managers&id=679140

Whyte, G., & Bytheway, A. (1996). Factors affecting information systems' success. *International Journal of Information Management, 7*(1), 74–93.

Williamson, O. E. (1981). Calculativeness, trust and economic organisation. *The Journal of Law & Economics, 26,* 453–486.

World Information Society Report. (2006). Digital opportunity index 2005. Retrieved February 13, 2007, from http://www.itu.int/osg/spu/publications/worldinformationsociety/2006/World.pdf

Xiong, J. A. (2006). *Current status and needs of Chinese e-government users.* Carbondale, IL: Southern Illinois University.

Zeithaml, V. A., Parasuraman, A., & Berry, L. L. (1999). *Delivering quality service: Balancing customer perceptions and expectations.* New York: The Free Press.

KEY TERMS AND DEFINITIONS

Confidence: Conviction or belief.

E-Governance: A noun for e-government. It means the use of ICTs to facilitate processes of government such as service delivery, political participation of citizens, electoral process, etc.

E-Government: The use of ICTs to improve the functions of government.

E-Government Interaction: Refers to different forms of e-government such as business to business, government to citizens, business to government, etc.

Ethics: Moral principles defining acceptable behaviour.

Trust: Belief that somebody means good and can be relied upon to deliver what is expected without being put under any scrutiny.

Trust Building: Process of inculcating confidence in a people.

User Satisfaction: Consumer feeling of contentment that a service provider is meeting user needs.

Chapter 3
Adopting Mobile Devices in Classroom:
An Empirical Case Study from Indonesian Teachers

Chockalingam Annamalai
SEAMEO RECSAM, Penang, Malaysia

Wahyudi Yososutikno
SEAMEO RECSAM, Penang, Malaysia

Ng Khar Thoe
SEAMEO RECSAM, Penang, Malaysia

ABSTRACT

Worldwide, Mobile phone technology is increasing at a remarkable rate. Mobile technologies (m-technologies) allow students to gather access and learn beyond the classroom. They are an integral part of human's lives today, including children in some societies. This chapter addresses issues related to the mobile phone evolution in Indonesia, pedagogical effectiveness of using it in educational context with empirical study on Indonesian teachers' perception of the aforementioned in the classroom for science and mathematics learning. The useful research evidence of the value of ICT in education thereby serves as empirical basis for the formulation of ICT policy in educational institutions, particularly ICT integration in instruction.

INTRODUCTION

Mobile technologies (m-technologies) are integral part of human's lives today in many parts of the world as they are powerful tools for communication and learning in the technology era. Learning via mobile technologies or mobile learning

DOI: 10.4018/978-1-61692-012-8.ch003

(m-learning) is increasingly gaining attention especially in the societies with limited access to Internet facilities such as Indonesia and South Africa. According to literature, not only is the Internet subscriber base in Indonesia significantly lower than in its Association of Southeast Asian Nations (ASEAN) counterparts (see Figure 1), it is also lower than the ASEAN average. Internet growth in Indonesia is also lagging behind as can

Figure 1. Internet subscriber in ASEAN countries (From ICT infrastructure in emerging Asia. Policy and regulatory roadblocks, Sage Publications (2008). Retrieved from http://www.idrc.ca/openebooks/378-2/#page_141).

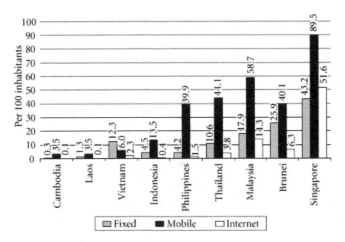

be seen in Figure 2 (Samarajiva & Zainudeen, 2008). Thus it is not surprising that m-learning is defined as "Knowledge in hand" by Brown & Metcalf (2008) as it facilitates the delivery of education or learning, foster communication, conduct assessments and provide access to performance support or knowledge.

This chapter deliberates on the broad definitions of m-learning with identification of various controversial issues and problems of practice as reviewed from literature. Elaboration will be made on the mobile phone evolution in Indonesia and the pedagogical effectiveness of using it in educational context. The findings from the empirical case study on Indonesian teachers' perception towards mobile learning in the science and mathematics classroom will also be reported with discussions on future trends. The useful research evidence of the value of ICT in education thereby serves as empirical basis for the formulation of ICT policy in educational institutions, particularly ICT integration in instruction.

The organization of this book chapter is divided into six sections: Section 2 (Background), Section 3 (Literature review), Section 4 (Research methodology), Section 5 (Finding and Discussion)

and Section 6 (Conclusion). Section 2 describes the background of related research on mobile phones and its pedagogical effectiveness. Section 3 discusses literature review. Section 4 discusses research methodology such as sample size, instruments, data collection and analysis. Section 5 elaborates findings discussions & conclusion drawn. Section 6 offers research implications and limitations. Finally, this chapter comes up with recommendation on formulation of ICT policy towards the usage of mobile devices in classroom.

BACKGROUND

Mobile learning (m-learning) is defined by Quinn (2000) as the intersection of mobile computing and e-learning: Accessible resources wherever you are, strong search capabilities, rich interaction, powerful support for effective learning, and performance-based assessment. Wei & Lin (2008, p.5251) defined m-learning as all forms of education in which the teacher and the learner are physically separated from one another by space and by time. A new m-learning architecture will support creation, brokerage, delivery and tracking of

Figure 2. Internet subscribers: Indonesia vs. India (1998-2005) (From ICT infrastructure in emerging Asia. Policy and regulatory roadblocks, Sage Publications (2008). Retrieved from http://www.idrc.ca/openebooks/378-2/#page_141).

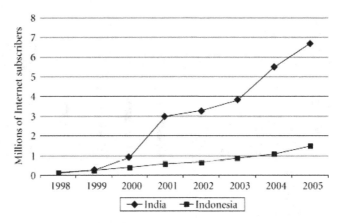

learning and information contents, using ambient intelligence, location-dependence, personalization, multimedia, instant messaging (text, video) and distributed databases (Mobilearn, 2003). Three ways in which learning can be considered for mobile "learning" are that it is mobile in terms of space; it is mobile in different areas of life; it is mobile with respect of time" (Kukulska-Hulme, Sharples, Milrad, Arnedillo-Sanchez, & Vavoula, 2009; Vavoula & Sharples, 2002)

REVIEW OF RELATED RESEARCH

Thanks to mobile phone technologies and personal digital assistants (PDAs), with the boundaries expanded beyond the classroom learning. With an increasing attention and concerns over the applicability of M-technologies, several researches have been conducted on M-learning environment in classroom activities for teaching and learning. One could use it for learning purposes beyond the classroom, create quizzes, surveys, questionnaire forms and evaluate them as well (Divjak, 2008; Kukulska-Hulme, 2008; Lindquist, 2007; Sung, Gips, Eagle, Madan, Caneel & DeVaul et al., 2005)

Technology tools such as mobile devices, wireless communications, and network technology have recently advanced significantly, and have been integrated into various wireless learning environments that attract many individuals' attention and expectations (Liu, 2007; Norris & Soloway, 2004; Roschelle & Pea, 2002). The tools contain semiautomatic and automatic guidance for learners' learning process, enabling automatic individualization of a learning process.

There are several controversial issues raised over the past decade about the use of m-learning. For example, Hoppe, Joiner, Milrad and Sharples (2003) wanted to shift the focus of M-learning from content delivery to interpersonal relations, as asserted "there is an imperative to move from a view of e- and m- learning as solely delivery mechanisms for content"(p.255).

According to Vogel (2007, p.2), adapting technology into learning is not new, and as times goes by, it is part of students' day-to-day's life. Mobile devices, e.g., PDAs and "smartphones," are categorically different forms of technologies with different behavioral consequences. The ubiquitous nature of these mobile technologies in terms of being constantly within reach of the

Figure 3. Asia Pacific: Top 10 markets by customers (Source: From Cellular News, 2008. Retrieved from http://www.itu.int/ITU-D/ict/newslog/content/binary/27-04-2009.gif).

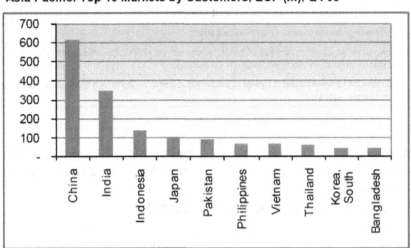

users and continuously connected to a broader communications network give them a unique status in the realm of technology support for education and learning.

Empirical studies have reported the advantages of using wireless technologies in learning environments, including supporting group work on projects, engaging learners in learning-related activities in diverse physical locations, and enhancing communication and collaborative learning in the classroom (Barker, Krull & Mallinson, 2005; Liu, Wang, Liang, Chan, Ko, & Yang, 2003)

While a great deal of studies had been done around the globe as aforementioned, however, none or practically only very few such study was done within Indonesian context, in relation to its pedagogical effectiveness in science/mathematics education. Therefore, this study was initiated to explore Indonesian mathematics and science teachers' views towards mobile learning with review of mobile phone evolution in Indonesia. It is hoped that the empirical findings will provide an overview of M-learning classroom practices that contributed to the intended formulation of information and communication technology (ICT) policy.

Mobile Phone Evolution and Pedagogical Effectiveness

Analysts and industry experts have previously predicted that the total number of cellular phone customers in Indonesia would climb to 120 million by the end of this year, from around 90 million at the end of 2007 (Reuters News, 2009). During 2008, Indonesia became the Asia Pacific region's third largest market with a total of 140.2 m connections (refer Figure 3) (Cellular News, 2008). The number of mobile phone users in Indonesia could more than double to 100 million by 2010 as a low penetration rate in the world's fourth most populous country offers huge potential. However, the mobile penetration rate in Indonesia is low compared with Malaysia, where around two-thirds of the population has access to mobile phones. Thailand also has a penetration rate of around 46% (The Edge Daily, 2006). -

The ultimate goal of this study is to see whether handheld device such as mobile phones

Table 1. Mobile phone evolution. Source: Soccio, D (2007). Victorian Adult Literacy and Basic Education Council (VALBEC) Conference. Retrieved from http://www.valbec.org.au/05/conf07/docs/Please_switch_your%20mobile_on.ppt.

Types	Description
1G	Used analogue technology for the transmission of voice traffic
2G	Encompasses competing standards or personal communications service (PCS). It uses digital technologies that enabled better audio quality in voice transmissions, increased capacity on networks
3G	Extension of the main 2.5 G standard and designed to enable rapid transmission of very large quantities of data, video clips and Multimedia Short Messages (MMS)

enhance teaching pedagogically and technically in the classroom activities. The first author tried with Nokia 6300 mobile device to show some applications on Science/Math Applications. The following Table 1 describes the evolution of mobile phones in general.

Handheld devices have the potential to effectively "push" and "pull" information and deliver learning whenever/wherever employee needs arise (Brown & Metcalf, 2008). Mobile devices themselves cannot be used effectively unless they meet pedagogical effectiveness criteria. According to Robson (2003), there are two basic requirements for the adoption of new technology by formal educational systems:

1. The technology must be pedagogically effective and viewed as an improvement
2. The technology must be available and accessible.

In mathematics and science, at least, there is widespread belief that microprocessor based technology is pedagogically effective and can lead to improvements in learning. Indeed, technology in the classroom has been *mandated* by curriculum standards for over a decade.

RESEARCH METHODOLOGY

This study employed questionnaire survey as main data collection method. The questionnaire was carefully developed and administered to the purposively chosen sample as described below.

Sample

A total of forty five science and mathematics teachers from Aceh, Indonesia with various years of science teaching experience and with at least a basic degree participated in this study. These respondents are the participants who attended a three months' in-service course on science and mathematics pedagogy. Prior to coming for this course, the majority had access to computers and were familiar in using it. During formal classes they had the opportunities to observe facilitators integrating ICT into their teaching mainly through presentation tools as well as exercises which involved the use of word processors to produce reports, presentation tool such as PowerPoint and exploration using the Internet. During the course they had been exposed to the use of mobile device as teaching tools in mathematics and science classroom.

Instruments

After the demonstration of using mobile device in teaching mathematics and science, a validated questionnaire *"My Perception on Mobile Learning"* (henceforth being referred to as MPML) adapted from Singh & Zaitun (2006) was administered. The questionnaire was intended to explore mathematics and science teachers' view

Table 2. Description of scales in MPML and representative items

Scale Name	Scale Description	Example of the item
Flexibility	Extent to which mobile learning device facilitates students learning in any condition	Learning can be done anytime and anywhere (*Item No.2*)
Effectiveness	Extent to which mobile learning helps the teachers to facilitate teaching	Students did not have to waste time copying what the instructor wrote on the whiteboard. (*Item No.12*)
Collaboration	Extent to which collaboration among stakeholders are needed to conduct mobile learning	Collaboration is a critical element to successful mobile devices integration in education. (*Item No.31*)
Restriction	Technical matter that limit teacher and students in doing mobile learning	Connectivity costs are kind of expensive for mobile devices (*Item No.18*)
Teacher Readiness	Extent to which teachers are able to engage in mobile learning classroom	Educators need to be trained on how to apply mobile devices in their practices (*Item No.29*)

towards the use of mobile devices in mobile learning classroom. The questionnaire consists of 36 items which required the respondents to respond to a 5-point Likert scales ranging from strongly disagree to strongly agree. The questionnaire is provided in the Appendix.

Data Collection Procedure

After experiencing the use of mobile device in science and mathematics classroom, the participants were asked to respond to the questionnaire that was put online [refer URL http://www.questionpro.com/akira/TakeSurvey?id=1185083]. The ethical concern of anonymity was abided by to ensure that the participants provided the most genuine response. The participants were told that the researchers were available if they have any question or clarification towards the items in the questionnaire. They were also informed that their responses will not affect their performance or grade in the course. The administration of questionnaire was followed up with some participants were randomly selected for interview to triangulate the findings.

Data Analysis

Data analysis included validation of the questionnaire with descriptive statistics on the analysis of responses for each scale in the questionnaire.

The Statistical Package for Social Science (SPSS) software version 13.0 was used to explore the internal consistency of scale and the reliability of questionnaire was followed up (see Table 3). Moreover, the teachers' views towards mobile learning, descriptive statistics via analysis of teachers' responses to the questionnaire about their views towards mobile learning was made. The average item mean, or the scale mean divided by the number of items in a scale and its standard deviation, was used as the basis of comparison among teachers' views on the items in different scales of the questionnaire.

FINDINGS AND DISCUSSION

Validity and Reliability of the Questionnaire

The 'My Perception on Mobile Learning' (MPML) questionnaire consists of 36 items which were unevenly distributed into five scales, namely *Flexibility* (a total of 7 items, i.e. No. 2, 3, 4, 5, 6, 26 and 35), *Effectiveness* (a total of 10 items, i.e. No. 7, 8, 9, 10, 12, 13, 20, 21, 23 and 24), *Collaboration* (a total of 5 items, i.e. No. 22, 26, 27, 30 and 31), *Restriction* (a total of 7 items, i.e. No. 16, 17, 18, 19, 32, 34 and 36), *and Teacher Readiness* (a total of 7 items, i.e. No. 1, 11, 14, 15, 28, 29, and 33) (Refer Appendix). The description

Table 3. Cronbach alpha (α) and mean correlation of scales of the MPML Questionnaire (N=45)

Scale	Number of Items	Criterion	
		Cronbach α	Mean Correlation
Flexibility	7	0.78	0.60
Effectiveness	10	0.78	0.45
Collaboration	5	0.72	0.49
Restriction	7	0.74	0.59
Teacher Readiness	7	0.69	0.57

Table 4. Teacher perceptions of mobile learning devices and classrooms (N=45)

No	Scale	Number of Items	Mean	SD
1.	*Flexibility*	7	4.07	0.45
2.	*Effectiveness*	10	3.82	0.47
3.	*Collaboration*	5	3.97	0.42
4.	*Restriction*	7	3.82	0.46
5.	*Teacher Readiness*	7	4.05	0.39

of these scales accompanied by examples of an item from each scale is provided in Table 2.

The criteria to ensure the validity and reliability of the questionnaire included scale internal consistency reliability, and factor structure. In this study, analysis was done to check whether or not each item in a scale of the questionnaire assessed a common construct. Therefore, Cronbach's alpha coefficient was calculated. On the whole, the statistics obtained were acceptable. Cronbach alpha coefficients of these five scales ranged from 0.69 (*Teacher Readiness*) to 0.78 (*Flexibility* and *Effectiveness*) (Table 3). These results suggest that the internal consistency for the MPML questionnaire is acceptable.

Furthermore, mean correlation of a scale with other four scales in the MPML questionnaire was calculated and used as the criterion of discriminant validity. The lower mean correlation infers greater discriminant validity for that scale. The results are presented in Table 3.

Table 3 shows that the discriminant validity, which measured the mean correlation of a scale with other four scales, ranged from 0.45 (*Effectiveness*) to 0.60 (*Flexibility*). These results suggest that the MPML used in this study possesses relatively satisfactory scales construct, although a degree of overlap still exists.

Teachers' Perceptions of Mobile Learning

To depict the teachers' perceptions of mobile learning, the average item mean (the scale mean divided by the number of items in that scale) and average item standard deviation of each scale were calculated and the results are shown in Table 4 below.

Generally, the teachers hold relatively positive view of mobile learning in which their average score for all scales ranged from 3.82 to 4.07 with standard deviation from 0.39 to 0.47.

With regard to the 'flexibility' of mobile learning classroom, teachers' perceptions towards the items in the questionnaire are in between agree and strongly agree (average mean score 4.07). It indicates that the teachers are in agreement that mobile learning will be able to provide students with learning activities regardless of time and place. Similarly, teachers also responded either 'agree or strongly agree' on their 'readiness' (average mean score of 4.05) which is important for the success of mobile learning delivery.

On the other hand, teachers are about in agreement on their perceptions towards the 'effectiveness' of using mobile learning for helping students, on 'collaboration' needed among stakeholder for the success of mobile learning delivery, and of 'restriction' they faced in implementing mobile devices in mobile learning classrooms.

RESEARCH IMPLICATIONS AND LIMITATIONS

The findings suggest a promising future and revealed the possibility on the use of mobile devices in delivering mobile learning for science and mathematics classroom in Indonesian context. At first, it was assumed that teachers' positives views on all scales in the questionnaire were due to the fact that the teachers felt refreshed with the opportunity to learn something new that they never knew and like to know about. It was observed that there were eagerness and enthusiasm during the presentation on mobile learning classroom by the facilitator as part of the course activities. However, this assumption was later rejected as it was found that the teachers hold positive perception and responded to the questionnaire as they had given their best knowledge as much as they could. Through interview we found that mostly participants were IT savvy, were familiar with mobile devices and did understand what mobile learning should be. It is obvious that they really knew the items in the questionnaire and had strong belief as what they

had expressed in the questionnaire. Therefore, this study implied that Indonesian science and mathematics teachers should be empowered to start implementing mobile learning to teach science and mathematics.

However, due to the fact that the survey was only conducted with samples from one country, the results of this study may not be generalizable to other nations. Time constraints faced during the conduct of this study within the one month in-service course also did not permit the researcher to conduct pilot testing in MPML in local schools before the real study although it is a validated instrument. Moreover, no follow up research activities or observational studies were made with the sample teachers participated in this study. Hence positive perception as evidenced from this case study does not necessarily guarantee the real practices.

FUTURE TRENDS

The National Information and Communication Technology Council (DeTIKNas) recently announced its plan to accelerate the implementation of Indonesia's ICT Blueprint. The announcement was part of the government's commitment to eliminate digital gap and build an e-Indonesia through the development of telecommunication network in 43,000 villages, 31,173 junior and senior high schools, 2,428 universities and 28,504 health centers in Indonesia by 2025 (ICT-Indo, 2007). As revealed from the literature, a backbone network will be developed across Indonesia, connecting 440 cities, in all 33 provinces of Indonesia. It consists of about 35,000 kilometers of submarine fiber optic cables. Indonesia also seeks to promote Asia regional interests in international and global forums, not to be merely target markets for every model and technology associated with the implementation of e-government (Sofyan, 2007 in Obi, 2007).

Table 5. Asia mobile penetration rates (2005-2010): Business Monitor International (BMI)'s report on 14 key Asia market. [Source: Cellular-News (2007). Retrieved from http://www.cellular-news.com/ story/17162.php].

	Mobile Penetration (2005)	BMI Forecast Mobile Penetration (2010)	Forecast Average Annual Growth
Hong Kong	118.5%	111.9%	-0.3%
Singapore	97.7%	100.7%	2.1%
Australia	96.1%	100.6%	2.0%
Taiwan	92.4%	92.6%	1.3%
Korea	79.1%	85.9%	2.1%
Malaysia	74.1%	91.2%	5.8%
Japan	70.3%	85.6%	4.2%
Thailand	46.9%	78.5%	14.4%
Philippines	42.7%	75.2%	16.9%
China	30.2%	58.9%	21.9%
Indonesia	**22.3%**	**41.5%**	**19.5%**
Pakistan	14.1%	37.7%	38.5%
Vietnam	10.3%	35.9%	62.3%
India	7.0%	32.8%	80.1%

According to Goswami (2007), for the year 2006, we may see an investment of more than $2.5 billion dollars made in the mobile infrastructure as the existing operators gear up to face the challenge from Hutch and Maxis who are rapidly rolling out their infrastructure. Literature revealed that more than 20 million Indonesians now own mobile phones, roughly 8 percent of the country's 220 million population. Research and Markets expects the country to have 60 million mobile subscribers by 2007 (The Gale Group, 2008). The recent report by Business Monitor International (BMI) on 14 key Asia market (refer Table 5) (Cellular-News, 2007) also revealed that mobile phone penetration in Indonesia was 22.3% in 2005. According to the BMI Forecast mobile penetration in the aforementioned country by 2010 is 41.5%. This means there is an increase of about 19.5% forecast average annual growth of mobile penetration in Indonesia. The implication is that there will be higher opportunity of mobile

technologies to be used in educational settings by wider population in future.

In view of the increasing attention and interest in mobile technologies, especially the trends to implement mobile learning in science and mathematics classrooms, it is imperative that some studies could be made with formulation of ICT policy to ensure effective implementation of mobile learning in science and mathematics classrooms. The future research that may be worth of doing can be focused on the investigation of how and to what extent these teachers use mobile devices in teaching science and mathematics through mobile learning in classroom.

IMPLICATIONS FOR ICT POLICY FORMULATION

As revealed in Table 4 of the research findings, though the Indonesian teachers hold positive view of mobile learning, their average scores for 'Effectiveness, Collaboration and Restriction' scales ranged from 3.82 to 3.97, which were less than the average scores for 'Flexibility and Teacher Readiness' scales that are 4.07 and 4.05 respectively. This means that some forms of measures and efforts need to be made so that teachers are more confident and can work 'collaboratively' (in terms of e.g. communication, networking, accommodating diverse learner-centred environment, stakeholders' collaboration and sharing) to introduce mobile learning. These types of learning should also be 'effectively' (in terms of e.g. feedback, assessment, experimentation, constructive learning support, balanced intellectual development) implemented in the classrooms with less 'restriction' (in terms of e.g. cost, functionality, potential to expand, cost effectiveness) (refer Appendix for the constructs identified for the questionnaire items).

The research implications show that some forms of ICT policy should be formulated. To enable the teachers adopt and implement mobile learning as the perceived, both necessary technical and pedagogical support need to be provided. Other educational stakeholders such as school principals, curriculum developer, education officers and mobile technologies company also need adequate support so that they can create conducive environment for teacher to implement mobile learning in classroom. Some forms of control need to be made to types of mobile technologies that could be implemented in classrooms with students of diverse education and sociocultural background. For example, the simpler and low cost mobile phone with no camera option may be allowed for use in schools as mobile learning is also feasible without camera. School management with financial viability should be encouraged to provide at least 10 low cost mobile phones per classroom. The following are some more suggestions that should also be considered when considering formulating ICT policy for mobile learning:

- Due to various ethical issues, mobile technology such as mobile phone was actually banned in secondary schools of some countries. Hence, there is a need to take into consideration the ethical issues of using mobile technologies such as mobile phone in classrooms so that students do not misuse its application to make unethical video recordings that create controversial issues if it is to be implemented at various levels of educational settings.
- Other management of learning via mobile devices or technologies may be considered, including sending students progress reports, parents-teachers meeting announcements.
- The integration of the existing independent system within the government by implementing E-government as has been promoted by Indonesia also need to be considered with some forms of policy guidelines formulated. Highly committed leadership with technical expertise to combine systems into an integrated process is essential. Effectiveness and transparency of E-government processes is also very important (Sofyan, 2007 in Obi, 2007).

CONCLUSION

This study was designed to tap Indonesian science and mathematics teachers' perception towards mobile learning classroom. Using online questionnaire, the data was collected and analysed using descriptive statistics with reliability and validity tests. The findings from the empirical case study revealed that generally Indonesian teachers hold positive views towards mobile learning in the science and mathematics classroom in terms of 'flexibility, effectiveness, collaboration, restriction and readiness'. The research evidence of the value of ICT in education also serves as empirical basis for the formulation of ICT policy in educational institutions, particularly ICT integration in instruction as aforementioned. It is hoped that this Chapter has initiated discussions towards empowering more science and mathematics teachers to use mobile devices in classroom environment.

REFERENCES

Alexander, B. (2004). Going nomadic: Mobile learning in higher education. *EDUCAUSE Review*, *39*(5), 29–35.

Cantoni & C. McLoughlin (Ed.). *Proceedings of World Conference on Educational Multimedia, Hypermedia and Telecommunications2004*, (pp. 4729-4736).

Csete, J., Wong, Y., & Vogel, D. (2004). Mobile devices in and out of the classroom. In L.

Divjak, S. (2008). Mobile phones in the classroom. *International Journal on Hands-on Science*, *1*(2), 92–94.

Hoppe, H. U., Joiner, R., Milrad, M., & Sharples, M. (2003). Guest editorial: Wireless and mobile technology in education. *Journal of Computer Assisted Learning*, *19*(3), 255–259. doi:10.1046/j.0266-4909.2003.00027.x

Kukulska-Hulme, A. (2008). Mobile Usability in Educational Contexts: What have we learnt? *International Review of Research in Open and Distance Learning*, *8*(2), 1–15.

Kukulska-Hulme, A., Sharples, M., Milrad, M., Arnedillo-Sánchez, I., & Vavoula, G. (2009). Innovation in mobile learning: A European perspective. *International Journal of Mobile and Blended Learning*, *1*(1), 13–35.

Lindquist, D., Denning, T., Kelly, M., Malani, R., Griswold, W. G., & Simon, B. (2007). Exploring the potential of mobile phones for active learning in the classroom. *ACM SIGCSE Bulletin*, *39*(1), 384–388. doi:10.1145/1227504.1227445

Liu, T. C. (2007). Teaching in a wireless learning environment: A case study. *Journal of Educational Technology & Society*, *10*(1), 107–123.

Liu, T. C., Wang, H., Liang, T., Chan, T., Ko, W., & Yang, J. (2003). Wireless and mobile technologies to enhance teaching and learning. *Journal of Computer Assisted Learning*, *19*(3), 371–382. doi:10.1046/j.0266-4909.2003.00038.x

Massey, A. P., Ramesh, V., & Khatri, V. (2006). Design, development and assessment of mobile applications: The case for Problem-based Learning. *IEEE Transactions on Education*, *49*(2), 183–192. doi:10.1109/TE.2006.875700

Mcconatha, D., Praul, M., & Lynch, M.J. (2008). Mobile learning in higher education: An empirical assessment of a new educational tool. *The Turkish Online Journal of Educational Technology (TOJET)*, *7*(3).

Metcalf, D. (2005). *mLearning: Mobile learning and performance in the palm of your hand.* Amherst, MA: HRD Press.

Montero, F., Córcoles, J. E., & Calero, C. (2006). *Handhelds and mobile phones to manage students and resources in classroom: A new handicap to the teacher?* (pp.876-880).

Naismith, L., et al. (2005). *Report 11: Literature review in mobile technologies and learning. Future lab series.* Bristol, U.K.

Norris, C., & Soloway, E. (2004). Envisioning the handheld-centric classroom. *Journal of Educational Computing Research*, *30*(4), 281–294. doi:10.2190/MBPJ-L35D-C4K6-AQB8

Pearson, J. (2004). Current policy priorities in information and communication technologies in education. In Trinidad, S., & Pearson, J. (Eds.), *Using information and communication technologies in education*. Singapore: Pearson Prentice Hall.

Roschelle, J., & Pea, R. (2002). A walk on the wild side: How wireless handhelds may change computer-support collaborative learning. *International Journal of Cognitive Technology, 1*(1), 145–168. doi:10.1075/ijct.1.1.09ros

Saipunidzam, M., Mohammad, N. I., Mohamad, I. A. M. F., & Shakirah, M. T. (2008). Open source implementation of M-learning for primary school in Malaysia. In *Proceedings Of World Academy Of Science, Engineering And Technology* (pp.752-756). 34.

Singh, D., & Zaitun, A. B. (2006). Mobile learning in wireless classrooms. [MOJIT]. *Malaysian Online Journal of Instructional Technology, 3*(2), 26–42.

Sofyan, D. (2007). ICT policy in Indonesia . In Obi, T. (Ed.), *E-governance: A global perspective on a new paradigm* (pp. 48–50). Amsterdam: IOS Press.

Sribhadung, P. (2006). Mobile devices in elearning. In *Proceedings of the Third International Conference on eLearning for Knowledge-Based Society* (pp.35.1-35.4), Bangkok, Thailand.

Sung, M., Gips, J., Eagle, N., Madan, A., Caneel, R., & DeVaul, R. (2005). Mobile-IT education (MIT.EDU): M-learning applications for classroom settings. Blackwell Publishing Limited. *Journal of Computer Assisted Learning, 21*, 229–237. doi:10.1111/j.1365-2729.2005.00130.x

Vavoula, G. N., & Sharples, M. (2002). KleOS: A personal, mobile, knowledge and learning organisation system. In M Milrad, HU Hoppe & Kinshuk (Eds), *IEEE International Workshop on Wireless and Mobile Technologies in Education* (pp.152–156). Los Alamitos, CA: IEEE Computer Society.

Vogel, D., Kennedy, D. M., Kuan, K., Kwok, R., & La, J. (2007). Do mobile device applications affect learning? In *Proceedings of the 40TH Annual Hawaii Conference on Systems Sciences*, 1-4.

Wagner, E. D. (2005). Enabling mobile learning. *EDUCAUSE Review, 40*(3), 40–53.

Wei, J., & Lin, B. (2008). Development of a value increasing model for mobile learning. In *Proceedings of the 39th Annual Meeting of the Decision Sciences Institute* (pp. 5251-5256).

Zhao, X., & Okamoto, T. (2008). A personalized mobile mathematics tutoring system for primary education. *The Journal of the Research Center for Educational Technology, 4*(1), 61–67.

Zurita, G., & Nussbaum, M. (2004). A constructivist mobile learning environment supported by a wireless handheld network. *Journal of Computer Assisted Learning, 20*(4), 235–243. doi:10.1111/j.1365-2729.2004.00089.x

WEBSITES

Barker, A., Krull, G., & Mallinson, B. (2005). A proposed theoretical model for M-learning adoption in developing countries. *Proceedings of the 4th World Conference on Mobile Learning*. Retrieved March 29, 2009 from http://www.mlearn.org.za/CD/papers/Barker.pdf

Brown, J., & Metcalf, D. (2008). *Mobile learning update*. Masie Center's Learning Consortium. Retrieved March 20, 2009 from http://www.masie.com

Cellular-News. (2007). *Asia phone penetration to reach 50% by 2010 – Report*. Retrieved October 9, 2009 from http://www.cellular-news.com/story/17162.php

Cellular-News. (2008). *Asia Pacific: Top 10 markets by customers*. Retrieved October 12, 2009 from http://www.itu.int/ITU-D/ict/newslog/content/binary/27-04-2009.gif

Goswami, D. (2007). *Evaluating ICT policy in Indonesia: Interview with LIRNEasia researcher*. Retrieved October 9, 2009 from http://lirneasia.net/2007/01/evaluating-ict-policy-in-indonesia-interview-with-lirneasia-researcher/

ICT-Indo. (2007). *ICT2007 Indonesia earmarked as a DeTIKNas flagship event to set ICT milestones for Indonesia*. Retrieved October 12, 2009 from http://www.ictindonesia.com/2008/PressRelease-Feb07.pdf

Kearns, P. (2002). *Towards the connected learning society. An overview of trends in policy for information and communication technology in education*. Canberra: Department of Education, Science and Training. Retrieved October 6, 2009 from http://www.dest.gov.au/sectors/higher_education/publications_resources/summaries_brochures/towards_the_connected_learning_society.htm

Merrill Lynch. *Bullish on mobile learning*. Retrieved March 12, 2009, from http://www.clomedia.com/index.php?pt=a&aid=2135

mLearnopedia. Retrieved March 12, 2009 from http://mlearnopedia.com

Mobile learning at IADT Online. Retrieved January 16, 2009, from http://online.academy.edu/iadtmobile/

Mobilearn (2003). *The Mobilearn project vision*. Retrieved 20 March 2009, from http://www.mobilearn.org/vision/visiton.htm

Quinn, C. (2000). *mLearning*. Mobile, Wireless, In-Your-Pocket Learning. Linezine. Fall 2000. Retrieved April 29, 2009 from http://www.linezine.com/2.1/features/cqmmwiyp.htm.

Retrieved June 11, 2009 from http://www.eduworks.com/ Documents/Publications/Mobile_Learning_Handheld_Classroom.pdf Samarajiva, R., & Zainudeen, A. (2008). Does regulation stifle or enable ICT connectivity? In R. Samarajiva, & A., Zainudeen (Eds), *ICT infrastructure in emerging Asia: Policy and regulatory roadblocks*. International Development Research Centre (IDRC), Sage Publications India Pvt Ltd. Retrieved October 12, 2009 from http://www.idrc.ca/openebooks/378-2/

Reuters News. (2009). Retrieved 20 March 2009, from http://www.reuters.com/article/rbssTech-MediaTelecomNews/idUSJAK10826820080606

Robson, R. (2003). *Mobile learning and handheld devices in the classroom.*

Silvernail, D. L., & Lane, D. M. M. (2004). *The impact of Maine's one-to-one laptop program on middle school teachers and students*. Maine Education Policy Research Institution, University of Southern Main. Retrieved October 6, 2009 from http://maine.gov/mlti/articles/research/MLTIPhaseOneEvaluationReport2004.pdf

Staudt, C. (2001). *Handheld computers in education*. Concord, MA: The Concord Consortium. Retrieved July 11, 2009 from http://playspace.concord.org/Documents/Learning from Handhelds.pdf

The Edge Daily. (2006) (Malaysia Online Newspaper) Retrieved March 12, 2009 from http://www.indonesia-ottawa.org/information/details.php?type=news_copy&id=2310

The Gale Group. (2008). *Indonesia's mobile penetration to triple by 2007, according to Research and Markets*. Retrieved October 9, 2009 from http://goliath.ecnext.com/coms2/gi_0199-2376036/Indonesia-s-mobile-penetration-to.html

Wei, J., & Lin, B. (2008). Development of a value increasing model for mobile learning. *Proceedings of 39th Annual Meeting of the Decision Sciences Institute*. Retrieved June 11, 2009 from www.decisionsciences.org/Proceedings/DSI2008/docs/525-2426.pdf

KEY TERMS AND DEFINITIONS

1G: First Generation mobile phones

BMI: Business Monitor International

E-Government: Electronic Government

ICT: Information and Communication Technology

MMS: Multimedia Short Messages

MPML: 'My Perception on Mobile Learning' (survey questionnaire)

M-learning: Mobile learning

M-technologies: Mobile technologies

PCS: Personal Communications Service

PDA: Personal Digital Assistant

Smartphones: Mobile phones offering advanced capabilities, often with PC-like functionality

SPSS: Statistical Package for Social Science

APPENDIX

Items in 'My Perceptions on Mobile Learning' (MPML) Questionnaire [An instrument adapted from Devinder Singh & Zaitun A.B (2006)]

No.	Questionnaire Items in MPML Questionnaire	Construct
1	Instructors can incorporate multimedia demonstrations in their lecturers and receive real-time feedback from their students using quizzes or surveys.	*Readiness*
2	Learning can be done anytime and anywhere. It supports continuous learning.	*Flexibility*
3	Able to collaborate with instructor's notebook during class.	*Flexibility*
4	Communication and teaching support while outside the classroom.	*Flexibility*
5	Mobile learning is able to synchronous team member's appointments and schedules.	*Flexibility*
6	Classroom seating does not have to have a fixed seating arrangement.	*Flexibility*
7	An instructor can get immediate feedback on the lesson being taught.	*Effectiveness*
8	Student's can be assessed on multiple choices, true/false questions in the classroom.	*Effectiveness*
9	An instructor can get immediate feedback on the lesson being taught.	*Effectiveness*
10	Real-time experiments can take place in classrooms.	*Effectiveness*
11	Instructors can provide examples such as simulations and web based documents that can be accessed at specific time to improve retention.	*Readiness*
12	Students did not have to waste time copying what the instructor wrote on the whiteboard.	*Effectiveness*
13	Students can be used as soon as they are turned on the device and don't have to be booted up.	*Effectiveness*
14	(Many, but not all) significantly cheaper than desktops or portable machines.	*Readiness*
15	Closed to youth lifestyle – part of their social and cultural life.	*Readiness*
16	Their functionality is limited compared to more powerful devices.	*Restriction*
17	They have limited potential for expansion and upgrade.	*Restriction*
18	Connectivity costs are kind of expensive for mobile devices.	*Restriction*
19	Concerns about security issues.	*Restriction*
20	It is considered as a learning supportive tool.	*Effectiveness*
21	Students are allowed the recording and maintenance of the lessons takes place.	*Effectiveness*
22	Mobile devices facilitate communication between faculty members and students through file sharing capabilities.	*Collaboration*
23	Mobile devices can be used as instructional tools to constructive learning.	*Effectiveness*
24	Mobile devices can be treated as tools that help students execute their tasks and promote the balanced development of their mental abilities by functioning as intellectual partners to the instructor and the learner.	*Effectiveness*
25	Activities within the curriculum can be designed to take place in classroom (deskwork) or mainly outside the classroom (fieldwork.).	*Flexibility*
26	There is a need for a shared, progressive pedagogy for mobile learning that will provide the scientific basis for networked and collaborative learning in both a virtual and a virtual-augmented environment.	*Collaboration*
27	It must accommodate different teacher- and learner perspectives, promote learner-centered environments and collaboration among learners and between learners and educators.	*Collaboration*
28	Educators need to be trained on how to apply mobile devices in their practices.	*Readiness*
29	In order to minimize that risk and increase the success probabilities, it is important to be proactive and apply a systemic, holistic approach to mobile technology integration.	*Readiness*
30	Collaboration (i.e. various stakeholders such as educators, students/ learners, computer scientists and engineers) is a critical element to successful mobile devices integration in education.	*Collaboration*

No.	Questionnaire Items in MPML Questionnaire	Construct
31	Stakeholders (i.e. Various Collaborators) need to communicate, coordinate their actions, transfer and share their knowledge and experiences, as well as align their needs and goals.	*Collaboration*
32	The screen size is one of the issue is mobile learning.	*Restriction*
33	The cost of the mobile is one of the issue is mobile learning.	*Readiness*
34	The battery life of the mobile is one of the issue is mobile learning.	*Restriction*
35	One can bring Science/Math Contents and Concepts beyond the classroom.	*Flexibility*
36	Mobile Bandwidth may degrade with large number of users.	*Restriction*

Chapter 4

Analysing an ICT4D Project in India Using the Capability Approach and a Virtuous Spiral Framework

Helena Grunfeld
Victoria University, Australia

Sriram Guddireddigari
Monash University, Australia

Benita Marian
The East West Foundation of India, India

John Peter
The East West Foundation of India, India

Vijay Kumar
The East West Foundation of India, India

ABSTRACT

The field research covered in this chapter represents the first wave of a longitudinal study, aimed at testing a framework for evaluating the contribution to capabilities, empowerment and sustainability of information and communication technology for development (ICT4D) projects. Key features of the framework are: it is conceptually informed by Amartya Sen's capability approach (CA), uses a participatory methodology and longitudinal timeframe, and considers the micro-, meso-, and macro- levels in understanding the role of ICT in development. Despite the longitudinal nature of the framework, each stage of the research is designed to be a case study in its own right. The research, conducted at a computer centre in the Indian state of Tamil Nadu, centred on the perception of participants with respect to whether the centre had played a role in any improvements in the community and whether they could see a role for it in changes they would like to see, or aspirations they may have for their communities.

DOI: 10.4018/978-1-61692-012-8.ch004

A key finding of the field research was that participants valued the centre mainly for its contribution to education of their children. Education was appreciated beyond its instrumental utility and included intrinsic value, i.e. value that exceeds its potential as a path to higher incomes. Participants frequently referred to how a higher level of literacy would empower them to deal with government officials without intermediaries. This finding is consistent with the CA's emphasis on development as a process facilitating capabilities that enable people to lead lives they have reason to value.

INTRODUCTION

When allocating scarce development resources, governments like to see hard evidence of the benefits. In the absence of such evidence, there is a risk that decision-makers misallocate resources, either through ineffective ICT deployments or no deployments. An appreciation of the environment in which ICT infrastructure is considered or deployed will "make us wary of blindly following the technological imperative and alert to situations where there is a trade-off between efficiency and human well-being" (Sawhney, 1996, p.311). It will also contribute to awareness of many benefits of ICT that are not necessarily quantifiable in economic terms, e.g. in the exchange of ideas and for governance processes.

As recognised by many researchers in this field, research aimed at understanding environments in which ICTs have been deployed and their impacts has not kept pace with the significant investments in ICT4D initiatives (e. g. Alampay, 2006a; Batchelor & Norrish, 2004; Gagliardone, 2005; Harris & Rajora, 2006; Hudson, 2006; Nielsen & Heffernan, 2006; O'Neil, 2002; Sciadas, (Ed.) 2005; Souter, Scott, Garforth, Jain, Mascarenhas, & McKemey, 2005; Torero & von Braun, 2006; Warschauer 2003). Torero & von Braun (2006) recommended investigations of the conditions required for ICT to contribute positively to sustainable development. Gagliardone (2005) argued that problems arise when localised experiences are scaled and identified the absence of an innovative culture, capabilities and links between ICT enclaves and the rest of society as factors impeding the beneficial use of ICT. Noting that ICT can contribute to

inequalities, some authors have called for further research to improve knowledge of this aspect of ICT4D (Forestier, Grace, & Kenny, 2002; Kumar & Best, 2006; Souter, Scott, Garforth, Jain, Mascarenhas, & McKemey, 2005; Torero & von Braun, 2006; van Dijk & Hacker, 2003).

ICT4D project evaluations exhibit diversity in frameworks, methodologies, methods and focus. They can be analytical, descriptive and/or prescriptive. Case studies represent a common approach (e.g. Batchelor & Sugden, 2003; Evans & Ninole, 2004; Falch & Anyimadub, 2003; Harris, 2001; Meera, Jhamtani, & Rao 2004; Overå, 2006; Talyarkhan, Grimshaw, & Lowe, 2005) and, apart from a few macro-level studies, they provide much of the evidence of the benefits of telecommunications in rural development (Hudson, 2006).

Whereas case studies indicate the importance of telecommunications in different sectors, such as agriculture, education and health, and functions such as marketing, they do not in general include any systematic analysis and are not undertaken within a specific theoretical framework. Some of the case studies include assertions based on varying levels of analysis, concluding with recommendations for authorities and other implementers of future projects. In commenting on the inadequate theoretical depth in ICT4D research, Heeks (2006), noted that while there has been reasonable theoretical underpinnings related to the first three letters of the ICT4D acronym: 'I' (library and information sciences), 'C (communication studies), and 'T' (information systems), this is not the case for 'D' (development studies), which in his view have been meagre.

The case study presented in this chapter represents the initial field research in a longitudinal study aimed at providing an input into the debate on suitable frameworks for understanding how ICT can contribute to capabilities, empowerment and sustainability from a development perspective. Despite the longitudinal nature of the framework, each stage of the research is designed to be a case study in its own right and be useful on a standalone basis, as it enables others to learn as the study progresses, even without an opportunity of reading results from subsequent research phases.

The conceptual framework is outlined in the next section, which includes an introduction to the capability approach (CA), as this has informed the conceptual framework. The case study includes sections on the micro-, meso-, and macro environments, methodology and key findings of the field research data, which was analysed through the lens of the conceptual framework.

BACKGROUND AND CONCEPTUAL FRAMEWORK

This chapter presents and tests a conceptual framework, designed to understand how and under what circumstances an ICT4D project can contribute to capabilities, empowerment and sustainability, taking into account factors that facilitate and inhibit their achievement.

The conceptual framework, the capabilities, empowerment, and sustainability virtuous spiral framework (CESVSF) is informed by the CA and applies a forward-looking longitudinal perspective to the micro-, meso- and macro-levels, using participatory methodologies. It recognises that many impacts are not direct, but are similar in nature to the concept of spillovers in economics, i.e. their influences may extend in unpredictable ways, even to those not directly involved in an activity.

"Static, one-shot, cross-sectional studies" identified by Orlikowski & Baroudi (2002, p.54) as the main form of IS research, are also predominant in

ICT4D, where most evaluations are undertaken upon the completion of a project. Although it may take some time before new technologies are accepted at a community level and even longer for them to have an impact, longitudinal studies are sparse (Gaved & Anderson, 2006). The study by Ramirez (2001) on community-based networks in Canada is one of the exceptions.

The micro-, meso- and macro-levels refer primarily, but not exclusively, to the geographic dimension. The micro-level is the smallest unit under consideration. In this case, it comprises the Vicki Standish e-Education Centre (VSeEC), the focus of this study, villages in its catchment area, and possibly the next institutional layers above, the panchayat. The central government, with its policies and practices are at the macro-level. The boundaries between these two and the meso-level are more difficult to define, as there are several nested hierarchies within this somewhat simplified three-tier scale. In addition to the central government, the state government can be considered to operate at the macro-level, although it has some meso-level characteristics, as do mediating organisations, such as NGOs. The boundary of the meso- level extends to the micro-level, in the geographic domain. It is not the exact definition of these layers that is of relevance in this context, but rather an understanding of the dependencies and information flows between them and the impact they have, can, or should have on each other. Analysis of these interdependencies is akin to systems theory, an approach advocated by Ramirez (2003) and Andrew & Petkov (2003) to enhance understanding of ICT and its contexts, particularly in rural environments.

In a conceptual dimension, the three tiers can be thought of in terms of the extent to which it is possible to generalise. From this and the geographic perspective, the meso-level is useful when considering scalability and replicability, as it is "less sweeping than macro concepts, without claiming that everything is different" (Bebbington, 2004, p. 348), i.e. it is expected that some degree of generalisation will be possible. Focussing on either the macro-level with studies on the relationship between ICT, the policy and regulatory environments and/or economic growth, or on implementation of specific projects at the micro-level, most ICT4D research overlooks the meso-level, despite the likelihood that it is at this level service provision can be most responsive. In addition to government authorities, this level is inhabited by infomediaries and a range of organisations that can facilitate effective use of ICT (Duncombe, 2006; Ramirez, 2001).

In terms of ICT deployments, there are several illustrations from India of three-tier ICT4D initiatives. The three-tier n-Logue (Jhunjhunwala, Ramachandran, & Bandyopadhyay, 2004) structure and the model used by the Informatics unit of the M.S. Swaminathan Research Foundation (MS-SRF) illustrate the diversity in the application of the three levels, with private sector involvement in n-Logue and a community based model in MSSRF. At the time of the study, MSSRF, headquartered in Chennai, had 18 Village Resource Centres (VCRs) spread across several states, supporting 94 Village Knowledge Centres with functions such as technical assistance, information gathering and coordination of training. A three-tier approach is not a guarantee for success, as indicated by Jain & Raghuram (2005) in their study of Community Information Centres in Nagaland, which, despite the involvement by central, state and local authorities, did not accomplish their objectives.

In contrast to classical economic theory, where motivations and perceptions of participants are not relevant, these often play a major role in the CA and in participatory frameworks (Anyaegbunam, Mefalopulos, & Moetsabi, 1999; Ramirez 2001; Robeyns, 2001), where local participation in assessing how ICT can contribute to capabilities would be important. The CESVSF envisages that participants would identify basic capabilities that are common to different places and cultures, such as literacy and employment as well as capabilities that are specific to each environment. Some capa-

bilities assumed to be essential for development and likely to be required for and facilitated by a sustainable ICT infrastructure are illustrated in Figure 1, which depicts a 'virtuous spiral' that is expected to emerge for appropriately implemented ICT4D projects.

The CESVSF assumes that a minimum set of capabilities are required to establish, manage and use a basic ICT infrastructure - shown as 'obtain initial funding' in Figure 1. Alternatively, this function can be performed by a mediating organisation such as the VSeEC, in this case study. Using a basic level of IT artfulness, individuals can gain confidence and increase the control over their own lives (Corea, 2007). This in turn can strengthen the IT infrastructure in their communities, whether in the form of better skills, equipment and/or services. With each cycle of this spiral, there are new insights and improvements in capabilities, which strengthen community members and expose them to new realities that can improve their livelihood conditions to enable them to do and to be what they have 'reason to value' in a sustainable way.

The Capability Approach

"Instead of asking about people's satisfactions, or how much in the way of resources they are able to command, we ask, instead, about what they are actually able to do or to be" (Nussbaum, 2000, p. 12). This question is central to the CA and stands in contrast to questions about utility, preference satisfaction, and/or access to resources, indicators that characterise the utilitarian and welfare approaches to development.

The seminal book by the 1998 Nobel Laureate in Economics, Sen (2001), 'Development as Freedom' ('DAF'), first published in 1999, was the culmination of considerable work carried out by Sen at least since the 1980s to develop a framework for development that is grounded in human development as an alternative to the prevailing focus on economic development. Oth-

Figure 1 The CES virtuous spiral conceptual framework (Source: Grunfeld, 2007)

ers (e.g. Alkire, 2005; Comin, 2001; Corbridge, 2002; Gasper, 1997; Robeyns, 2001; Stewart (2005), Stewart & Deneulin, 2002; Nussbaum, 2000, 2006) have contributed to the development of this framework both before and following the publication of DAF. The versatility of the CA is illustrated by the wide range of research to which it has been applied, e.g definition by children of their capabilities in an endeavour to understand appropriate dimensions of children's well-being (Biggeri, et al. 2006), analysis of poverty alleviation programmes in New Zealand and Samoa (Schischka, Dalziel & Saunders (2008) and addressing a river water dispute between different Indian states (Anand, 2007).

Rather than focussing on economic growth and income to evaluate outcomes of development initiatives, adherents of the CA have argued in favour of using capabilities of individuals to 'lead the lives they have reason to value', as an informational base for evaluations. This approach has had considerable influence on welfare and development economics and is reflected in the UNDP human development index (Sen, 2000).

At the heart of the CA lies the importance of the "expansion of freedom … both as the primary end and as the principal means of development" (Sen, 2001:xii). Development is considered to be an extension of freedoms, which are viewed as the basic building blocks to development, as well as "the expansion of 'capabilities' of persons to lead the kinds of lives they value — and have reason to value" (Sen, 2001, p.18). This focus on freedom, which distinguishes the CA from frameworks advocating growth at any price, including doctrines justifying that the end justifies the means, does not mean that economic variables such as income and access to commodities are considered irrelevant. They are, however, inadequate for measuring quality of life and livelihoods. In the CA framework, certain political and social freedoms, such as the freedom to participate in political activities and to receive basic education are considered to be constitutive of development (i.e. they are relevant whether or not they contribute to economic development and/or growth). Certain capabilities are required to achieve and enjoy freedom. There is also a link between capabilities and human equality, as discrimination of any kind is considered a "failure of associational capability" (Nussbaum, 2000, p. 86).

As individuals, according to the CA, are responsible for their own well-being, it is up to them to decide which capabilities are important to them and how, subject to external constraints, these should be translated into functionings. These describe what a person is actually doing with his or her capabilities or the state of being resulting from this.

As many who have tried to effect change in a community would be able to attest to, there are many barriers, of an institutional nature, between somebody's capability of doing something and actual achievements. Considering the CA to be too focused on individualism, Stewart & Deneulin (2002) suggested an extension of the concept to include "valuable structures of living together" (p.68), i.e. structures that can impact positively on people's well-being. They specifically enumerated "functional families, cooperative and high-trust societies and social contexts" (p. 68) and argued that this inclusion would be important from a policy and research perspective. From the policy angle, more attention would be placed on structures that facilitate or inhibit development.

This critique of the CA did not take sufficient account of the CA's recognition of external constraints, including institutional constraints, involved in obtaining and using capabilities. Furthermore, the CA recognises reciprocity between individuals and institutions in that a person's capabilities not only depend on social arrangements and institutions but also influence others, as described by Sen (1985a): "Given the intrinsic importance of well-being, and indeed of agency, it is not credible that a person can morally evaluate his or her actions without taking note of their effects on the well-being and agency aspects of others (including their well-being freedom and agency freedom)" (p.216). This means that the CA accounts for impacts at a wider community level, without defining geographic or other limitations in the definition of a community.

Another way of illustrating the importance attached to institutions in the CA framework is the recognition that, although poverty "entails a lack of basic capabilities to lead full, creative lives" (UNDP 2003, p.27), capabilities represent only one of four dimensions identified by Sen (2001) as essential for poverty alleviation. The others are: opportunity (access to markets and employment), security/vulnerability to economic risk and to all forms of violence, and empowerment, external to as well as within households, all of which depend on institutional frameworks at the macro-, meso-, and micro-levels. The importance of these domains played out is several ways in the research, whether in the form of absence of a broadband infrastructure, potential security issues when accessing VSeEC after dark, and education as a tool for empowerment.

Developed as a critique against the more prevalent utilitarian approach to evaluation and with its emphasis on the importance of capabilities as the basis for evaluations, Comin (2001) described the CA as "a framework for evaluating and assessing social arrangements, standards of living, inequality, poverty, justice, quality of life or well-being" (p. 4). However, a main difficulty of applying this framework is the lack of operationalisation of the concept (Alampay, 2006a; Comin, 2001; Gasper, 2002).

One way of overcoming the lack of operationalisation, thereby making the concept more useful to development policy could be to list basic capabilities. Nussbaum has been a vocal advocate of doing this and has also developed a tentative list, which she admits must be subject to review over time and in different contexts. Despite insisting that capabilities should be formulated through democratic processes, in his practical work, Sen has nevertheless assumed that there would always be democratic support for capabilities of being healthy, well nourished, and educated: "expansion of health care, education, social security, etc., contributes directly to the quality of life and to its flourishing" (Sen, 2001, p. 144). Sen has also recognised the role that ICT can play in contributing to these basic capabilities.

ICT4D and the Capability Approach

In the CA, access to physical ICT infrastructure would not be a sufficient determinant of how individual preferences, capabilities and choice would influence the use of and benefits derived from it. Inherent in the CA is a reciprocal relationship between ICT and capabilities in that individuals require certain capabilities to be able to benefit from ICT, which in turn facilitates the free flow of information: vital to democratic freedom. As expressed by Sen (2005), "… access to the web and the freedom of general communication has become a very important capability that is of interest and relevance to all Indians" (p.160). 'Access' to physical infrastructure alone is not sufficient and the concept of access must be extended to include capabilities, (e.g. [computer] literacy), to actually use the infrastructure.

Several studies have applied the CA to ICT, or at least referred to a relationship between these (Alampay 2006a, 2006b; Barja & Gigler, 2005; Byrne & Sahay, 2007; Garnham, 1999; Gigler, 2004; Madon, 2004; James, 2006; Mansell, 2006; Musa, 2006; Thomas & Parayil, 2008; Walsham & Sahay, 2006; Warschauer, 2003, Zheng & Walsham, 2008).

A common thread in the literature linking the CA and ICT is the attention given to the capabilities of the user to benefit from technology in ways that will achieve the desired functionings. In this context the CA can be useful in actually shaping the design of ICT interventions, taking into account interests and perspectives of individuals. For example, rather than treating ICT4D interventions as just infrastructure, Musa's (2006) modified version of the technology acceptance model (TAM), refers to the relevance of CA in its focus on the intrinsic value to individuals of such initiatives. Garnham (1999) analysed the contribution made by communication media to enhancing a range of

functionings, using the capabilities approach to highlight that both the type of infrastructure and the ability of people to use it should be taken into account when evaluating impacts of ICT on human development. This is similar to the concept of 'effective use', which has been of concern in the community informatics approach. Calling for more resources to be allocated to the development of applications and support, rather than just infrastructure, to benefit users in developing countries, Gurstein (2003), defined 'effective use' as "the capacity and opportunity to successfully integrate ICTs into the accomplishment of self or collaboratively identified goals".

Other researchers with an interest in the CA have gone a few steps further and applied the framework to specific countries or projects. Alampay (2006b) used the CA for explanatory purposes in an investigation of ICT ownership and access in two locations in the Philippines and concluded that in order to contribute to human development; those who are marginalised must first be made aware of opportunities inherent in new ICTs. In their case study on a community health information system, Byrne & Sahay (2007) referred to DAF when advocating in favour of a participatory methodology for establishing the informational base for the project and determining how the information collected should be used, taking into account capabilities of community members to actually use the information. Informed by the CA, Barja & Gigler (2005) suggested a conceptual framework for measuring information poverty in Latin America that recognised the role of ICTs in the advancement of human freedoms. They drew attention to the requirement for new capabilities for the exchange of information about the economy, politics, and society in addition to the need to strengthen the capabilities of the poor with respect to ownership and use of economic assets.

Applying the CA, or 'functionings' approach, as he referred to the CA, James (2006) explored the relationship between the Internet and poverty, focussing on what occurs after the 'point of purchase', i.e. the usage, contrasting this to traditional welfare economics, where the focus is on the point of purchase. In doing so, he critiqued the 'telecentre' model in favour of initiatives that blend the Internet with radio, using the example of Kothmale in Sri Lanka. Several single initiatives have in fact deployed a mix of technologies and a range of different methods to reach a wider audience. Others are planning to do so.

Comparing research results from one village in Kerala and one in Andra Pradesh, Thomas & Parayil (2008) found better capabilities to use ICTs and convert information to useful knowledge in Kerala and attributed this to the more equitable socio-economic development in that state. They concluded that access to ICTs does not in itself lead to development, but requires social and political intervention to achieve progress in this area. These findings are similar to those made by Niles & Hanson (2003), in their analysis of the key role played by the social context in the process of conversion from physical Internet infrastructures to useful knowledge. Conditions existing prior to the deployment of ICT infrastructure shape both constraints and capabilities to use this medium.

Despite this reasonable body of work related to ICT and CA, the literature is nevertheless sparse when it comes to applying the CA to participatory evaluation of ICT4D initiatives. Unlike writers who refer to the CA as a way forward for future research in general terms, but without including reference to in-depth field studies based on this approach, Mansell (2006), suggested that "one way of ensuring greater participation of the poor in ICT4D initiatives could be an evaluation of priorities in the light of entitlements as outlined in DAF ..." (p.903).

This research accepts the challenge presented by Mansell. Attempting to fill a knowledge gap in the relationship between ICT and the CA at the empirical level, the research aims at developing a framework that can explore the reciprocal influences between ICT and capabilities in a systematic, forward looking longitudinal manner

through a participatory approach. In doing so, the significance of having contrasted the CA with the more traditional development economics lies in the method of choosing the informational base for exploring the impact on individuals and communities of an ICT4D initiative in India. Consistent with the CA, the impact will be assessed in areas that are relevant for the population, rather than using pre-defined indicators. Outcome Mapping (Earl, Carden, & Smutylo, 2001) and Most Significant Change (Dart & Davies, 2003) are two approaches that, although not making specific reference to the CA, have adopted participatory methods in identifying factors that are relevant for evaluative purposes in different contexts. These methods have influenced the design of this research and some aspects of them may be useful for operationalising the CA in general. However work on trying to synthesise the CA and these approaches is beyond the scope of this chapter, but could be an important research topic in its own right in an endeavour to contribute to operationalising the CA.

CASE STUDY: VICKY STANDISH E-EDUCATION CENTRE (VSEEC)

In this section we present the first phase of research conducted at the VSeEC in the form of a case study. Although this research forms part of a three-year longitudinal study of VSeEC's influence on communities in its coverage area, we have designed the research to enable each phase to also be analysed and presented on a standalone basis. The section starts with describing the centre and the environments within which it operates. We then report on the field research, including methodology and key findings.

Micro-Level

The East West Overseas Aid Foundation (TE-WOAF), an Australian based NGO and its Indian partner organisation, The East West Foundation of India (TEWFI) fund and operate the VSeEC, the Uluru Children's Home (UCH), with which it is associated, a health clinic, and other community development initiatives in the Alamparai-Kadapakkam area in the Indian state of Tamil Nadu. The funding comes from private donations to these organisations. Engineers Without Borders Australia (EWB), was responsible for designing and has been instrumental in operating the centre. No charges had been applied for use of the VSeEC when the research was conducted. Financial sustainability of VSeEC, an issue that has plagued many ICT4D initiatives, is not addressed in this study. VSeEC was established as a non-profit centre, funded through donations to TEWOAF. However, as the centre expands from its current nine computers to full capacity of 36 (through donation of additional funds to cover this cost), the operational expenses, including staff, maintenance and power, could increase to such an extent that it may be necessary to pay greater attention to financial aspects. There was one full time systems administrator, and one part time employee at VSeEC, when the study was conducted, but this may not be sufficient when the centre is fully equipped.

The TEWFI complex is located at the Bay of Bengal backwaters, in a relatively isolated rural coastal community, five km east of the East Coast Road (ECR), between Chennai (125km) and Pondicherry (46km). The catchment area of TEWFI's activities is defined as having a radius of approximately five km. Seven villages are located within this distance, three of which are defined as fishing villages and the others as farming villages. The main village of Kadapakkam, on the ECR and the market one km inland are included within this radius. Administratively, the area is in the Edaikalinadu Panchayat, Cheyyur Taluk, Kancheepuram district. There are also informal leadership structures within the villages. The 2001 census showed a population of 25,793 in

Edaikalinadu, the lowest administrative level for which census data could be found.

Consistent with the inscription on the inauguration plaque at the entrance of the VSeEC building: '…..with intent to serve the local community', the aim of the centre is to serve as a resource for the local communities as well as for the children living at UCH. Inaugurated in 2008, the VSeEC had been operational for only eight months when the research reported in this chapter was conducted in September/October 2008.

Although no loss of life was suffered when the fishing villages were affected by the December 2004 tsunami, many homes, fishing boats and other fishing equipment were damaged or lost. The villages were subsequently relocated further from the coastline. As part of tsunami rehabilitation projects there has been an influx of foreign NGOs. The economic characteristics of the area are atypical because of the influence of TEWFI's presence, which has contributed considerably to employment levels. In addition to TEWFI's permanent and casual staff, the purchasing power of TEWFI and its overseas volunteers has boosted employment in the private sector. To the extent possible, TEWFI sources food, for approximately 50 people, and other requirements locally. The transport, primarily in the form of auto-rickshaws (three-wheelers), and construction sectors have also been stimulated by the requirements of TEWFI. Other employment opportunities in the area are limited and remittances from family members working elsewhere represent another income source. There is limited scope for higher paid employment in the area, which however has some potential for tourism (Lummen & Ruiter, 2008).

There are several castes in the area, but in to order avoid bringing up an issue that could potentially be divisive, the issue of caste was not addressed during the focus groups. Caste is not relevant to the activities of TEWFI. The government's affirmative action policies for scheduled castes and tribes have contributed to overcoming some of the disadvantages that previously plagued their lives. TEWFI has also been open to employing persons of all caste backgrounds and especially those from disadvantaged communities.

A key obstacle confronting VSeEC has been the lack of broadband connectivity throughout the area and Internet was not available when the research was conducted. Access via dial-up was not an option, as the location of the centre was serviced with narrowband wireless local loop lacking data capability. Some of the mobile operators had limited data coverage (GPRS and CDMA) of questionable quality and high prices. This wireless connectivity was primarily used for administrative purposes and had, on occasions provided villagers with information, e.g. for examination results. The local telephone exchange in Kadapakkam, despite being located along a main road and serving a reasonable population size, had not been equipped with DSL at the time of the study. While regional telecommunications infrastructure initiatives and the introduction of competition into the market may have improved network access in some rural regions of India, this area was still under-served and lacking reasonable broadband infrastructure. And this despite the fact that the normal reasons for lack of adequate infrastructure in rural areas, viz mountainous topography and/or thinly distributed population, do not apply for the Alamparai-Kadapakkam area. There were no public Internet facilities in the area. A Drishtee outlet in Kadapakkam (at ECR), where services, including computer courses and access to computers for typing and other functions were available on a commercial basis, did not have Internet access either.

From a CA perspective, the absence of broadband infrastructure is an illustration of a barrier inhibiting both the development of capabilities and the conversion of capabilities into functionings.

Meso- and Macro-Level

There are rarely clear boundaries between macro- and meso-environments. For the purpose of this study, India with its central government constitutes the macro-environment. This is where the telecommunications policy is formulated and administered. With almost 70 million citizens, the state of Tamil Nadu can sometimes be too remote from a village to be considered as a meso-level. The boundary between the macro and meso-levels is therefore somewhat blurred in this analysis.

On paper, India has a regulatory framework, including a Universal Service Obligation (USO) Fund which, through a levy of 5% on gross revenue on licensed carriers, is intended to support broadband connectivity and a progressive transformation of village public telephones into ICT community centres (Thomas, 2007). Several reasons have been suggested for the lack of progress in implementation of these policies. According to Rao (2008) various operational, procedural, legal and regulatory issues have impeded progress, whereas Thomas (2007) argued that consideration of access as a basic right is no longer a public policy priority. These reasons describe what has happened, without advancing any explanations as to why the ICT infrastructure momentum has abated. For example, it would be useful to know whether the central government re-evaluated the policies in light of any research findings and whether it consulted with meso- and micro-level authorities and mediating organisations on the usefulness of ICT projects for rural development. If so, what was the reaction? A key objective of undertaking research of the nature covered in this chapter is to provide authorities with a relevant informational base for deciding between competing interests and to provide the public with evidence that is useful for influencing government policies.

In the meantime, for whatever reason(s) and despite the government's promising framework, villagers have been deprived of a tool they could have used to obtain capabilities for facilitating implementation of changes they want to be able to effect in their communities.

The macro-level also has an important role in the provision of e-government services and to that end the Indian government and many of the state governments have developed many innovative e-government applications. At the central government level, the portal (http://india.gov.in/) provides access to a range of services, including the Right to Information Act, 2005, referred to in some of the focus groups as potentially being of great significance in their dealings with the government.

Despite the resources devoted to these many initiatives, the Indian e-readiness index, according to the 2008 UN e-government study (UN, 2008) declined from 0.4001 in 2005 to 0.3814 in 2008. This resulted in India dropping from 87th place in 2005 to 113th out of 192 countries in 2008. The UN study (UN, 2008) noted the paradox that, whereas India has witnessed significant growth of ICT use in urban centres, this has not been the case in rural locations, despite the initial view that e-government would be a means of overcoming distance. The study hypothesised that the lack of progress was, to a large extent, due to underestimation of the infrastructure challenges.

There also appeared to be an absence of institutions at the meso-level that could influence the government to ensure that such infrastructure be provided. One 'official' path for villagers to approach the government is via the panchayat ward councilor, who can raise issues with the panchayat. We understand that this had not occurred, so there is scope for more action by elected representatives at the meso-level.

Despite having embraced the development of the IT sector from an early stage with a comprehensive IT policy and the implementation of several initiatives (Mitra, 2000), Tamil Nadu seems to be losing its foothold at the top of the Indian e-readiness index league table. Although still ranked among the eight e-readiness leaders in the 2006 survey, Tamil Nadu's position dropped from 2nd to 7th place between 2005 and 2006. Its composite index declined from 0.99 to 0.82.

While the e-readiness and e-usage indices both increased, the "environment" index fell from 0.97 to 0.69. This index reflects infrastructure and associated regulatory and market environments (DIT & NCAER, 2006).

The meso-level also includes non-government organisations, many of which can and have played a very important role for the diffusion of ICT, particularly as infomediaries (Ramirez, 2001; Ramirez & Richardson, 2005). The VSeEC already had a working relationship with some mediating organisatons; it was using Aid India's lesson plan material for training and was making extensive use of primary school education material produced by the Azim Premji Foundation. Partnerships also extended to local institutions. One computer had been made available to one of the primary schools in the area. An environmental studies teacher employed by the NGO Pitchandikulam Forest/Bio Resource Centre used another VSeEC computer in his work at the government secondary school in the area. India in general, and the state of Tamil Nadu in particular have a long tradition of involvement by innovative mediating organisations in the provision of ICT services to rural areas. Tamil Nadu is home to several pioneering organisations, well known in the ICT4D community, including the Telecommunications and Computer Networks (TeNeT) Group at the Indian Institute of Technology, Madras (Chennai) and MSSRF. At the time of the study the VSeEC was actively seeking co-operation with these and other organisations.

The situation in the Kadapakkam-Alamparai region fits with the general trend in many developing countries in Asia, where limited connectivity in rural areas has tempered improvements in broadband penetration at the national level. WSIS (2008) attributed this to "inadequate and restrictive policy environments, lack of focus on technological R&D innovations, and limited understanding of the effects of ICTs on communities" (p.7). The research described in this chapter has been conducted in an endeavour to improve this understanding.

Research Design

Before describing the research methodology, we first elaborate on what we mean with the term 'participatory', which is used to define one dimension of the evaluation framework. Parfitt (2004) differentiated between participation as a means and participation as an end. The former, unlike the latter, is politically neutral in that it does not deal with any power relationships. It simply invites participants to present their views about what is important to them and may, by doing so in a group situation, be able to draw on the strength of others. This is participatory in contrast to the utilitarian approach, which would typically not involve participants in defining the scope of the research. However this type of participation does not occupy the top rungs of Arnstein's (1969) ladder of citizen participation, where Arnstein placed partnership, delegated power, and citizen control. Although the methodology used in the field research was not designed to transfer any power, there was also no cynical intent behind the participation, as in the middle rung–"tokenism"–of Arnstein's ladder (informing, consultation, and placation). Whereas the objective of the research was learning and understanding, the participatory research process could well have contributed to empowerment, although this was not defined as an objective for the research.

Falling within the definition of the interpretivist or naturalistic paradigm, the research methodology adopted for this study was of a qualitative nature, with focus groups, each consisting of participants from specific stakeholder groups: leaders from fishing and farming villages, parents from fishing and farming villages, a women's self help group including members from fishing and farming communities, a youth group and employees of an NGO. A separate focus group with teachers had

been planned, but due to practical difficulties this did not take place. Three teachers participated in the focus group with leaders from farming villages.

As this was an exploratory study without any claims of being representative, there was no attempt to use probabilistic sampling for the invitations to participate in focus groups. Invitations were instead to a large extent based on existing contacts and some snowballing. No information on socio-economic, caste or any other status (except for the village leaders) of the participants was obtained.

The seven focus group sessions were conducted over a period of 20 days in September-October 2008. The intention was to conduct additional focus groups, but for practical reasons that was not possible during this wave of research. The seven focus groups are considered sufficient for a meaningful, albeit limited analysis.

As the lead researcher did not speak Tamil, the local language, the research was somewhat complicated by the need for interpretation, which was undertaken by a social worker employed at TEWFI's health clinic and the systems administrator at VSeEC. They were also responsible for recruiting participants.

It is reasonable to question the objectivity, representativity, and reliability of the information provided by participants, when considering that the sessions took place in the presence of staff, with what could be perceived as vested interests in the outcomes of the research. Their presence could have influenced participants to express views considered favourable to them and to avoid critical comments. There was also a risk of bias arising from the lead researcher's reliance on interpretation by staff members and that the views could reflect what the participants expected the researcher wanted to hear. Despite these potential biases, some of the responses were critical of the existing arrangements associated with the VSeEC, e.g. participants in two groups expressed the view that the VSeEC needed publicity to attract more users and the remoteness of the centre in relation to where people live was raised in three groups. These views indicate that participants were not afraid to voice any criticism. In any case, this was not an evaluation of the performance of staff, so whatever bias there may have been is not considered to have seriously influenced the results. It was also in the interest of staff for VSeEC to provide services that are useful for the intended target groups and that maximum use is made of the centre.

There were also some benefits associated with the role of staff members as interpreters in this study. As they had not previously been exposed to this type of research, the study provided capacity building opportunities for them, particularly in the form of learning the importance of listening to perceptions of community members about services provided by them. Their presence also enabled the focus group sessions to take on the form of exchange of information and impressions as well as research, as they were able to respond to questions from participants. This would not have been possible by an interpreter without knowledge about VSeEC. Such exchanges are consistent with a naturalistic form of inquiry (Guba & Lincoln, 1981).

In some cases, the views expressed by participants were simply noted and in others some probing was applied for the purpose of clarification and verification when bias was suspected. For example, when the fishing village parent group suggested longer opening hours and more computers, we probed why this was necessary, as VSeEC was not always fully occupied.

The research centred on the perception of participants with respect to whether the VSeEC had played a role in any improvements in the community and whether they could see a role for it in changes they would like to see, or aspirations they had for their communities. Using a strength-based approach, the structure of the question framework and conduct of the focus groups

were organised around first identifying strengths in the communities. This was followed by asking participants to refer to recent improvements and consider contributing factors to these improvements. This questioning aimed at exploring the role of the VSeEC in any improvements. We then asked participants whether there were any changes and/or enhancements they would like to see in their communities, how they would go about implementing these changes and the role VSeEC could play for this purpose. With those questions we intended to capture the essence of capabilities and functionings they would value and the role VSeEC could play in contributing to their achievements. The framework thus encouraged participants to identify their own solutions to development and to think about ICT as a potential tool. In concluding, participants were invited to provide general comments, views, and suggestions about the VSeEC.

Cognisant of the difficulties and dangers associated with attribution, the focus of the inquiry was on contribution, rather than attribution (Ramirez, 2007) of ICT to communities. Any outcome is likely to be attributable to several factors, some of which are within and others beyond the control of VSeEC's management, e.g. government policies. The study revealed an insight among villagers that there are multiple sources of attribution to different outcomes, including education.

Consistent with the CA's views on the informational base for evaluating outcomes and impacts, participants were encouraged to suggest possible indicators that would be useful in measuring the extent to which the suggested changes have been achieved. Participant involvement in defining community based indicators is not limited to the CA. One illustration of where this has been done in the context of ICT is the KNET (2001) project in indigenous communities in Canada.

As the information sought was related to perceptions of the groups, rather than facts, triangulation using other methods for obtaining data was

Table 1. Key findings from the first phase of research at the VSeEC

	Leaders		Parents		NGO group	Women	Youth
Major strengths/assets in the community	**Fishing**	**Farming***	**Fishing**	**Farming**			
Unity	✓		✓	✓	✓		✓
Farming		✓			✓	✓	
Fishing	✓		✓		✓	✓	✓
Education			'			✓	✓
Recent improvements							
Education	✓	✓		✓		✓	✓
Tsunami relief programmes, particularly women's empowerment					✓	✓	
Factors contributing to improvements							
Standard of education	✓						✓
Government policies relating to education	✓						
Greater interest in and "thirst" for education	✓		✓				
Other TEWFI activities (incl UCH and activities related to education)				✓			✓
Tsunami relief programmes, particularly women's empowerment					✓	✓	
Priority areas for further improvements							
Improved education, particularly in English	✓	✓	✓	✓		✓	✓
More employment opportunities	✓	✓		✓		✓	
Local vocational school and other higher education institutions		✓			✓		✓
Sports and entertainment facilities	✓		✓				✓
New industries in the area	✓	✓					
24 hour health service (Multi-specialty hospital with pathology, etc)		✓			✓		
Maternity care	✓	✓					
Possible indicators for measuring impact of VSeEC and other changes							
Changes in employment levels and type of employment	✓		✓		✓		
Education in general		✓	✓				
* This group included three primary school teachers							

not undertaken. However, the conduct of several focus groups could be considered as some form of triangulation, as it enabled views to emerge from different contexts. The results were verified by participants at a meeting called to discuss the outcomes of the research.

Key Findings

This section summarises the general direction of responses during the focus group sessions. Where any of the groups did not include the issues included in this section, neither did these groups express views that contradicted any of these findings.

Table 1 captures themes that arose in at least two different focus groups. There are some apparent holes in the table, e.g. farmers did not mention farming as a major strength. This does not mean that they would disagree with farming being a major strength, only that they did not raise it. As shown in Table 1, to be educated emerged as far the most important capability, both in terms of defining strengths, identifying recent improvements in the community and as an area where further improvement is required.

The Importance of Education

A striking feature of most focus group discussions was the enthusiasm about the use of computers by children at the centre and the words expressed in one of the sessions that the VSeEC is the "Gift of God" reflected a widespread perception. The children's ability to use computers had strengthened their, and thereby their parents' confidence and self-esteem.

It did not matter for the participants, all of whom were over the age of 18, that they did not fully understand what their and other children were doing at the VSeEC. There was trust that the activities were useful and an understanding that the children used programmes with cartoons (Azim Premji software) to learn different subjects.

A prevailing perception was that the new capability acquired by children to master computers was valued as a capability for empowerment, and not only because of any immediate increase in prosperity that would result from this new skill. This does not preclude that the greater confidence can later expand to economic aspects, (e.g. by improving their ability to find employment or start their own businesses). The capability of finding work outside the farming and fishing sectors was mentioned as a high priority. Participants recognised the importance of narrowing the gap between current skill levels and the requirements

for employment in the 'formal' sector, through investment in education.

Education was a recurring and dominant theme throughout the sessions, whether in the context of identifying strengths or expressing priorities for improvements and the role of the VSeEC in this. There was general agreement that the VSeEC has boosted the prospects for good education outcomes for children. Education also appeared to serve functions other than being a ticket for better employment opportunities. It is also a tool for empowerment, as illustrated by a description in one of the sessions of how community members felt stigmatised and embarrassed when having to rely on others to complete forms and accompany them to visit government authorities due to their low or non-existent literacy skills. This is reminiscent of the concept of being ashamed to appear in public, used by Sen (1985b) to define one aspect of deprivation. Another example of deprivation was the reliance on interpreters when dealing with English speaking NGO representatives in the aftermath of the tsunami. In addition to the material nature of this deprivation in the form of suspicions that they may not have received their full entitlements, there was a sense of powerlessness in not understanding discussions about their lives and livelihoods.

The empowering features of education for normal village life implied in these examples could be one reason for the high importance attached to informal as well as formal education, where the former refers to acquiring specific skills outside the formal education system and the latter to education within the formal system. Participants nevertheless considered certificates to be of significant value, whether or not these are issued by accredited institutions. They represent a sense of achievement. The VSeEC thus has an important role to play in facilitating supported and self-directed learning about topics that are of relevance for community members. High on the list of priorities were computer and English language skills and there was a widespread perception that the use of computers was helping the children with their English skills, (e.g. by getting used to the alphabet and a few English words on keyboards). The importance attached to these skills is consistent with the CESVSF, as these capabilities are fundamental to being able to exploit the full potential of ICT.

Diversifying Livelihoods

Despite the inherent value of education, the capabilities of being educated and employable were nevertheless so intertwined in the minds of the participants that it is difficult to refer to these in separate sub-sections of this paper. Participants in several groups, particularly from fishing villages, were adamant that they wanted their children to find employment in other sectors. Education was considered a pre-requisite for reasonable employment conditions, particularly better English language and computer skills. The level of English language skills was quite low in the area.

The wish for a local vocational school and other higher educational institutions were suggested in the context of education for livelihood diversification. Several participants considered this particularly important for young women, many of whom, due to cultural and financial barriers, do not continue their education after completing high school in the local area. They commented that women just sit at home, waiting to get married, but marriage often does not solve their livelihood situation. There were several cases of women having been abandoned by their husbands after giving birth.

The VSeEC, in partnership with accredited vocational education providers, could play an important role in developing capabilities that would enable both genders to diversify their livelihoods.

The Illusion of and Search for Unity

One unexpected finding was the frequent reference to 'unity' and co-operation, in response to the topic dealing with strengths and assets. Having expected something more directly related to livelihoods, 'unity' came as a surprise. It was particularly astonishing against the background of what appeared to be common knowledge of instances of disunity. There were considerable tensions within at least one of the villages, between some of the fishing villages, and between the fishing and farming communities. Participants in the youth group explained that they refer to community members helping each other as unity and illustrated this with how members of their group were funding tertiary education for one of their members who suffers from physical disabilities. Hidden behind the term 'unity' could also be a wish to portray the community as being united, a valued capability that could relate to previous dealings with government authorities and NGOs. It is understood that villagers were told that they must speak with one voice in order to obtain certain benefits for their communities.

The ability to unite, form and join groups is another essential capability, which in turn can also be a source of empowerment and a manifestation of the meso-level. There are many inspiring case studies of what can be achieved through groups, rather than by individuals acting alone, in terms of building sustainable livelihoods. However, according to Stewart (2005), disadvantaged persons often lack assets such as networks and human capital required to form groups. In an endeavour to overcome this disadvantage, the Indian government has facilitated the establishment of women's self-help groups.

The extent to which the capability of forming groups (particularly at the regional level where they could add strength to individual communities) can be facilitated by ICTs could not be tested. Whereas ICT in education can be useful to a limited extent without any communications capability, this is not the case when considering ICT as a tool for establishing groups. Without Internet access, it was not possible to explore whether the VSeEC could be useful in extending the geographic coverage of groups, e.g. to facilitate communication between the different villages in the area and between the different panchayats in the taluk, etc. This will be explored in further detail in the next research phase, when it is expected that the VSeEC will be equipped with Internet access. But Internet access in itself is unlikely to be sufficient, particularly with the low level of literacy and inexperience with this medium. In order to maximise the benefits of the Internet, its introduction will be accompanied with involvement by infomediaries (mediating individuals and organisations) that can overcome barriers of low levels of literacy and skills to facilitate effective use of the Internet in a way that can improve livelihoods of the communities. The importance of infomediaries has emerged as a key success factor for ICT4D initiatives (Duncombe, 2006; Fillip & Foot, 2007; Ramirez, 2001; Ramirez & Richardson, 2005; Schilderman, 2002; Warren, 2007). MSSRF's use of village knowledge workers, with access to facilities provided by the village resource centres, is one example of how infomediaries can synthesise information from different sources and facilitate contacts between villagers and the outside world.

Dealing with the Government

None of the groups presented positive views about dealing with government authorities. The reluctance to deal with these was reflected in reactions when asked how participants would achieve desired changes. The most common response was to approach NGOs. However, in considering whether and how the VSeEC would be useful in achieving changes, many participants appeared keen on using the Internet for contacting the government or obtaining information on different government (funding) schemes. For someone who has had limited, if any, exposure to the Internet, many displayed surprisingly high awareness of the Internet's potential in dealing with government authorities. This included knowledge about the

e-government portal and the Rights to Information Act, both of which enable citizens to deal directly with more senior government officials via e-mail, and access information about government grants and subsidies. There was a perception that community members missed many opportunities for funding due to ignorance of available grants.

Interestingly, while aware of the possibilities arising from access to the Internet, the knowledge about what is available through this medium did not translate directly into an eagerness to learn to use the computers themselves. Many of the participants seemed to be satisfied with using their children as intermediaries between themselves and information. This may well change when Internet becomes available at the centre and adults can experience its power, without intermediaries. Such access could become a catalyst for wider interest in using the Internet, although this would require a certain level of keyboard literacy, which, with its 247 characters, is not easily attainable in Tamil. Other options, including English and Hindi were not viable alternatives, as skill levels in these languages were very low.

There is a role for staff at VSeEC to act as intermediaries for those who cannot rely on family members or acquaintances to access information.

Indicators

The question about what indicators to use when measuring progress in achieving desired changes, gave some credence to the CA's critique of the emphasis on growth and GDP by utilitarian and welfarist approaches to development. Most participants focused on indicators for education and employment. Against this, it can be argued that although the importance attached to indicators on employment levels and types was not explicitly linked to income, such linkage could nevertheless have been implied, at least to some extent. It is interesting to note though that income was raised by participants in one group only and material items arose only in the youth group.

Similarly, education was not explicitly linked to employment and thereby income. In the minds of the participants, there could have been flow-on effects from education to income via employment. But when explaining why education was so important to them, it was clear that it had an intrinsic value associated with empowerment.

Simply knowing the type of indicators the participants would find useful to measure is not enough. As government agencies do not collect this type of statistics, the challenge is now to develop the informational base to enable this analysis to be undertaken. One option being considered for this is to involve the local school community in this activity as part of the curriculum for high school students, possibly in conjunction with tertiary institutions in Tamil Nadu.

Policy Implications

The main policy implication to emerge from the meso- and macro-level perspectives is that an ICT initiative such as VSeEC requires a supportive regulatory environment that can deliver adequate infrastructure to the area. The lack of broadband infrastructure has made it difficult for VSeEC to operate as intended. While the absence of infrastructure is a reflection of the telecommunications policy and its administration, a central government responsibility, the different formal meso-level institutions also have a role to play, (e.g. by influencing the government). There was no evidence of elected representatives in the area doing this, so another policy implication is the importance of collaboration between the micro- and meso-levels in representing the interests of villagers. Inadequate representation also manifested itself in the lack of trust in governments, indicated by participants when suggesting they would turn to NGOs, rather than a government authority for support in realising aspirations for their communities. An NGO is limited in what it can achieve in an environment without active government involvement, particularly in providing services that are ultimately the responsibility of the government.

By more actively co-operating with initiatives such as the VSeEC, meso-level institutions could become major resources for their communities, amplifying the contributions made by NGOs, without necessarily incurring expenses. The type of research presented in this chapter could provide input into negotiations with authorities that have power and resources to address the inadequate infrastructure situation.

The results also suggest that the many indicators developed at a global level to measure the information society and which show levels of uptake and activity based on countable items, such as uptake of computers, Internet and mobile services, are not sufficient in themselves. They do not necessarily reflect what is important for the population. Governments and international organisations involved in developing and measuring ICT related indicators should therefore pay more attention to factors associated with effective use and contribution to improved livelihoods.

FUTURE TRENDS AND RESEARCH

Although the research conducted so far has provided important insights into the dynamics surrounding the application of the CESVSF to an ICT4D initiative, it would be necessary to undertake similar studies in other areas in order to assess whether this framework is a useful contribution to understanding how ICT can contribute to capabilities, empowerment and sustainability. A more extensive test of the framework would also include other types of ICT4D initiatives to compare the results, including mobile based initiatives. There has been a shift in focus on ICT4D from shared access facilities to mobiles, as these have become increasingly ubiquitous and affordable, over the past few years (Howard, 2008; McNamara 2008; Sey & Fellows, 2009). This may influence methods of incorporating ICT into rural livelihoods programmes. Comparisons of CESVSF results between public access centres and individually owned mobile services would provide an important input into policy reviews dealing with what type of ICT facilities to be encouraged and supported in the future to achieve development objectives.

CONCLUSION

With reference to the CESVSF, this research has positioned the VSeEC situation at the time of the research at the starting line of the spiral. The VSeEC was building skills that will be useful in mastering the next stage of deployment, which still awaited broadband Internet access. Participants in the focus groups demonstrated an understanding of the potential empowering qualities inherent in these skills.

The importance attached to education and employment provided empirical evidence to support the validity of the CA, as a framework for understanding what is important to villagers. It is not just a theoretical framework, but an approach that is consistent with the ways participants were thinking about their and their children's lives.

While there were implied links between education-employment-income, education also emerged as having intrinsic value to participants. The intrinsic value was associated with dignity and empowerment, attributes that would enable individuals to "effectively shape their own destiny and help each other [and] need not be seen as passive recipients of the benefits of cunning development programs" (Sen 2001, p.11). Such attributes, important as they are for human development, are ignored in utilitarian and welfare economics, which tend to limit its focus to variables such as income and growth.

In this case, at this time, great importance was attached to measuring employment and education, but other factors may emerge in other places or in the next phase of the VSeEC study. These results indicate that, rather than focusing informational

efforts relating to livelihoods solely on economic indicators, data collection should be channeled towards developing an informational base that incorporate indicators that participants consider important for leading lives they have reason to value. Collection of such statistics need not be expensive, as they could be conducted by community members themselves with some support on methodological aspects from the government and other entities, (e.g. tertiary education institutions).

This VSeEC case study has confirmed the limit to what local communities can achieve without supportive meso- and macro-level environments. The lack of Internet access had, at least until the time when the research was conducted, deprived communities of their entitlement to an important source of knowledge and a communication tool, preconditions for being propelled along the CES virtuous spiral.

The study supports the value of a longitudinal perspective when trying to understand the impact of an ICT initiative on communities. For this evaluation to have followed the common approach, which is to study the impact a short time after 'project completion', defined as installation of infrastructure, would have given a distorted view of the impact of VSeEC, as it will take some time for the community to understand and realise the potential of this facility. A report prepared to TE-WOAF, TEWFI and EWB on this initial research has already encouraged some new initiatives. Had the study been undertaken at the 'end' of the project period, usually defined as the expiry of the funding period, it could have been too late to use the constructive input from participants.

The field research methodology, in the form of focus groups, was also found to be useful; both in providing insights that are unlikely to have surfaced in structured survey forms, and in actively contributing to greater awareness and promotion of the VSeEC among its constituency.

The next phases of the research at the VSeEC and other ICT4D initiatives are designed to populate the CESVSF with more empirical evidence to gain a better understanding of the reciprocal relationships between ICT and capabilities.

ACKNOWLEDGMENT

Numerous people have contributed to the research presented in this chapter. Many thanks go to the trustees, foundation members and the many staff members and volunteers at TEWOAF, TEWFI and EWB, as well as the many community members who, in their different ways enabled this study to be conducted. We wish to express special appreciation to Dr Natteri V. Chandran, Founder and Chair, Board of Trustees of TEWOAF and TEWFI for facilitating this research and for his support. The contributions by Professor John Houghton for his helpful input to the formulation of the conceptual framework and by Adjunct Professor Ricardo Ramirez for his many inputs, insightful comments and suggestions on an earlier draft are also gratefully acknowledged.

REFERENCES

Alampay, E. A. (2006a). Beyond access to ICTs: measuring capabilities in the information society. *International Journal of Education and Development Using Information and Communications Technology, 2*(3), 4–22.

Alampay, E. A. (2006b). The capability approach and access to information and communications technologies. In Minogue, M., & Cariño, L. (Eds.), *Regulatory governance in developing countries* (pp. 183–205). Cheltenham, UK: Edward Elgar.

Alkire, S. (2005). Why the capability approach? *Journal of Human Development, 6*(1), 115–133. doi:10.1080/146498805200034275

Anand, P. B. (2007). Capability, sustainability, and collective action: an examination of a river water dispute . *Journal of Human Development, 8*(1), 109–132. doi:10.1080/14649880601101465

Andrew, T. N., & Petkov, D. (2003). The need for a systems thinking approach to the planning of rural telecommunications infrastructure. *Telecommunications Policy, 27*(1-2), 75–93. doi:10.1016/S0308-5961(02)00095-2

Anyaegbunam, C., Mefalopulos, P., & Moetsabi, T. (1999). *Participatory rural communication appraisal (PRCA) methodology, the first mile of connectivity*. Rome: FAO. Retrieved May 29, 2006, from http://www.fao.org/sd/CDdirect/CDan0024.htm

Arnstein, S. R. (1969). A ladder of citizen participation. *Journal of the American Planning Association. American Planning Association, 35*(4), 216–224. doi:10.1080/01944366908977225

Barja, G., & Gigler, B.-S. (2005). The concept of information poverty and how to measure it in the Latin American context . In Galperin, H., & Mariscal, J. (Eds.), *Digital poverty: perspectives from Latin America and the Caribbean*. Ottawa: IDRC.

Batchelor, S., & Norrish, P. (2004). *Framework for the assessment of ICT pilot projects: beyond monitoring and evaluation to applied research*. Washington, DC: InfoDev. Retrieved July 21, 2006, from http://infodev.org/en/Publication.4.html.

Batchelor, S., & Sugden, S. (2003). *An analysis of InfoDev case studies: lessons learned*. Washington, DC: InfoDev. Retrieved May 26, 2006, from http://www.sustainableicts.org/infodev/infodevreport.pdf

Bebbington, A. (2004). Social capital and development studies 1: critique, debate, progress? *Progress in Development Studies, 4*(4), 343–349. doi:10.1191/1464993404ps094pr

Biggeri, M., Libanora, R., Mariani, S., & Menchini, L. (2006). Children conceptualizing their capabilities: results of a survey conducted during the first Children's World Congress on Child Labour. *Journal of Human Development*, *7*(1), 59–83. doi:10.1080/14649880500501179

Byrne, E., & Sahay, S. (2007). Participatory design for social development: a South African case study on community-based health information systems. *Information Technology for Development*, *13*(1), 71–94. doi:10.1002/itdj.20052

Comin, F. (2001 June). *Operationalising Sen's capability approach*. Paper prepared for the conference Justice and Poverty, examining Sen's capability approach, Cambridge, UK. Retrieved August 20, 2006 from http://www.st-edmunds. cam.ac.uk/vhi/sen/papers/comim.pdf.

Corbridge, S. (2002). Development as freedom: the spaces of Amartya Sen. *Progress in Development Studies*, *2*(3), 183–217. doi:10.1191/1464993402ps037ra

Corea, S. (2007). Promoting development through information technology innovation: The IT artifact, artfulness, and articulation. *Information Technology for Development*, *13*(1), 49–69. doi:10.1002/itdj.20036

Dart, J., & Davies, R. (2003). A dialogical, story-based evaluation tool: the most significant change technique. *The American Journal of Evaluation*, *24*(2), 137–155.

DIT & NCAER (Department of Information Technology & National Council of Applied Economic Research). (2006). *India: e-readiness assessment report 2006 - for States/Union Territories*.

Duncombe, R. (2006). Using the livelihoods framework to analyse ICT applications for poverty reduction through microenterprise. *Information Technologies and International Development*, *3*(3), 81–100. doi:10.1162/itid.2007.3.3.81

Earl, S., Carden, F., & Smutylo, T. (2001). *Outcome mapping: building learning and reflection into development programs*. Ottawa, Canada: IDRC.

Evans, J., & Ninole, M. (2004). Minding the information gap in Papua New Guinea: a view from Wewak. In R.F. Garcia, R.F. (Ed.), *Divide and connect: perils and potentials of information and communication technology in Asia and the Pacific*. (pp. 49-60). Mumbai: Asian South Pacific, Bureau of Adult Education (ASPBAE).

Falch, M., & Anyimadub, A. (2003). Tele-centres as a way of achieving universal access: the case of Ghana. *Telecommunications Policy*, *27*(1-2), 21–39. doi:10.1016/S0308-5961(02)00092-7

Fillip, B., & Foote, D. (2007). *Making the connection: scaling telecenters for development*. Washington, DC: Information Technology Applications Center of the Academy for Education Development. Retrived July 10, 2008, from http://connection.aed.org/pages/MakingConnections.pdf

Forestier, E., Grace, J., & Kenny, C. (2002). Can information and communication technologies be pro-poor? *Telecommunications Policy*, *26*(11), 623–646. doi:10.1016/S0308-5961(02)00061-7

Gagliardone, I. (2005). Virtual enclaves or global networks? The Role of information and communication technologies in development cooperation. *PsychNology Journal*, *3*(3), 228–242.

Garnham, N. (1999). Amartya Sen's capabilities approach to the evaluation of welfare: its application to communications . In Calabrese, A., & Burgelman, J.-C. (Eds.), *Communication, citizenship and social policy: rethinking the limits of the welfare state* (pp. 281–302). Oxford, UK: Rowan and Littlefield.

Gasper, D. (1997). Sen's capability approach and Nussbaum's capabilities ethics. *Journal of International Development, 9*(2). doi:10.1002/(SICI)1099-1328(199703)9:2<281::AID-JID438>3.0.CO;2-K

Gasper, D. (2002). Is Sen's capability approach an adequate basis for considering human development? *Review of Political Economy, 14*(4), 435–461. doi:10.1080/0953825022000009898

Gaved, M., & Anderson, B. (2006). *The impact of local ICT initiatives on social capital and quality of life*. Chimera Working Paper 2006-6, University of Essex, Colchester, UK. Retrieved October 30, 2006, from http://www.essex.ac.uk/chimera/content/pubs/wps/CWP-2006-06-Local-ICT-Social-Capital.pdf

Gigler, B.-S. (2004 September). *Including the excluded – can ICTs empower poor communities? Towards an alternative evaluation framework based on the capability approach*. Paper for the 4th International Conference on the capability approach, University of Pavia, Pavia. Retrieved May 26, 2006, from http://www.its.caltech.edu/~e105/readings/ICT-poor.pdf

Grunfeld, H. (2007). *Framework for evaluating contributions of ICT to capabilities, empowerment and sustainability in disadvantaged communities*. Paper presented at the CPRSouth2 (Communication Policy Research) Conference, 'Empowering rural communities through ICT policy and research,' Chennai, India, 15-17 December.

Guba, E. G., & Lincoln, Y. S. (1981). *Effective Evaluation*. San Francisco: Jossey-Bass.

Gurstein, M. (2003). Effective use: a community informatics strategy beyond the digital divide. *First Monday, 8*(12).

Harris, R., & Rajora, R. (2006). *Empowering the poor. Information and communications technology for governance and poverty reduction: a study of rural development projects in India*. UNDP Asia-Pacific Development Information Programme.

Harris, R. W. (2001). *A place of hope, connecting people and organisations for rural development, through multipurpose community telecentres (MCTs) in selected Philippine Barangays; a learning evaluation*. Report for International Development Research Centre (IDRC), Ottawa. Retrieved October 5, 2006, from http://www.idrc.ca/IMAGES/ICT4D/PanAsia/HarrisPhilippineReport.pdf

Heeks, R. (2006). Theorizing ICT4D research. *Information Technologies and International Development, 3*(3), 1–4. doi:10.1162/itid.2007.3.3.1

Howard, I. (2008). *Unbounded possibilities: observations on sustaining rural information and communication technology (ICT) in Africa*. Retrieved on October 29, 2008, from http://www.apc.org/en/system/files/SustainingRuralICTs_0.pdf

Hudson, H. E. (2006). *From rural village to global village: telecommunications for development in the information age*. Mahwah, NJ: Lawrence Erlbaum Associates.

Jain, R., & Raghuram, G. (2005). *Study on accelerated provisions of rural telecommunication services (ARTS)*. Ahmedabad, India: Indian Institute of Management. Retrieved December 22, 2006, from http://www.iimahd.ernet.in/ctps/pdf/Final%20Report%20Edited.pdf

James, J. (2006). The Internet and poverty in developing countries: welfare economics versus a functionings-based approach. *Futures, 38*(3), 337–349. doi:10.1016/j.futures.2005.07.005

Jhunjhunwala, A., Ramachandran, A., & Bandyopadhyay, A. (2004). n-Logue: the story of a rural service provider in India. *Journal of Community Informatics, 1*(1), 30–38.

KNET (Keewaywin Local Government and Economic Day). (2001). Retrieved on November 15, 2008 from http://smart.knet.ca/keewaywin/governance.html

Kumar, R., & Best, M. (2006). Social impact and diffusion of telecenter use: a study from the sustainable access in rural India project. *Journal of Community Informatics*, *2*(3).

Lummen, M., & Ruiter, E. (2008). *Product report: Edaikazhinadu project*. Dissertation, Saxion University of Applied Sciences, Hospitality Business School, Education: Tourism Management, Retrieved on November 11, 2008 from: http://dms01.saxion.nl/C12574DB0035EDCF/All+documents/52C462DA81331CA4C12574DB00630312/$File/Afstudeerscriptie%20E.%20Ruiter%20en%20M.%20Lummen.pdf

Madon, S. (2004). Evaluating the development impact of e-governance initiatives: an exploratory framework. *The Electronic Journal on Information Systems in Developing Countries*, *20*(5), 1–13.

Mansell, R. (2006). Ambiguous connections: entitlements and responsibilities of global networking. *Journal of International Development*, *18*(6), 901–913. doi:10.1002/jid.1310

McNamara, K. (Ed.). (2008). Enhancing the rural livelihoods of the poor: recommendations on the use of ICT in enhancing the livelihoods of the rural poor. *Infodev*. Retrieved on April 1, 2009, from http://www.infodev.org/en/Publication.510.html

Meera, S. N., Jhamtani, A., & Rao, D. U. M. (2004). *Information and communication technology in agricultural development: a comparative analysis of three projects from India*. Agriculture Research and Extension Network, ODI Paper No.135. London: Overseas Development Institute. Retrieved August 1, 2006, from http://www.odi.org.uk/agren/papers/agrenpaper_135.pdf

Mitra, R. (2000). Emerging state-level ICT development strategies . In Bhatnagar, S., & Schware, R. (Eds.), *Information and communication technology in development* (pp. 195–205). New Delhi: Sage Publications.

Musa, P. F. (2006). Making a case for modifying the technology acceptance model to account for limited accessibility in developing countries. *Information Technology for Development*, *12*(3), 213–224. doi:10.1002/itdj.20043

Nielsen, L., & Heffernan, C. (2006). New tools to connect people and places: the impact of ICTs on learning among resource poor farmers in Bolivia. *Journal of International Development*, *18*(6), 889–900. doi:10.1002/jid.1321

Niles, S., & Hanson, S. (2003). A new era of accessibility. *URISA Journal*, *15*, I.

Nussbaum, M. (2000). *Women and Human Development: The Capabilities Approach*. Cambridge, UK: Cambridge University Press.

Nussbaum, M. (2006). Capabilities as fundamental entitlement . In Kaufman, A. (Ed.), *Capabilities equality: basic issues and problems* (pp. 44–70). New York: Routledge.

O'Neil, D. (2002). Assessing community informatics: a review of methodological approaches for evaluating community networks and community technology centres. *Internet research: electronic networking applications and policy 12* (1), 76-102.

Orlikowski, W. J., & Baroudi, J. J. (2002). Studying information technology in organizations: research approaches and assumptions. (2002. In M.D Myers. & D. Avison, (Eds.), *Qualitative research in information systems,* (pp. 51-77). London: Sage Publications.

Overå, R. (2006). Networks, distance and trust: telecommunications development and changing trading practices in Ghana. *World Development, 34*(7), 1301–1315. doi:10.1016/j.worlddev.2005.11.015

Parfitt, T. (2004). The ambiguity of participation: a qualified defence of participatory development. *Third World Quarterly, 25*(3), 537–555. doi:10.1080/0143659042000191429

Ramirez, R. (2001). A model for rural and remote information and communication technologies: A Canadian exploration. *Telecommunications Policy, 25*(5), 315–330. doi:10.1016/S0308-5961(01)00007-6

Ramirez, R. (2003). Bridging disciplines: the natural resource management kaleidoscope for understanding ICTs. *The Journal of Development Communication, 1*(14), 51–64.

Ramirez, R. (2007). Appreciating the contribution of broadband ICT with rural and remote communities: stepping stones toward an alternative paradigm. *The Information Society, 23*(2), 85–94. doi:10.1080/01972240701224044

Ramirez, R., & Richardson, D. (2005). Measuring the impact of telecommunication services on rural and remote communities. *Telecommunications Policy, 29*(4), 297–319. doi:10.1016/j.telpol.2004.05.015

Rao, S. S. (2004). Role of ICTs in India's rural community information systems. *Info, 6*(4), 261–269. doi:10.1108/14636690410555663

Robeyns, I. (2001). *Understanding Sen's capability approach*. Cambridge, UK: Wolfson College. Retrieved on August 20, 2006, from http://www.ingridrobeyns.nl/Downloads/Under_sen.pdf.

Sawhney, H. (1996). Information superhighway: metaphors as midwives. *Media Culture & Society, 18*, 291–314. doi:10.1177/016344396018002007

Schilderman, T. (2002). *Strengthening the Knowledge and Information Systems of the Urban Poor*. London: DFID.

Schischka, J., Dalziel, P., & Saunders, C. (2008). Applying Sen's capability approach to poverty alleviation programs: two case Studies. *Journal of Human Development, 9*(2), 229–246. doi:10.1080/14649880802078777

Sciadas, G. (Ed.). (2005). *The digital guide to digital opportunities: measuring infostates for development*. Montréal: Claude-Yves Charron.

Sen, A. (1985a). Well-being, agency and freedom: the Dewey Lectures: 1984. *The Journal of Philosophy, 82*, 169–221. doi:10.2307/2026184

Sen, A. (1985b). *Commodities and capabilities*. Amsterdam: North-Holland.

Sen, A. (2000). A decade of human development. *Journal of Human Development, 1*(1), 17–23. doi:10.1080/14649880050008746

Sen, A. (2001). *Development as freedom*. London: Oxford University Press.

Sen, A. (2005). Human rights and capabilities. *Journal of Human Development, 6*(2), 151–161. doi:10.1080/14649880500120491

Sey, A., & Fellows, M. (2009). *Literature review on the impact of public access to information and communication technologies*. CIS Working Paper No. 6. Seattle: University of Washington Center for Information & Society. Retrieved on May 9, 2009, from http://cis.washington.edu/depository/publications/CIS-WorkingPaperNo6.pdf

Souter, D., Scott, N., Garforth, C., Jain, R., Mascarenhas, O., & McKemey, K. (2005). *The economic impact of telecommunications on rural livelihoods and poverty reduction: a study of rural communities in India (Gujarat), Mozambique and Tanzania*. Commonwealth Telecommunications Organisation for UK Department for International Development. Report of DFID KaR Project 8347. London.

Stewart, F. (2005). Groups and capabilities. *Journal of Human Development, 6*(2), 185–204. doi:10.1080/14649880500120517

Stewart, F., & Deneulin, S. (2002). Amartya Sen's contribution to development thinking. *Studies in Comparative International Development, 37*(2), 61–70. doi:10.1007/BF02686262

Talyarkhan, S., Grimshaw, D. J., & Lowe, L. (2005). *Connecting the first mile: investigating best practices for ICTs and information sharing for development*. Rugby, UK: ITDG Publishing.

Thomas, J. J., & Parayil, G. (2008). Bridging the social and digital divides in Andhra Pradesh and Kerala: a capabilities approach. *Development and Change, 49*(3), 409–435. doi:10.1111/j.1467-7660.2008.00486.x

Thomas, P. (2007). Telecom musings: public service issues in India. *info, 9*(2/3), 97-107.

Torero, M., & von Braun, J. (Eds.). (2006). *Information and communication technologies for development and poverty reduction: the potential of telecommunications*. Baltimore: Johns Hopkins University Press.

UN – United Nations. (2008). *e-Government Survey 2008: From e-government to connected governance*. New York: United Nations.

UNDP – United Nations Development Programme. (2003). *Human Development Report 2003: Millennium Development Goals: a compact among nations to end human poverty*. New York: Oxford University Press.

van Dijk, J., & Hacker, K. (2003). The digital divide as a complex and dynamic phenomenon. *The Information Society, 19*(4), 315–326. doi:10.1080/01972240309487

Walsham, G., & Sahay, S. (2006). Research on information systems in developing countries: current landscape and future prospects. *Information Technology for Development, 12*(1), 7–24. doi:10.1002/itdj.20020

Warren, M. (2007). The digital vicious cycle: links between social disadvantage and digital exclusion in rural areas. *Telecommunications Policy, 31*(6-7), 374–388. doi:10.1016/j.telpol.2007.04.001

Warschauer, M. (2003). Demystifying the digital divide. *Scientific American, 289*(2), 42–48. doi:10.1038/scientificamerican0803-42

WSIS (World Summit on the Information Society). (2008). WSIS follow up Report 2008. Advanced unedited draft (for comment). Note by the Secretariat. Retrieved on August 12, 2008, from http://www.unctad.org/en/docs/none20081_en.pdf

Zheng, Y., & Walsham, G. (2008). Inequality of what? Social exclusion in the e-society as capability deprivation. *Information Technology & People, 21*(3), 222–243. doi:10.1108/09593840810896000

KEY TERMS AND DEFINITIONS

Capability and Capability Approach (CA): Capability in the capability approach means the freedom to achieve valuable beings and doings, i.e. diverse combinations of functionings that a person can achieve. The capability approach is an economic and philosophical paradigm initiated by the Nobel Laureate in Economics, Amartya Sen.

Capability, empowerment, and sustainability virtuous spiral framework (CESVSF): is the conceptual framework underlying the research presented in this chapter.

Empowerment: is the process through which individuals and communities take charge of their multiple environments of which they form part, e.g. family, economic, physical, and cultural, in a way that gives them influence and a high degree of control over decisions affecting them.

Engineers Without Borders Australia (EWB): works with disadvantaged communities to improve their quality of life through education and the implementation of sustainable engineering projects.

Micro-, meso-, and macro: in the conceptual framework, there is no precise definition of these terms. Geographically, micro refers to the village and possibly the next layer, the panchayat and macro refers to the central government and possibly the state government. The importance is to recognise the interplay between the three layers. At the conceptual level, the terms refer to the degree of generalisation. At the extremes micro means that everything is different, so it is impossible to generalise and macro refers to sweeping generalisations that do not take into account any local differences. The boundaries between the meso- and the other two levels are blurred and dynamic, both from a geographic and conceptual perspective.

The East West Foundation of India (TEWFI): is the Indian partner organisation of TEWOAF.

The East West Overseas Aid Foundation (TEWOAF): is an Australian volunteer driven organisation, with operations in Alamparai, near Kadapakkam, a small fishing village in the southern Indian state of Tamil Nadu. Its main projects are the Uluru Health Care Centre, the Uluru Children's Home, a coastal eco-education centre, community development programs concentrating on health, the environment and education in the region, and the Vicki Standish e-Education Centre (VSeEC).

Vicki Standish e-Education Centre (VSeEC): is TEWOAF's computer centre located adjacent to the Uluru Children's home in Alamparai, Kadapakkam, Tamil Nadu, India.

Chapter 5
Analysis of Speedy Uptake of Electronic and Digital Signatures in Digital Economy with Special Reference to India

Swapneshwar Goutam
Hidayatullah National Law University, India

ABSTRACT

This chapter focuses on the issues evolved out of the Indian Information Technology Act of 2000; the key subject related to authentication of digital signatures with special reference to India based on case studies; the benefits of strong information technology infrastructure in India for advancement of future technologies and expansion of domestic market worldwide as well as the vital suggestions on advantages of electronic and digital signatures in enriching and ensuring swiftness in business desires and security.

INTRODUCTION

Information technology (IT) sector added diverse advantages to the contemporary face of India. Digital signatures and electronic signatures are one of the sophisticated means for authentication of electronic records in advancement of electronic commerce in digital world. Information communication technology (ICT) constitutes a threshold requirement for e-commerce adoption. International e-commerce transactions are based on the sophisticated catalog from end to end which is used for companies to advertise their products that result in import and export of their goods and services.

This chapter discusses recent and emerging legal complexities regarding the use of digital signature and electronic signatures in India. It also discusses basic evidentiary questions related to digital and electronic signatures testimony and authenticity issues related to civil and commercial disputes. Moreover, it explores benefits of digital Signature in uplifting the best use -in-class technologies and shares best practices within the digital economy.

DOI: 10.4018/978-1-61692-012-8.ch005

BACKGROUND

Digital Signature is defined as means of authentication of any electronic record by a subscriber by means of an electronic method or procedure. In other words, it is as an electronic identifier, created by computer, intended by the party using it to have the same force and effect as the use of a manual signature.

Electronic signature means authentication of any electronic record by a subscriber by means of electronic technique and it also includes digital signature. Also it can be understood as electronic sound, symbol or process attached to or logically associated with a record and executed or adopted by a person with the intent to sign the record.

Digital and Electronic signatures attracts various technical issues; the major concern lies in legal validity and authentication issues, in determining the security concern over business profitability. The utility of digital or electronic signatures facilitate trade and secures transactions over the Internets which enhance rapid growth of electronic commerce.

The prospect of electronically concluding contracts and other legally significant transactions raises a number of technical and legal questions about how to establish the genuineness of electronic documents. The term electronic document is understood as "electronic message," or "electronic record" which is often used interchangeably. In general, "electronic" should not be taken to mean exclusively electrical, but may also include other forms of document preparation, transmission, and storage, including fiber optic transmission lines.

As used in this paper, "electronic document" refers to a *digital* representation of information, where the human-readable characters and images have been reduced to a unique set of binary digits or bits—ones and zeros—which represent those characters.

The difference between a digital, electronic document and an analog image of the same document is that the digital document has effectively captured the raw keystrokes used, to create it, whereas, in case of a written document it is the *image* of the document that has been captured. In a sense, digital electronic documents are normally stored and transmitted in computer-readable form only. The term "written document" will be used when it is necessary to differentiate between a traditional written document, whether recorded on paper or carved in stone, versus an electronic document in digital form, even if such a document were recorded on some semi-permanent medium such as a writeable CD-ROM. Digital documents consist solely of streams of binary digits or "bits" a seemingly endless series of ones and zeroes-they lack the distinctive, semi-permanent physical attributes of a written document.

Difference between Electronic Signature and Digital Signature

There is no universally accepted meaning of e-signatures (Sneddon, 1998). When trying to explore the definition of e-signatures, another term 'digital signatures' will have to be mentioned and differentiated as well. Actually, these two terms have created considerable confusion and sometimes they refer to the same meaning (Finocchiaro, 2002). However, digital signatures are developed to specially refer to one kind of e-signature technologies, that is, the e-signature employing asymmetric encryption. Digital signature is a specific term of art within the technical community that has been used consistently since the landmark publication describing public key cryptography (Diffie & Hellman, 1976).

The digital signature algorithm is based on the use of "public key cryptography" and involves the use of two codes known as "keys" that are used by the signer to authenticate the source and content of his electronic documents, and by the recipient to validate their correctness. One of a pair of keys which are generated at the same time, the private key, is kept solely in the possession of the signer of an electronic document and is used to encode the text of the document into the digital signature (ABA, 1996).

The public and private keys are mathematically related, but the relationship is so complicated that it is "computationally infeasible" to deduce the private key solely from knowledge of the public key or to create a signed message which can be verified by application of the public key without the knowledge of the private key. Hence, the relying party, having trustworthy access to the public key, can validate the documents as having been signed by someone who had the knowledge of the corresponding private key, but cannot deduce the private key from the public key, nor create such a signature without the private key (ABA, 1996).

The digital signature thus provides very reliable algorithmic evidence of the source of an electronic document, assuming that the relying party has a reliable way of verifying the identity of the person with whom the private key is associated and assuming that the secrecy and control over the private key has been maintained. In order to secure such electronic transactions, encryption technologies are used. Such encryption technologies, which are supported by the appropriate legal mechanisms, have the potential to expand global electronic commerce (Swindells & Henderson, 1998).

Digital signatures on electronic records are of imperative significance to guarantee the authenticity and integrity of the contents (Aalberts & van der Hof, 1999). The Government of India has adopted sound techniques in promoting and affiliating international trade by enhancing sound information technology structures which will facilitate in authentication and recognition of new techniques before the court of law in resolving vital important issues around business risk as well as security.

Complex Issues of Authentication around E-Commerce Deal

Major issues of dispute over authentication and recognition of electronic and digital signatures revolve around e-commerce before the court in civil and criminal cases. Normally, electronic commerce is defined as the use of online facilities for doing business. It also includes the internet, private networks, and any other facility that enables buyers to communicate with suppliers (Fellenstein & Wood, 2000). To resolve such dispute of authentication and recognition, the legislation should be construed in a way that accommodates scientific changes; in particular, the meaning to be given to 'form ... of storage' must keep abreast with the digital age (Abichandani, 2004). It has been recognised worldwide that digital signatures are basis to facilitate authentication toward corporate assets, encryption of sensitive information; also the use of digital signatures en route aid to verify identities (Reed, 2007).

It is also argued that, government should frame the policies welcoming new technological changes; which will result in pumping up the economy of country. The new technology makes globalisation a process of wider prospective for the changing world (Michael, 2002).

Legal Requirement of Digital Signature

This section deals with corroboration techniques with respect to electronic and digital signatures authentication issues in commercial disputes, major issue involved with regard to corroboration technique. How much will it be helpful for courts of law?

Where the Indian Information Technology Act of 2000 is silent over authentication issues, and recognition of any new emerging technology which is challenged before the Indian court of law, such issues have always been treated with "UNICTRAL model law" before the enactment of the enactment of the information technology act, 2000 as tool of interpretation in determining the dispute, but even after the enactment of the act of 2000, the courts in India has relied and referred, the model laws as a major source of law in resolving the disputes regarding electronic commerce . In numerous cases subjected to international arbitrational and commercial disputes before courts of law in India, the courts have noticed that UNCITRAL Model Law is suitable to Indian industrial climate under such circumstance where Indian Information Technology Act, 2000 is silent (Sundaram Brake Linings Ltd *v.* Kotak Mahindra Bank Ltd, 2008). Indian Judiciary has taken a broad view in welcoming sound and sophisticated technologies over the issue of authenticity and recognition in commercial and civil disputes. The most significant question is about legitimacy of valid digital signature before courts. In India, such disputes are covered under Information Technology Act, 2000 - under section 8 of IT Act, 2000 under the head of "authentication of electronic record". In order to further facilitate, our courts by and large refer to "the common law" (Cloud Corp, 2002) and the "Universal Commercial Code" (Indian Petrochemicals Corporation Ltd *v.* Union of India, 2006), or other international source and treaties in case of divergence to fill the absence of adequate laws to facilitate adequate infrastructure.

American Bar Association's drafted Digital Signature guidelines over the period 1992 to 1995, with final publication in August 1996 (Jueneman & Robertson, 1998) states that *Digital signatures* are created and verified by cryptography, the branch of applied mathematics that concerns itself with transforming messages into seemingly unintelligible forms and back again. Digital signatures use what is known as "public key cryptography", which employs an algorithm using two different but mathematically related "keys" one for creating a digital signature or transforming data into a seemingly unintelligible form, and another key for verifying a digital signature or returning the message to its original form.

Generally, public key infrastructure is understood as an authentication technology. Using a combination of secret key and public key cryptography (PKI) enables a number of other secu-

rity services including data confidentiality, data integrity, and key management (Portable Mass Storage Device with Virtual Machine Activation, 2003, Paragraph 29).

Today, public private participation is a key element to develop infrastructure in our economy (Delhi Electricity Regulatory Commission, 2007).

Winn (1998) states "cryptography is the process of taking some information called the plaintext and passing it through an encryption process to produce an encrypted copy of the information (called the ciphertext) that can be decrypted and restored to the original plaintext through the application of the cipher key"(p.1198).

Law recognizes 'Asymmetric Crypto System' which can be understood as system that creates a secured key-pair consisting of a private key creating a digital signature and a public key to verify the digital signature (Tulsian, 2006).

The Private Key is explicitly accessible and used by those that need to validate the signer's digital signature. PKI [Public Key Infrastructure] encompasses different components which include a "Certificate Authority", (Basu & Jones, 2002) end-user enrollment software, and tools for organization, renewing, and revoking keys and certificates (Carlisle & Lloyd, 2002).

Mainly certification authorities may create special policies and procedures intended for associating individual persons through online identities. Any person by the use of a public key of the subscriber can verify the electronic record it is a one of legal requirement for authentication. "UNCITRAL MODEL LAW ON ELECTRONIC SIGNATURES (2001)" intends to develop uniform legislation that can facilitate the use of both digital signatures and other forms of electronic signatures in the electronic environment (Chan & Gligor, 2002). Many countries have adopted the Model law or introduced legislation related to electronic facilitation issues (Panagariya, 2000). Working Group recalled that, alongside digital signatures and certification authorities, future work in the area of electronic commerce might also need to address issues of technical alternatives to public-key cryptography; In digital signature "public-key cryptography" is the emerging electronic-commerce practice, general issues of functions performed by third-party service providers; and electronic contracting (United Nation, 2002), the regulations put forward the PKI system and are limited to public-key cryptography, (i.e., digital signatures) (Cerina, 1998 and Aalberts & van der Hof, 1999).

'Information security' is now a familiar concept at the highest levels of corporate structures. The security consultant is taking his place as an advisor along with the legal and accounting experts that are essential to conducting business today. Information security, when approached from a corporate perspective, is an enabler of traditional business goals in an electronic environment (Nortel, 2002). The Information Technology Act, 2000 widely covers the major issues over legitimacy of digital signature when challenge before the courts. There is no doubt that emerging efforts carried out by government in providing adequate base to the emerging technology. I believe there is urgent need to endorse alertness towards the escape.

Legal Issues Involved in Electronic and Digital Signatures Laws in India

This section analyzes issues involved in legal assessment and framework on digital signature and its utility towards international trade, commerce and present regime adopted by India courts to promote the authentication and recognition of electronic and digital signatures as tool of modern trend in development of market. Information Technology Act, 2000 was enacted on the basis of UNCITRAL Model Law on Electronic Commerce 1996; the bill passed by the Indian Parliament in May 2000; notified on 17 October 2000. Under Section 2(d) of Act, defines "*affixing digital signature*" with its grammatical variations

and cognate expressions means adoption of any methodology or procedure by a person for the purpose of authenticating an electronic record by means of digital signature.

Authentication of the Electronic Record

The authentication of the electronic record shall be affected by the use of 'asymmetric crypto system' and 'hash function' which envelop and transform the initial electronic record into another electronic record. A written document can be introduced as an evidence at trial but it must be 'authenticated', a term of art in the law of evidence. Under section 22 A of Indian Evidence Act, 1872; oral admission as to content of electronic records are not relevant, unless the genuineness of the electronic record produced are in question. Genuineness of document under the Act under section 47-A whenever the court has to form opinion as to digital signatures of any person, the opinion of the certifying authority which has issued the Digital signature certificate is a relevant fact. Admissibility of electronic records is to be under Section 65-(B).

In (Stovekraft Private Limited, 2007) Karnataka High Court held that a secure digital signature of any subscriber is alleged to have been affixed to an electronic record of fact that such digital signature is the digital signature of the subscriber must be proved.

Documentary evidence under Indian Evidence Act, 1872 under sections 85-B, 85-C where it specifically states the court may take presumption as to electronic records and digital signature along with court may also take presumption as to digital signature certificates (Batuk Lal, 2006). The details of type of digital signature, manner and format of affixing, and the procedure related to authentication of electronic records are handed over to secondary legislation which has provided a swift in the era of dynamic trade.

IT Security Procedure Rules, 2004

Special rules are framed for securing digital signature as per Rule 4, which states that digital signature shall be deemed to be a secure digital signature for the purposes of the Act if the following essential elements are fulfilled: (1) smart card or hardware token, with cryptographic module in it, is used to create the key pair; (2) that the private key used to create the digital signature; (3) that the hash of the content to be signed is taken from the host system to the smart card or hardware token and the private key is used to create the digital signature, (4) that the information contained in the smart card or hardware token, is solely under the control of the person who is purported to have created the digital signature; (5) that the digital signature can be verified by using the public key listed in the Digital Signature Certificate issued to that person; (6) that the standards referred to in rule 6 of the "Information Technology (Certifying Authorities) Rules, 2000" have been complied relate to the creation, storage and transmission of the digital signature; and (7) that the digital signature is linked to the electronic record in such a manner that if the electronic record was altered the digital signature would be invalidated.

Recent Development

In the year 2004, Business Rules Pertaining to Department of Information Technology were enacted to pro-actively motivate, facilitate, promote and spread IT to masses and ensure speedy IT led development. One of the main vital issues in cyber infrastructure protection is to promote the use of digital signatures in the financial sector, judiciary and education.

Amendments to Information Technology Act 2000

In the year 2006 ("Information Technology (amendment) Bill 2006"), further to amend IT Amendment Act 2000, a bill was introduced in Lok Sabha during winter session 2006. and the hon'ble speaker referred the bill to standing committee on Information Technology. It was presented to Lok Sabha on 7 September, 2007. The bill is yet to be approved by the parliament. Sanghi (2008) points that Reports of the Department-related Parliamentary Standing Committee on Information Technology as well as expert committee "Reports on amendments to IT (amendment) Act 2000" submitted by *Mr. T D Maran,* Hon'ble Minister for Communications and Information Technology has noted that the field of cyber laws, being a nascent area, experience of its formulation and implementation are still evolving worldwide and more so in India. The Act is being made technology neutral with minimum change in the existing IT Act, 2000. This has been made by amendment of the relevant sections of Information Technology Act to provide for electronic signature with digital signature as one of the types of electronic signature and by enabling the details of other forms of electronic signature to be provided in the Rules to be issued by the Central Government from time to time. This is an enabling provision for the Central Government to exercise as and when the technology other than digital signature matures. Certain related amendments to this effect have been made. For a country like India where we are trying to enhance the positive use of Internet and working towards reducing the digital divide, there is need to ensure that new users do not get scared away because of publicity of computer related offences. At the same time, it must be ensured that offenders do not go unpunished. This balancing spirit has been incorporated in the proposed amendments in relevant sections.

This bill calls for substitute of word digital signature with electronic In Chapter II of the principal Act, for the heading, the heading "Digital signature And Electronic Signature" shall be substituted. Insertion of new section 6(A) which deals with delivery of services to the public through electronic means authorise, by order, any service provider to set up, maintain and upgrade the computerized facilities and perform such other services as it may specify, by notification in the Official Gazette. Insertion of new section 10(A) after section 10 of the principal Act, it states validity of contracts formed through electronic means. The Central Government may, for the purposes of sections 14 and 15, prescribe the security procedures and practices regard to the commercial circumstances, nature of transactions and such other related factors as it may consider appropriate. Insertion of new section 40A; will lay down duties of subscriber of Electronic Signature Certificate (Rajesh Saini *v.* State of Himachal Pradesh, 2008)

Digital Signature Mandatory for Filing of Statutory Documents

Recently, mandatory use digital signatures have been launched by the Government of India for filing of statutory documents and other transactions of commercial documents by the help of electronic mode to facilitate e-commerce. The Ministry of Corporate Affairs, Government of India, has made it mandatory for registered companies to use digital signatures for e-filing of documents. All the company authorised signatories or the companies professionals who file documents on behalf of the companies are required to obtain digital signature certifications to enable e-filing have to acquire a digital signatures.

One of the major benefits of acquiring digital signatures is that it reduces cumbersome paper work, enhanced observance management, overall intelligibility from end to end e-governance, customer centric approach, building digitalized corporate operating and monitoring system. Digital signatures make the process swifter for filing of statutory documents.

Importance of Digital Signature in Business and Technical Issues

Digital signatures have emerged as a sophisticated mode in promotion of corporate sectors; especially with regards to business and technical issues which help in promotion of e-governance with regards to business and technical issues. Under section 48 of Act of 2000; provides for the establishment of 'Cyber Appellate Tribunal' hereinafter referred to as CAT; any person aggrieved by an order made by controller or an adjudicating officer under this Act may make an appeal to a CAT having jurisdiction in the matter. The IT Act also clearly mentions under section 61 that civil court do not have jurisdiction. Any person aggrieved by any decision or order of the CAT may file an appeal to the High Court within sixty days from the date of communication of the decision or order of the CAT to him on any question of fact or law arising out of such order.

CASE STUDY

If any person who dishonestly or fraudulently causes, sign's, seals, executes, and alters a document an electronic record or affixes his digital signature on any electronic record shall be liable for penal consequences. (M.K. Razdan *v.* The State, 2008)

Justice S.B. Sinha, (Hon'ble Judge of the Supreme Court of India) strongly argued: "Various new developments leading to various different kinds of crimes unforeseen by our legislature come to immediate focus. Information Technology Act, 2000 although was amended to include various kinds of cyber crimes and the punishments therefore, does not deal with all problems which are faced by the officers enforcing the said Act."

In a changing cyber world, the information may reside in several systems and can be deciphered using the process of computer. Under Indian Evidence Act and the IT Act, there are provisions relating to digital signatures, which make the digital records and digital signatures admissible. Once they are treated as documents, the action of the will falls under Sub-sections (3) and (4) of Section 110 of the Act and not under Sub-sections (1) and (2) of Sect 110 of the Act held by Supreme Court of India in (*State of Punjab and Ors. v. Amritsar Beverages Ltd. and Ors.,* 2006, para 17).

The IT Act, that provides legal recognition for transactions carried out by means of electronic data interchange and other means of electronic communication is commonly referred to as 'electronic commerce' (Diebold Systems Pvt. Ltd. *v.* Commissioner of Commercial Taxes, 2006). The digital signatures are the main degree to any transactions and it depends on its authenticity. The courts while deciding such critical issues must look that are unique to the signer, and are under the signer's sole control.

Proposed Amendment over Admissibility of Digital and Electronic Signatures

If the proposed bill becomes a law it will bring vital changes under the present regime of information technology sector in India as well as more clear direction over the admissibility of digital and electronic signatures before the court of law in a dispute related to electronic commerce.

The above said bill shall enhance monitoring cyber crimes and it shall also lay down harsh and rigorous penal consequences for disclosure of information with regards to security of national, important deals financial transitions and internal security against e- crimes. A bill proposes to establish the national nodal agency in respect of any critical information infrastructure protection is also proposed under a bill before the parliament of India. Under the proposed bill the power to issue directions for blocking for public access of any information through any computer resources which is not state friendly is provided to the Government of India,

Digital Signature and Future of Digital Economy of India

It cannot be denied that digital Signature uprising has brought about far-reaching transformation to the financial industry/institutions, such as fund intermediation. In present time, the financial transactions and settlements have been digitalized; many cyber technologies have been acquired by many national and international financial institutions and financial portal services which have evidentially resulted in boosting the faster methods of business transactions which are less cumbersome and are sound in their nature. It has also resulted in sound security and transparency concerns in regular transactions and promoting electronic commerce. A rapid uptake of digital signatures has supported the major concern of securities in electronic based transactions. The nations should craft their legislations and regulatory framework in such a manner which is supportive to the growth of the nation's economy.

Generally, corporate sector of any developing or developed nation is encouraging the utility of digital signatures in their main stream of transaction and promotion of economy. India has also mandated the use of digital signatures for filling mandatory documents and transactions. The government of India has taken strong initiative in mandating the digital signature which result the market to do good business transactions throughout the nation and international market.

Digital signature used in e-filing has facilitated networked business. India has recently witnessed 30 per cent growth but that is due to the mandatory requirements imposed by the MCA; digital signatures is a method to achieve e-governance which will lead to transparency systematic risk-management based e-business security policy (Otuteye, 2008).

The Security Exchange Board of India has also promoted the utility of digital signatures used in primary and secondary markets; brokers under such markets are allowed to issue contract notes authenticated by means of digital signatures provided that the broker has obtained digital signature certificate from Certifying Authority under the IT Act, 2000.

Other *Depository institutions in India* "National Security Depository Limited 'NSDL' and Centre Depository Service Limited 'CDSL', have also decided to expand the infrastructure of SPEED-e for electronic delivery of digitally signed contract notes by brokers to custodians / fund managers in pre-defined, uniform formats. Major benefit is to avoid hard cumbersome paper work in to modern sophisticated digital trends; digital signature has strengthened security issues in such deals. The key impact of digital signature on financial sector is evident in accuracy of IT infrastructure blended with internet technologies. Overall accuracy is the major apprehension in the financial transactions which are based on verified signature, certified servers and authenticated digital signatures.

The two big stock exchanges in India (i.e., 'National Stock Exchange of India' and 'Bombay Stock Exchange') are the leading stock exchanges of India, promoting the use of digital signatures in stock trading. Trading members to these stock exchange use digital signatures on contract notes; it is compulsory that digital signatures certificate should be obtained from certifying authority under the IT Act, 2000. The Ministry of Corporate Affairs of India has taken vital steps implementing mandatory system of digital signatures, for strengthening to meet the changing requirements of the market.

The major concern lies on international investments issues where any foreign investor interested in investing in any particular nation has to opt for right kind of business or corporate entity which best suits its purposes and takes care of liability issues and tax planning issues. It is argued that

there should be mandatory implementation of digital signatures which will be less burdensome, secured and faster in making investment in to a particular deal. Recently, government of India has mandated the directors of Indian companies (both Indian and foreigner directors) to obtain Digital Signature Certificate. It is one of the mandatory requirements for directors, of Indian companies, both Indian and foreigners, to engage in business transactions in the country.. Digital Signature Certificate (DSC) is required for all directors or authorized representatives of any company and professional who will require signing to Registrar of Companies forms or documents. Digital signature must be adopted by company representatives, professionals and others who are required to affix digital signatures for submitting an electronic Form (Hand Book MCA21, Pg. 37). Indian banking sector has taken vital step in mandating the use of digital signatures in normal day's transactions. Digital signature under Indian banking sector has emerged as major protection during transmission of cash within banks to banks and banks to costumers.

The Ministry of Corporate Affairs of India, has recently implemented the, information systems security guidelines for the 'Banking' and 'Financial Sectors', which will help in protecting the audit trail information's from deletion, modifications, fabrications or resequencing by use of digital signatures. Moreover, the guidelines also highlights the major use of digital signature in controls against malicious software, protection during transmission, repudiation and specifically spell out legal the advantages, roles and use of digital signature for banking and financial sector India. Digital signatures play a major role in addressing the issues relating to the maintenance of the keys, used for encryption, authentication of contracts.

It is hoped that the legal provision for mandatory adoption of digital signature in business transaction in India would have tremendous impact in the digital economy of the country. It will build confidence in promote e-government, e-commerce and e-banking in India.

CONCLUSION

In India, bold attempts have been made by the government to promote the use of digital signatures in commercial transitions as well as in strengthening laws to implement best security practices in electronic commerce transactions. Digital signatures are also addressed as tools of building 'trust' between parties. Courts have also recognized and authenticated the utility of modern sophisticated, high-tech digital records and digital signatures in commercial transactions and building affirmative tools of security and 'trust'.

It is believed that the whole implementation exercise is designed to build trust. Digital signatures facilitate secure electronic transactions, authentication, confidentiality, and integrity. It can be said that 'digital signature plays a major role to create trust in electronic environment'.

A bold attempt made by the Indian government in promoting and mandating the use of digital and electronic signatures has provided "feel safe" attitude to the banking industry and corporate affairs, which led towards has swifter trading process and complete secured system place. If the proposed bill converts in to a law, it will provide more bases to the new security concept and authentication process will be more adaptable to shifting legal prerequisites.

ACKNOWLEDGMENT

I am grateful to Mrs. Aruna Hyde, MA. B. Ed, CIEFL (NELTS), Hidayatullah National Law University, Raipur, C.G, India for her assistance and Dr. Esharenana E. Adomi, who graciously supported by providing thoughtful comments and advised throughout; I am immensely grateful for his encouragement.

REFERENCES

Aalberts., B.P., & Van der Hof, S. (1999). *Digital Signature Blindness Analysis of Legislative Approaches to Electronic Authentication*, (pp. 32). Retrieved on Nov.29, 2008 from; http://www.buscalegis.ufsc.br/arquivos/Digsigbl.pdf

Abichandani, R. K. (2004). *Controvertial Copyright Issues*, National Judicial Academy, Bhopal, Retrieved on August, 21, 2009, from http://gujarathighcourt.nic.in/Articles/roundtable.htm.

American Bar Association. (1996). *Digital signature guidelines: Legal Infrastructure For certification Authorities and Secure Electronic Commerce*, (2nd Ed.). Chicago: Author. Retrieved on Oct. 30, 2008, from American Bar Association website; http://www.abanet.org/scitech/ec/isc/dsg.pdf

Atkinson, R. D. (2001). The New Growth Economics: How to Boost Living Standards through Technology, Skills, Innovation, and Competition. *Blueprint Magazine*. Retrieved on 22 Feb 2009 from http://www.dlc.org/ndol_ci.cfm?contentid=2992&kaid=107&subid=123

Bae, K. S. (2001). Korea's e-commerce: Present and future. *Asia-Pacific Review, 8*(1), 77.

Basu, S., & Jones, R. P. (2002). *Legal Issues Affecting E-Commerce A Review of the Indian Information and Technology Act 2000,* 17th Annual BILETA Conference, Amsterdam. Retrieved on Oct. 22, 2008; from http://works.bepress.com/subhajitbasu/33

Batuk lal. (2006). *The Law of Evidence,* Allahabad, India: Central Law Agency.

Carlisle, A. & Lloyd. (2002). *Understanding PKI: Concepts, Standards, and Deployment Considerations,* (pp. 11-17). London: Addison-Wesley.

Cerina, P. (1998). The New Italian Law on Digital Signatures. 6 *CTLR,* 193.

Chan, A. H., & Gligor, V. (2002). *Information Security, (LNCS)*. Berlin: Springer. doi:10.1007/3-540-45811-5

Chwee Kin Keong and Others v. Digilandmall. com Pvt. Ltd, (2004) 2 SLR 594.

Cloud Corp. v. Hasbro Inc, 314 F.3d 289 (7th Cri. 2002).

Diebold Systems Pvt. Ltd. *v*. The Commissioner of Commercial Taxes, (2006). 144STC59(Kar).

Diffie, W., & Hellman, M. E. (1976). New Directions in Cryptography. *IEEE Transactions on Information Theory, IT-22*(6), 644–654. doi:10.1109/TIT.1976.1055638

Fellenstein, C., & Wood, R. (2000). *Exploring e-commerce, global e-business, and societies, 30*. Upper Saddle River, NJ: Prentice Hall.

Finocchiaro, G. (2002). Digital Signature and Electronic Signatures: The Italian Regulatory Framework After The d.lgs 10/2002. *Electronic Communication Law Review,* (9), 127.

Indian Petrochemicals Corporation Ltd. *v*. Union of India, (Decided on: Sept.19, 2006). (2006). MANU/GJ/8490.

Indira, C., & Stone, P. (2005). *International Trade Law*. London: Routledge Cavendish.

Information Technology Security Procedure Rules. *(2004)*.

Jueneman., R R. & Robertson, Jr., R. J. (1998). Biometrics and Digital Signatures in Electronic Commerce 38 *Jurimetrics* 427.

Michael, D. (2002), *Globalization Of The World Economy: Potential Benefits And The Cost And A Net Assessment*. Governing Stability Across the Mediterranean Sea: A Transatlantic Perspective, Columbia International Affairs Online. Retrieved on Dec.3,2008 from http://www.ciaonet.org/coursepack/cp09/cp09c.pdf

Ministry of Corporate Affairs. (n.d.). *MCA21 Stakeholder Handbook*. New Delhi, India: Tata Consultancy. Retrieved on July. 12, 2009, from, http://www.mca.gov.in/MinistryWebsite/dca/help/ProcessHandbook.pdf

M.K. Razdan *v*. The State and Shri Indukant Dixit, Crl.Rev.P. No. 861-62/2005, (Decided on March.03, 2008), Delhi.

Nortel., Shashi Kiran., Lareau P., & Lloyd., S. (2002). *PKI Basics - A Technical Perspective*. Retrieved on Nov.19,2008.from http://www.oasis-pki.org/pdfs/PKI_Basics A_technical_perspective.pdf

Otuteye., E. (2008), *A Systematic Approach to E- Business Security: Ninth Australian World Wide Web Conference*. Retrieved on August. 21, 2009, from http://ausweb.scu.edu.au/aw03/papers/otuteye/

Panagariya, A. (2000). E-Commerce, WTO and Developing Countries. *World Economy, 23*(8), 959–978. doi:10.1111/1467-9701.00313

Portable Mass Storage Device With Virtual Machine Activation, U. S. Patent No. WO/2008/021682, PCT/US2007/074399 (2008, Nov. 26). fig.6 Retrieved on Oct. 29, 2008, from http://www.wipo.int/pctdb/en/wo.jsp?IA=US2007074399andwo=2008021682andDISPLAY=CLAIMS

Professional Horizon. (2006). *Important Messages For e-Filing Under MCA21, The Chartered Accountant* (p. 1806). ICAI.

Rajesh Saini *v*. State of Himachal Pradesh, (2008). Cr. LJ 3712

Reed, C. (2007). Taking a side on Technology Neutrality . *SCRIPTed, 4*(3), 263. doi:10.2966/scrip.040307.263

Report of Standing Committee on Information Technology. *Raja Sabha Parliamentary Bulletin*. (n.d.). Retrieved on 22 Oct 2008 http://164.100.47.5/Bullitensessions/session-no/214/221008.pdf

Reserve Bank of India. (2002). *Information Systems Security Guidelines for the Banking and Financial Sector.* Retrieved on Feb 22, 2009, from http://www.prsindia.org/docs/bills/1168510210/bill93_2007112393_Press_Release_for_IT_Act_Amendment.pdf

Russell, I. (2008, Nov.27). Sale of digital signature double in 18 months. *Business Standard.* Retrieved on 27 Jan 2009. Business-standard http://www.businessstandard.com/india/storypage.php?autono=297405.

Sharma, B. R. (2005). *Bank Frauds Prevention and Detection, 61.* New Delhi: Universal Law Publication.

State of Punjab and Ors. *v.* Amritsar Beverages Ltd. and Ors., AIR 2006 SC 2820. (2006). (Paragraph 7 & 17, pp. 3488-89).

Stovekraft Private Limited, rep. by its Managing Director Rajendra Gandhi v The Joint Director, Directorate of Revenue Intelligence and Ors. *(2007). 214 ELT179(Kar)*

Sundaram Brake Linings Ltd *v.* Kotak Mahindra Bank Ltd., A. No. 8078 of 2007 in C.S. No. 1072 of 2007, High court of Madras. (2008). MANU/TN/0938.

Swindells, C., & Henderson, K. (1998) Legal regulation of Electronic Commerce, 3. *The Journal of Information, Law and Technology,* 3. Retrieved on Oct. 29, 2008 from http://www2.warwick.ac.uk/fac/soc/law/elj/jilt/1998_3/swindells/#a3

Tulsian, P. C. (2006). *Business Law.* New Delhi, India: Tata McGraw-Hill.

United Nation. (2002). *56th Sess. General Assembly Resolution. A/RES/56/80.* Model Law on Electronic Signatures of the United Nations Commission on International Trade Law.

United Nations. (2002). *UNCITRAL Model Law on Electronic Signatures with Guide to Enactment.* New York: United Nations Publication. Retrieved on Jan.02, 2009, from UNICITRAL website http://www.uncitral.org/pdf/english/texts/electcom/ml-elecsig-e.pdf

United Nations Conference on Trade and Development. (2000). *A Positive Agenda for Developing Countries: Issues for Future Trade Negotiations,* (pp. 474). New York: UN Publication.

Ved Ram and Sons Pvt. Ltd *v.* Director, (2) AWC 2053 (UP), (2008).

Winn, J. K. (1998). Open Systems, Free Markets, and Regulation of Internet Commerce. *Tulane Law Review, 1177,* 72.

KEY TERMS AND DEFINITIONS

Cyber Security: Protecting information, equipment, devices computer, computer resource, communication device and information stored therein from unauthorised access, use, disclosure, disruption, modification or destruction.

Electronic Form Evidence: Any information of probative value that is either stored or transmitted in electronic form and includes computer evidence, digital audio, digital video, cell phones, and digital fax machines.

Electronic Record: An image or sound stored, received or sent in an electronic form or micro film or computer generated micro fiche.

Electronic Signature: Authentication of any electronic record by a subscriber by means of the electronic technique and it also includes digital signature. Or it can be understood as electronic sound, symbol or process attached to or logically associated with a record and executed or adopted by a person with the intent to sign the record.

Electronic Signature Certificate: A certificate issued by certifying authority and it includes 'Digital Signature Certificate' in such a form as prescribed by central government of India.

Intermediary: With respect to any particular electronic records, means any person who on behalf of another person receives, stores or transmits that record or provides any service with respect to that record and includes telecom service provid-ers, network service providers, internet service providers, web hosting service providers, search engines, online payment sites, online-auction sites, online-market places and cyber cafes.

Secured Electronic Signature: An electronic signature shall be deemed to be a secure electronic signature if the signature creation data, at the time of affixing signature, was under the exclusive control of signatory and no other person; and the signature creation data was stored and affixed in such exclusive manner as may be prescribed.

Chapter 6
Context for ICT's Role in South African Development

Udo Richard Averweg
Information Services, eThekwini Municipality and University of KwaZulu, South Africa

Geoff Joseph Erwin
The Information Society Institute (TISI), South Africa

ABSTRACT

This chapter discusses that information and communication technologies (ICTs) can (and should) be used to disseminate information and participation to disadvantaged communities in order to foster socio-economic development in South Africa. The objective of this chapter is twofold: (1) how should ICT policies and frameworks in South Africa be implemented (e.g. by a "top-down", "bottom-up" or "mixed approach" paradigm) in order for the South African government to achieve its socio-economic goals?; and (2) can socio-economic development in South Africa be effectively assisted by the use of ICT? A discussion of these points may assist in the formulation of national ICT policies in South Africa and thereby spawn the setting up of social appropriation of ICT advancement programs. Such programs are particularly relevant to the digital divide, for fostering socio-economic development and in promoting an inclusive information society in South Africa.

INTRODUCTION

South Africa's liberation struggle was led by the African National Congress (ANC) and in 1994 the first democratically elected State President was Nelson Rolihlahla Mandela. Since the earliest days of the post-apartheid era, the ANC-led government has adopted a position that information and communication technologies (ICTs) can

(and should) be used to disseminate information and participation to disadvantaged communities in order to foster socio-economic development. Socio-economic development "refers to continuous improvement in the well-being and in the standard of living of the people" (see www. nepad.org).

The impact of this development focus is evidenced by the fact that South African government officials (e.g. South Africa's Minister of Communications) accept that ICT can play an

DOI: 10.4018/978-1-61692-012-8.ch006

important role in accelerating development in rural areas (Snyman & Snyman, 2003). Furthermore, the South African government actively supports the promotion and realisation of the Millennium Development Goals (MDGs) through the use of ICT (UnitedNations, 2003). As part of its strategy to promote economic growth, the South African government has implemented a plan to promote the adoption of ICT, especially Internet technology (South Africa, 2005). However, the high cost of broadband access in South Africa and the limited access of this technology to all South Africa citizens, remains a problem (Masango, 2007). Micrososft South Africa (MSA) and Universal Services Agency in South Africa) (USASA), an ICT parastatal, announced plans for a partnership to spread access to technology to an increased number of people in South Africa (Masango, 2005). However, the mechanism, if any, by which increasing access to technology promotes socio-economic development was left open. The question thus arises whether a policy framework should be formulated which will guide future research in this regard.

The implications for socio-economic development policy and implementation from the South African context discussed in this chapter are:

- How should ICT policies and frameworks in South Africa be implemented (e.g. by a "top-down", "bottom-up" or "mixed approach" paradigm) in order for the South African government to achieve its socio-economic goals?; and
- Can socio-economic development in South Africa be effectively assisted by the use of ICT?

Discussion of these points may assist in the formulation of national (and provincial) ICT policies in South Africa and thereby spawn the setting up of social appropriation of ICT advancement programs. Such programs are particularly relevant to the digital divide, and the emergence

of the multi-disciplinary field of Community Informatics (CI) supporting and promoting an inclusive information society in South Africa. CI seeks to realize the social appropriation of ICT for local benefit. Participation in an inclusive information society is a current aim of the South African government, enabled by ICT and via the national Department of Communications (DOC), and reflected in national government structures and frameworks such as the Meraka e-Skills Institute, and the related National e-Skills Dialogue Initiative (Ne-SDI) (DOC, 2009; ITWeb, 2009).

The objectives of this paper are to set out the context in which a developing country such as South Africa is attempting to include all sectors of society in an inclusive information society, to position this within the overall activity for *socio-* as well as economic development, and to explore the relevance of mechanisms by which this can be achieved, potentially by a series of initiatives, both "top-down" from government, "bottom-up" from civil society, and "mixed".

BACKGROUND

With the democratisation of South Africa in 1994, coupled with its current level of technological development, the country is a beacon of technological hope for the rest of Africa. As declared by a previous South African Minister of Communications, Dr Ivy Matsepe-Casaburi, ICTs present Africans with an opportunity to leapfrog decades of development into becoming information societies. Government policies in South Africa (ICT Charter) are being established which attempt to ensure that all citizens have the opportunity to access and effectively use ICT in order to enable them to participate fully in educational, social and economic activities and democratic processes (Cullen, 2002). In this regard, Ne-SDI is an initiative to grow thought leadership and intellectual discourse among and between major sectors of society, namely Higher

Education, business / industry, government itself and civil society.

Various spheres of interest in the societal application of ICT exist, such as the digital divide, information gap, CI and the information society. All embrace the idea that ICT can be employed to promote economic growth. Technology plays a defining role in understanding the concept of the information society (Oyedemi & Lesame, 2005). There is no universally agreed definition for the term "information society". The South African Presidential National Commission on Information Society and Development (PNC on ISAD) defines "information society" as one that has

- Built the necessary capacity to use ICT maximally to accelerate social and economic development;
- Set goals for such development; and
- Formulated policy and legislative measures to realise these goals.

According to the PNC on ISAD, South Africa's ICT policy assumes that there can be no information society without proper ICT infrastructure and delivery mechanisms adapted to the global development needs of the people. The second (post 1994) South African President Thabo Mbeki, stated that in co-operation with international and regional organisations, government contributes to the information society through tax and other incentives. The South African government has introduced "managed liberalisation" of the telecommunications (and the communications sector in general), which is designed to encourage the entry of new telecommunications companies and to foster competition within the sector (Lesame, 2005). Managed liberalisation recognised that it was no longer reasonable for the South African government to own and operate huge telecommunications and broadcasting infrastructures indefinitely but to rather foster economic growth and development by opening these infrastructures to a wider ownership and usage by the private sector.

Furthermore, this infrastructure was seen as key to the development of previously disadvantaged and under-served communities within the country.

Moving from Policy Formulation to Implementation

The failure of some governments to achieve the goals of ICT policies and the recognition that ICT policies cannot be understood from the means of their execution, provide a rationale for policy implementation research. Policy implementation studies, predominantly in the form of documented cases studies do offer a variety of explanations, and these are commonly referred to as the "information gap." The arguments and rationale for failure for reducing such information gaps differ according to how the process of policy implementation was conceptualised by the different researchers and authors (Grossenbacher, 2000). From a review of the literature on policy implementation research, two main conceptual frameworks are identified: "top-down paradigm" and "bottom-up approach."

In the "top-down paradigm", Sabatier & Mazmanian (1979, 1980) perceived policy making as linear with a clear division between policy formulation and policy execution. Policy formulation is seen as political with value judgments while policy execution (i.e. policy implementation) is administrative or managerial. Policy implementation is therefore seen as primarily a technical process to be carried out by administrative agencies at the national and sub-national governments. In South Africa there are three levels of government: national central government, the nine (second-tier) provincial governments and the (third-tier) local government or metropolitan/district municipalities. In the case of ICT policies this will mean that decisions by politicians and bureaucrats within the Department of Communications are communicated to managers and planners at central levels who operationalise policies by designing appropriate programs (with guidelines), rules and monitoring systems. These are then cascaded to the provincial governments and municipalities to be put into practice.

In the "bottom-up approach," Barrett & Fudge (1981) and Hjern & Hull (1982) argue that implementation should be regarded as an integral part of the policy process rather than as administrative "follow on" from policy making. The political process by which policy is negotiated and modified during its formulation do not stop when initial policy decisions have been made but continue to influence policy through the behaviour of those responsible for its implementation and those affected by the policy, who act to protect and enhance their own interests (Grossenbacher, 2000). Instead of placing a focus on formal organisational hierarchies, communication and control mechanisms, these types of models place more emphasis on

- The multiplicity of actors and agencies involved in the linkages between them;
- The interactions taking between them (particularly in terms of negotiation and bargaining); and
- Their value systems, interests, power bases and relative autonomies.

In this view, implementers very often play a significant role in policy implementation. They do not act as downward-cascaders of policies (such as ICT policies) but act as active participants in a complex process that informs policy upwards as well. Thus implementers may change the way a policy is implemented, or even re-define the objectives of the policy because they are closer to the problem and the local situation (Grossenbacher, 2000). This approach facilitates the resolution of unintended consequences that may arise during the course of policy implementation. Since 2000 the new multi-disciplinary field of CI has arisen reflecting aspects of a "bottom-up" approach towards socio-economic development (Taylor, Erwin & Wesso, 2006; Erwin & Taylor, 2006; Gurstein, 2007).

CI is the discipline and systematic set of approaches to the research, design, development, implementation and evaluation that underpins these activities. The potential impacts of CI fall into five key areas:

- **Strong democracy:** democratic participation via a meaningful association of citizens within a community, including political discussion and online voting among others;
- **Social capital development:** civic engagement, social networks, interaction on matters of public concern, solidarity, altruism, loyalty and reciprocity;
- **Individual empowerment:** "the ability of people to gain understanding and control over personal, social, economic, and political factors in order to take action to improve their life situations";
- **Enhanced sense of community:** facilitating the transition from agrarian to urbanised societies, extending local cohesion;and
- **Economic development opportunities:** enabling shared and inclusive community socio-economic development, small business encouragement, economies of disaggregation.

ICT simultaneously increases the porosity of nation state boundaries and contributes to the growth of trans-national economies of scale. It also provides a means for South Africa to balance the disparities in the localised social and economic impacts of these developments through the use of ICT in locally relevant and beneficial ways. As part of this process communities are, through the use of ICT being involved in local, national and international applications to realise effective social cohesion, civic participation and equity.

CI is basically a framework and set of collaborative practices which begin with ICT as a set of resources and tools that individuals and communities can use to provide access, uptake and usage of

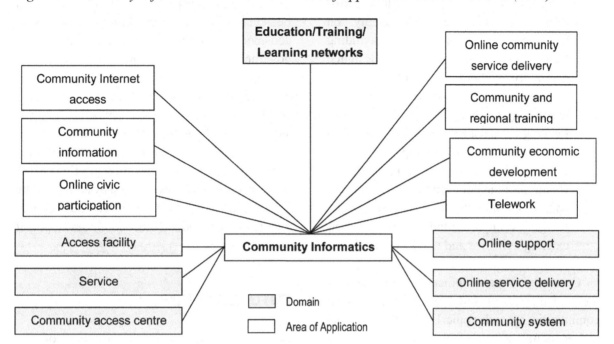

Figure 1. Community Informatics: domains and areas of application. Source: Mitrovic (2006).

information management, sharing and processing. Through this use of ICT it is possible to achieve community benefits and the electronically-enabled development of community practice (Gurstein, 1999, 2003; Taylor, 2004). This approach to local socio-economic development includes areas such as electronic commerce, community and civic networks, community Technology Centres, electronic democracy, cultural enhancement, and online participation (O'Neil, 2002) and social networks; combined with local economic development, cultural issues, civic activities, community based health and environmental initiatives, online support, and service delivery, training and learning networks, and "telework" (Gurstein, 2003), as shown in Figure 1.

CI includes an area of application encompassing "Education/Training/Learning networks", and the CI knowledge base has a rich body of experience (case studies) regarding the evaluation of community networks and community Technology Centres (also known as "telecentres") and in the development of indicators that can be used to

gauge the effects of such centres on local communities. Furthermore, this approach to socio-economic development is helpful in understanding other important related issues directly or indirectly, such as[1]:

- The introduction of ICT as a tool to help communities meet their needs and aspirations and in the targeting of effective uptake and usage of ICT, rather than the simple provision of *access* as the basis for establishing ICT priorities, funding, measurement and evaluation criteria;

- Supporting the community with the necessary inputs to seed initiatives and as a basis for the continuance of investment, not only in ICT but in the resultant development outcomes; resulting in broadly embedding the applications in the local community and enhancing sustainability;

- Ensuring a priority for initiatives to address the real community based needs of the target communities;

- Focusing on how to use ICT to improve community outcomes, not simply on bringing ICT to the poor and avoiding technology-driven approaches designed externally without a socially-embedded purpose i.e. needs-driven rather than technology-driven;
- Leveraging existing community resources–organisational, leadership, and skills– and involving community groups; and
- Using participatory approaches that encourage communities to define the need, to design the implementation based on their existing networks and to take ownership.

Currently, CI-based indicators for the assessment of telecentres and their impact on local communities are embryonic. Evaluating the social and economic effects of ICT interventions is a complex problem, demanding multi-disciplinary approaches (O'Neil, 2002; ITU 2006).

INFORMATION AND COMMUNICATION TECHNOLOGIES

The use of ICT is not widely socially dispersed in South Africa. The South African government realises this shortcoming and efforts are underway to "get" ICT to rural areas and to ensure that rural people have access to digital and online technologies. The USASA is an agency which has been set up by the government to facilitate the provision of universal access. Making ICT and the associated benefits available in most rural areas is a policy implementation challenge faced by many governments. South Africa, like many developing countries, still has a long way to go to develop and implement its information policy (Oyedemi & Lesame, 2005). Such policies must be determined according to the different users and their specific access needs to ICT. This raises the question of whether such policies in South Africa should be "universalised" and implemented by

using a top-down paradigm or whether it will be better, following certain kind of standards, to consider the features of each of the nine provincial governments (or municipalities) using a bottom-up approach? While we do not specify which approach should be adopted, we do contend that ICT policy frameworks (national and provincial) should be so aligned that they facilitate community engagement and participation towards local decision-making and social appropriation of ICT for local needs (bold style added by authors). The opportunities presented by the discipline of CI may augment ICT-based socio-economic development in South Africa.

FUTURE TRENDS

Access to ICT has dominated the agenda of telecommunications policies in South Africa since the aim is to redress past inequities (prior to 1994). It is through the implementation of current policies that a larger proportion of South African citizens can have access to telecommunications from which they were previously excluded. ICT are now a standard operating tool in today's information society (Jacobs & Herselman, 2006). The long-term goal for the South African government should be to lead the formulation and implementation of ICT policies and frameworks to obtain local benefits in the information society. ICT must be regarded as a pre-requisite for economic and social development in South Africa. ICT should be seen as *tools* for facilitating socio-economic development as technology *per se* does not stimulate socio-economic development.

CONCLUSION

Socio aspects of development introduce major issues not normally addressed or embraced by *economic* development approaches. Traditional projects and programs often measured success by quantifiable items, such as the number of computers installed, the number of people using the Internet, the amount of money spent on ICT-related infrastructure. The emergence of the World Wide Web allowed connectivity to be extended to participation in society by an ill-defined audience. This has threatened the previous paradigm of government providing information and services by giving this audience direct knowledge and access to information and process services of government, including collaboration on policy and practice. This socio-economic role builds a democratic form of active citizenship and is altering both the role and methods of all levels of government, civil society, Higher Education Institutions and corporations. Socio aspects, added to economic aspects, of development look towards an inclusive society collaborating on societal issues, including economic policy. Society, now with unbridled ability for participation between groups, including government, is undergoing a paradigm shift. Government should welcome this move to government of the people, for the people, and now, *by* the people.

REFERENCES

Barrett, S., & Fudge, C. (Eds.). (1981). *Policy and Action*. London: Methuen.

Cullen, R. (2002). Addressing the digital divide. *Online Information Review*, *25*(5), 311–320. doi:10.1108/14684520110410517

DOC. (2009). Launch of the National e-Skills Dialogue - From Agrarian, Industrial to Information Society. Retrieved July 2009, from http://www.doc.gov.za/index.php?option=com_content&task=view&id=309&Itemid=457.

Erwin, G. J., & Taylor, W. (2006). Assimilation by Communities of Internet Technologies: Initiatives at Cape Peninsula University of Technology, Cape Town . In Marshall, S., Taylor, W., & Yu, X. (Eds.), *Encyclopedia of Developing Regional Communities with Information and Communication Technology* (pp. 40–46). Hershey, PA: Idea Group Publishing.

Granqvist, M. (2005). Looking critically at ICT-4Dev: the case of Lincos. *Journal of Community Informatics*, *2*(1), 21–34.

Grossenbacher, K. (2000). *Implementing the District Health System in KwaZulu-Natal: A Systemic Inquiry into the Dynamics at the Policy/Practice Interface*. Unpublished masters dissertation, University of Natal, Durban, South Africa.

Gurstein, M. (1999). Flexible networking, information and communications technology and local economic development, *First Monday 4* (2). Retrieved March 15, 2003, from http://www.firstmonday.org/issues/issue4_2/gurstein/index.html.

Gurstein, M. (2003, December). Effective use: a community informatics strategy beyond the digital divide, *First Monday, 8*(12). Retrieved June 18, 2004, from http://firstmonday.org/issues/issue8_12/gurstein/index.html.

Gurstein. M. (2007). *What is Community Informatics (and Why Does It Matter)?* Retrieved February 2008, from http://eprints.rclis.org/archive/00012372/01/What is Community Informatics reading.pdf.

Hjern, B., & Hull, Ch. (1982). Implementation Research as Empirical Constitutionalism. *European Journal of Political Research, 10*(June), 105–115. doi:10.1111/j.1475-6765.1982.tb00011.x

ITU. (2006). Measuring ICT for Social and Economic Development. *World Telecommunication/Ict Development Report*. Retrieved January 2007, from http://www.itu.int/publications.

ITWeb. (2009). *Meraka boosts e-skills*. Retrieved July 2009, from http://www.itweb.co.za/sections/business/2009/0903241046.asp?S=IT%20in%20Government&A=ITG&O=google, 24 March 2009

Jacobs, S. J., & Herselman, M. E. (2006). Information Access for Development: A Case Study at a Rural Community Centre in South Africa. *Issues in Informing Science and Information Technology*, *3*, 295–306.

Krishna, A. (2002). *Active Social Capital, Tracing the Roots of Development and Democracy*. New York: Columbia University Press.

Lesame, Z. (2005). Bridging the digital divide in South Africa. In N. C. Lesame (Ed), *New Media. Technology and policy in developing countries* (pp. 17-29). Hatfield, Pretoria, South Africa: Van Schaik Publishers.

Masango, D. (2005). *Partnership to bring IT to MPCCs countrywide*. Compiled by the Government Communication and Information System, 3 Oct. Retrieved September 22, 2008, from http://www.buanews.gov.za/news/05/05100316451001.

Masango, D. (2007). *ICT Key for Economic Growth, development*. Compiled by the Government Communication and Information System, 23 Oct. Retrieved September 22, 2008, from http://www.buanews.gov.za/news/07/07102316151004.

Mitrovic, Z. (2006). *ICT and small business development: the importance of government support and community involvement*. Unpublished working paper.

O'Neil, D. (2002). Assessing community informatics: a review of methodological approaches for evaluating community networks and community technology centres. *Internet research: electronic networking applications and policy*, *12*(1), 76-102.

Oyedemi, T., & Lesame, Z. (2005). South Africa: an information Society? In N. C. Lesame (Ed) *New Media. Technology and policy in developing countries* (pp. 75-97). Hatfield, Pretoria: Van Schaik Publishers.

Sabatier, P., & Mazmanian, D. (1979). The Conditions of Effective Implementation. *Policy Analysis*, *5*(Fall), 481–504.

Sabatier, P., & Mazmanian, D. (1980). The Implementation of Public Policy: A framework of Analysis. *Policy Studies Journal: the Journal of the Policy Studies Organization*, *8*, 538–560. doi:10.1111/j.1541-0072.1980.tb01266.x

Schware, R. (Ed.). (2005). *E-Development: From Excitement to Effectiveness*. Washington, DC: World Bank Group.

Snyman, M., & Snyman, R. (2003). Getting information to disadvantaged rural communities: the centre approach. *South African Journal of Library and Information Science*, *69*(2), 95–107.

South Africa (2005). South Africa Millennium Goals. *Country Report 2005*.

Taylor, W. (2004). Community Informatics: The basis for an emerging social contract for governance, research and teaching. *Building & Bridging Community Networks: Knowledge, Innovation & Diversity through Communication*. Brighton, UK, 31 March and 1-2 April.

Taylor, W. J., Erwin, G. J., & Wesso, H. (2006). *New public policies for the emerging information society in South Africa - a strategic view. Governments and Communities in Partnership conference*, Centre for Public Policy, University of Melbourne, 25-27 September. Retrieved September 2007, from http://www.public-policy.unimelb.edu.au/conference 06/Taylor W.pdf.

UnitedNations (UN). (2003). *Tools for Development - Using Information and Communications Technology to Achieve the Millennium Development Goals*. Working Paper, United Nations ICT Task Force, August.

KEY TERMS AND DEFINITIONS

Community Informatics (CI): Refers to an emerging set of principles and practices concerned with the social appropriation of information and communication technologies for the personal, social, cultural or economic development of and within communities.

Digital Divide: Refers to the unequal access by some members of society to information and communication technologies and the unequal acquisition of related skills.

Information Society: Is a society in which the creation, distribution, diffusion, use, integration and manipulation of information is a significant economic, political, and cultural activity.

Policy Formulation: Is a process that typically includes an attempt to assess as many areas of potential policy impact as possible, to lessen the chances that a given policy will have unexpected or unintended consequences.

Social Network: Is a social structure made of nodes (which are generally individuals or organisations) that are tied by one or more specific types of interdependency (e.g. values, visions, ideas, etc).

Socio-Economic Development: Refers to continuous improvement in the well-being and in the standard of living of the people.

Telecentres: Are a public place where citizens can access computers, the Internet, and other digital technologies that enable them to gather information, create, learn, and communicate with others while they develop essential digital skills.

World Wide Web: (Commonly abbreviated as "the Web") is a very large set of interlinked hypertext documents accessed via the Internet.

ENDNOTE

¹   See, for example, Gurstein (2003); Krishna (2002); Schware (2005); Granqvist (2005) for a more detailed insight into these issues.

Chapter 7
Cybercrime Regulation:
The Nigerian Situation

Alex Ozoemelem Obuh
Delta State University, Nigeria

Ihuoma Sandra Babatope
Delta State University, Nigeria

ABSTRACT

This chapter discusses cybercrime and cybercrime regulation in the Nigeria. It gives the meaning to cybercrime, types of cybercrimes (of which advance fee fraud is the most prevalent in Nigeria), means of perpetrating cybercrimes, the current situation and efforts towards combating cybercrime in Nigeria.

INTRODUCTION

Advancements in information and communication technologies (ICTs) have led to the representation of different types of information in electronic formats. The consequence is that currently text, pictures and voice can all be digitised. Along with these geometric changes in information presentation and distribution are tandem demands in user expectations for more rapid, open, and global access to information than has been available in the past. However, this migration from traditional communication medium to the new mediat seems to constitute a threat to the existence of a number of traditional print institutions and has provided a platform for fraudulent (criminal) Internet activi-

ties. Computers are increasingly more affordable and Internet connectivity is also becoming commonplace. The introduction and embracement of the Global System for Mobile Communication (GSM) in Nigeria and the influx of digital and online services such as the MP3 players, Ipod, cell phones with internet access and blogs (instant news reporting on personal and corporate web pages) (Longe & Chiemeke, 2008)

Cybercrime is a major concern to the global community. The introduction, growth, and utilisation of information and communication technologies (ICTs) have been accompanied by an increase in criminal activities (Parker, 1998). With respect to cyberspace, the Internet is increasingly used as a tool and medium by transnational organised crime (Lyman & Potter, 1998). Cybercrime is an obvious form of international crime that has been

DOI: 10.4018/978-1-61692-012-8.ch007

affected by the global revolution in ICTs (Parker, 1998). As a recent study noted, cybercrimes differ from terrestrial crimes in the following four ways (McConnell, 2000):

- They are easy to learn how to commit;
- They require few resources relative to the potential damage caused;
- They can be committed in a jurisdiction without being physically present in it; and
- They are often not clearly illegal.

On such a basis, cybercrimes present new challenges to lawmakers, law enforcement agencies, and international institutions. This necessitates the existence of an effective supra-national as well as domestic mechanisms that monitor the utilisation of ICTs for criminal activities in cyberspace.

Nigerian 419 scam has become a major concern for the global community. The introduction, growth and utilization of information and telecommunication technologies (ICTs) have been accompanied by an increase in illegal activities. With respect to cyberspace, anonymous servers, hijacked emails and fake websites are being used as a tool and medium for fraud by cyber scammers. Nigerian advance fee fraud on the Internet is an obvious form of cybercrime that has been affected by the global revolution in ICTs. This form of crimes is not exclusive to advance sums of money to participate into business proposals but also covers romance, lottery and charity scams. The term '419' is coined from section 419 of the Nigerian criminal code (part of Chapter 38: Obtaining Property by false pretences; Cheating) dealing with fraud. Currently, the axiom '419' generally refers to a complex list of offences which in ordinary parlance are related to stealing, cheating, falsification, impersonation, counterfeiting, forgery and fraudulent representation of facts (Tive, 2006).

According to 2007 Internet Crime Report prepared by the National White Collar Crime Centre and the FBI, Nigeria currently ranks third in the world with 5.7 per cent of perpetrators of cybercrime (2007 Internet Crime Report). The Nigerian government has over the years enacted far-reaching laws aimed at checkmating transnational organized crime and punishing the perpetrators of these crimes. Efforts in regulating cybercrimes such as advance fee fraud and 419 are reflected in the Criminal Code Act, Economic and Financial Crimes Commission Act 2004, Computer Security and Critical Information Infrastructure Protection Bill 2005 and Advance Fee Fraud and other Fraud Related Offences Act 2006.

This chapter is aimed at explaining the concept of cybercrime especially as it relates to Nigeria, issues relating to cybercrime legislation and suggests ways of getting out of these problems in the present days of internet usage and applications.

BACKGROUND

Cybercrime is generally regarded as any illegal activity conducted through a computer. Cybercrime is any criminal activity employing an information system (which may not be computerized) as the channel through which it is committed (Parker, 1998). It is illegal computer -mediated activities which often take place in the global electronic networks (Thomas & Loader, 2000). Cybercrime is when criminals use computers or networks as a tool, place, or target for criminal activity and behavior. The evolvement of cybercrime has affected law enforcement agencies and society. Enforcement has led to the creation of laws, policies, and legislature. Law enforcement agencies must vigorously fight and prevent cybercrime in order to help create a society that is safer (Thomas, 2006).

Cybercrime is a major problem faced by businesses attempting to establish and maintain an online presence (Smith & Rupp, 2002), and cybercrime attacks can potentially be just as damaging to a nation's infrastructure as attacks by classical criminals. Computer-related crime includes theft of telecommunications services or computer services by using hacking techniques. Servers and websites could be targets of denial-of-service attacks, viruses and worms. Computers are also used as instruments to commit crime, such as modification of data, electronic vandalism, forgery and counterfeiting, information piracy, industrial espionage and copyright infringement. There are many types of computer-related crime involving attacks on banks or financial systems, as well as fraud involving transfer of electronic funds. Other problems involve telemarketing and "phishing" or spoofing spam. Existing offences such as extortion and harassment are also carried out online. In recent years, increasing attention has been devoted to the relation between terrorism and the Internet, as the Internet is being used to facilitate terrorist financing and as a logistics tool for planning terrorist acts (United Nations, 2005).

Cybercrime uses the unique features of the Internt: sending of e-mail in seconds, speedy publication/ dissemination of information through the web to anyone on the planet. Computer attacks can be generated by criminals from anywhere in the world, and executed in other areas, irrespective of geographic location. Often these criminal activities can be faster, easier and more damaging with the use of the Internet (Awe, n.d.).

Regulation is a rule or order prescribed for management or government; prescription; a regulating principle; a governing direction; precept; law; as, the regulations of a society or a school. According to the free dictionary (2009), regulation is the act or process of regulating or a rule, principle, or condition that governs procedure or behavior. Also according to Merriam-Webster dictionary (2009) defines regulation as a rule or order issued by an executive authority or a regulatory agency of a government and having the force of law.

Cybercrime Regulation therefore implies a regulating process which entails a rule, principle or conditions that governs behaviour or procedure as they relates to cybercrime. And such must be issued by an executive authority or a regulatory agency of a government and enforceable by law.

Wilson (2000) cites the need to combat computer crime, cyber terrorism and information warfare on parallel paths. Development of effective security countermeasures for each and every type of attack are needed to control potential threats.

CYBERCRIME IN NIGERIA

A nation with high incidence of crime cannot grow or develop. That is so because crime is the direct opposite of development. It leaves a negative social and economic consequence (Sylvester, 2001). For Nigeria, a nation in the process of saving her face regarding cybercrimes, efforts are now being directed at the sources and channels through which cybercrimes are being perpetuated: the most popular one being Internet access points.

Most of the cybercrimes perpetrated in Nigeria generally are targeted at individuals and not necessarily computer systems, hence they require less technical expertise. The damage done manifests itself in the real world. Human weaknesses such as greed and gullibility are generally exploited. The damage dealt is largely psychological and financial. These crimes are similar to theft, and the likes that have existed for centuries offline even before the development of high-tech equipment. Through the internet, the same criminals or persons with criminal intents have simply been given a tool which increases their potential pool of victims and makes them all the harder to trace and apprehend (Aghatise, 2006).

The challenge in fighting cybercrimes today relates to the fact that cybercrimes have been in existence for only as long as the cyber space exists. This explains the unpreparedness of society and the world in general towards combating them (Longe & Chiemeke 2008). Numerous crimes are committed daily on the Internet with Nigerians at the forefront of sending fraudulent and bogus financial proposals all over the world. Criminals involved in the advance fee fraud schemes (419) are now popularly referred to as "yahooboys' in

Nigeria. The nation has therefore carved a niche for herself as the source of what is now generally referred to as '419' mails named after Section 419 of the Nigerian Criminal Code (Cap 777 of 1990) that prohibits advance fee fraud.

The following categories of crime are the most common ones in the Nigerian Internet landscape (Longe & Chiemeke, 2008).

(a) **Hucksters:** Hucksters are characterized by a slow turnaround from harvest to first message (typically at least 1 month), a large number of messages being sent to each harvested spam-trapped addresses, and typical product based Spam (i.e. Spam selling an actual product to be shipped or downloaded even if the product itself is fraudulent). Email addresses are obtained from Internet access points using e-mail address harvesting software (web spiders) such as E-Mail Extractor Lite1.4. These tools can automatically retrieve e-mail addresses from web pages. They are therefore referred to as harvesters (Longe, & Chimeke, 2006)

(b) **Fraudsters:** The fraudsters are characterized by an almost immediate turnaround from harvest to first message (typically less than 12 hours), only a small number of messages are sent to each harvested addresses (e.g. phishing, "advanced fee fraud"-419 from the Nigerian perspective). Fraudsters often harvest addresses and send only a message to them all at a particular time. The tools for getting addresses are mailing address extractors (Longe & Chiemeke, 2006).

(c) **Piracy:** Piracy involves the illegal reproduction and distribution of software applications, games, movies and audio CDs. (Longe, 2004). This can be done in a number of ways. Usually pirates buy or copy from the Internet an original version of a software, movie or game and illegally make copies of the software available online for others to download and use without the notification of the original owner of the software. This is known as Internet piracy. Modern day piracy may be less dramatic or exciting but is far subtler and more extensive in terms of the monetary losses the victim faces. This particular form of cybercrime may be the hardest of all to curb as the common man also seems to be benefiting from it.

(d) **Hacking:** Young Nigerians can be observed on daily basis engaging in brainstorming sessions at cybercafés trying to crack security codes for e-commerce, Automated Teller Machine (ATM) cards and e-marketing product sites. The surprising thing is that even with their low level of education or understanding of the intricacies of computing techniques, they get results! Phishing is also becoming popular as criminals simulate product websites to deceive innocent Internet users into ordering products that are actually non-existent. Phishing refers to imitating product and e-commerce web pages in order to defraud unsuspecting users. This method is used mostly to obtain credit card numbers.

HAVENS FOR PERPETRATING CYBERCRIMES IN NIGERIA

Cybercafés provide havens for perpetration of cybercrimes in Nigeria (Adomi & Igun, 2008) Apart from the availability and usage of Internet facilities in cyber cafes for scam mails and other cybercrimes, the evolution of fixed wireless facilities in the Nigerian network landscape has added another dimension to the cybercrimes problem (Longe and Chiemeke, 2008). Fraudsters who can afford to pay for Internet connection via fixed wireless lines can now perpetrate their evil acts within the comfort of their homes.

In some cybercafés, a number of systems are dedicated to fraudsters (popularly referred to as "yahoo boys") for the sole purpose of hacking and sending fraudulent mails. Other cybercafés share their bandwidth (popularly referred to as home use) to some categories of customers who acquire systems for home use in order to perpetuate cybercrimes from their homes (Longe & Chiemeke, 2008).

Efforts at preventing financial Cybercrime in Nigeria are on both at entrepreneurial, private and public pedestal. For café owners, notices are pasted on walls warning of possible arrests of scammers who send fraudulent mails. Individuals can only take precautions within the limit of the knowledge of the dynamics of the Internet and the e-mail system. Users are generally learning not to respond to scam mails or mails presenting financial bogus proposals. For the government, the Economic and Financial Crimes Commission (EFCC) has been given powers to arrest and prosecute individuals and organisations suspected to be involved in the promotion of cybercrimes. The bill on cybercrime has also been passed by the National Assembly and it is not unusual to see bill boards donning Nigerian roads warning cybercriminals that the "hands of the law will soon get to them" (Longe & Chiemeke, 2008).

EFFECTS OF CYBERCRIME ON NIGERIA

According to Chawki, (2009), an Internet Crime Report in 2007 prepared by the National White Collar Crime Centre and the FBI, Nigeria currently ranks third in the world with 5.7 per cent of perpetrators of cybercrime (2007 Internet Crime Report). Though the perpetrator percentage of 5.7 from Nigeria appears low, we can regard it as rather high considering that less than 10 per cent of the 150 million population of Nigeria use the internet (Adomi, 2007). In Africa, though internet use is higher in South Africa, cybercrime perpetrators percentage is higher for Nigeria than South Africa. Cybercrime has a negative impact on Nigeria. It can be explained in the terms of the following statistics (WITFOR 2005):

- Annual global loss of $ 1.5 billion in 2002.
- 6% of global Internet spam in 2004.
- 15.5% of total reported FBI fraud in 2001.
- Highest median loss of all FBI Internet fraud of $ 5,575.
- VeriSign, Inc., ranked Nigeria 3rd in total number of Internet fraud transactions, accounting for 4.81% of global Internet fraud.
- American National Fraud Information Centre reported Nigerian money offers as the fastest growing online scam, up 900% in 2001.

Nigerian Cybercrime has the potential to impact technology growth which is a key requirement for productivity improvement, and ultimately for socio-economic growth because (WITFOR 2005; Adomi & Igun, 2007):

- International financial institutions now view paper-based Nigerian financial instruments with scepticism. Nigerian bank drafts and checks are not viable international financial instruments.
- Nigerian ISPs and email providers are already being black-listed in e-mail blocking blacklist systems across the Internet.
- Some companies are blocking entire Internet network segments and traffic that originate from Nigeria.
- Newer and more sophisticated technologies are emerging that will make it easier to discriminate and isolate Nigerian e-mail traffic.
- Key national infrastructure and information security assets are likely to be damaged by hostile and fraudulent unauthorized use.

Accordingly cybercrime has created an image nightmare for Nigeria. When one comes across phrases like 'Nigerian scam', the assumption which crosses one's mind is that virtually all scam e-mails originate from Nigeria or Nigerians, though this is actually not the case . Advance fee fraud has brought disrepute to Nigeria from all over the world. Essentially, Nigerians are treated with suspicious in business dealing. Consequently, the honest majority of Nigerians suffer as a result (Adomi & Igun, 2008).

CYBERCRIME REGULATORY EFFORTS IN NIGERIA

It has been argued that organized crime weakens the very foundation of democracy, as there can be no good governance without rule of law. This observation is quite apt for the situation in Nigeria. As the nation faces the challenges of nurturing a stable democracy, after many years of military dictatorship, organized crime poses a great threat to the survival of the country. Therefore, the Nigerian government has mapped out strategies to deal with cybercrimes (Chawki, 2009).

1. Legislative Approaches

The Nigerian government has, over the years, enacted far-reaching laws aimed at checkmating transnational organized crimes and punishing the perpetrators of these crimes. Under this subsection, we shall focus on the Criminal Code Act, Economic and Financial Crimes Commission Act 2004, Computer Security and Critical Information Infrastructure Protection Bill 2005 and Advance Fee Fraud and other Fraud Related Offences Act 2006.

- **Criminal Code Act**

Advance fee fraud scam under the Nigerian Criminal Code Act, qualifies as a false pretence (section 418), while a successful Internet scam would amount to a felony under section 419. This section provides as follows:

"Any person who by any false pretence, and with intent to defraud, obtains from any other person anything capable of being stolen, or induces any other person to deliver to any person anything capable of being stolen, is guilty of a felony, and is liable to imprisonment for three years. If the thing is of the value of one thousand naira or upwards, he is liable to imprisonment for seven years. It is immaterial that the thing is obtained or its delivery is induced through the medium of a contract induced by the false pretence. The offender cannot be arrested without warrant unless found committing the offence." (Laws of the Federation of Nigeria, 2009)

Furthermore, a suspect could alternately be charged under section 421 of the Criminal Code Act which provides as follows:

"Any person who by means of any fraudulent trick or device obtains from any other person anything capable of being stolen, or induces any other person to deliver to any person anything capable of being stolen or to pay or deliver to any person any money or goods, or any greater sum of money or greater quantity of goods than he would have paid or delivered but for such trick or device, is guilty of a misdemeanour, and is liable to imprisonment for two years. A person found committing the offence may be arrested without warrant." (Laws of the Federation of Nigeria, 2009)

The Criminal Code is a British legacy which predates the Internet era and understandably does not specifically address email scams (Oriola, 2005). Advance fee fraud methodology obviously falls within the remit of the Act for the following reasons: first, there is a false pretence to the existence of non-existent money; second, a solicitation for financial help to get the fictitious money released; and third, the fraudulent retention of various fees paid to the scammers to release the phony millions of dollars. The scammers' modus operandi fits snugly the elements of the offence under section 419 of the Criminal Code Act cited above, and have been used for years by the Nigerian law enforcement agencies for prosecuting alleged acquisition of property by false pretence (Oriola, 2005). However, the Criminal Code Act provisions on advance fee fraud are ill-suited for cyberspace criminal governance. Oriola (2005) argues that:

"Although section 419 of the Criminal Code Act deems advance fee fraud a felony, the provision that an advance fee fraud suspect cannot be arrested without a warrant, unless found committing the offence, does not reflect the crime's presence or perpetration in cyberspace. Only in rare circumstances could a suspect be caught in the act because most of the scam emails are sent from Internet café's. Aside from the fact that the country lacks the resources to police every known cyber cafe´, doing so could actually raise privacy or other rights issues. If found guilty, an advance fee fraudster is liable to a mere three years imprisonment or seven years if the value of stolen property exceeds 1000 naira. The punishment, to say the least, is paltry relative to the enormity of the crime and unjust rewards that characteristically run into millions of dollars. Thirdly, in criminal trials, the State is the complainant, and there is hardly any compensation for victims of crime under the Nigerian criminal justice system. The victims could no doubt resort to civil court for remedies. However, the prospects for success for the plaintiff in the typical advance fee fraud case scenario are extremely slim. For instance, a contract to assist in the transfer from Nigeria of millions of dollars illegally to a foreign account, or to pay bribes to certain government officials to ensure release of such moneys, or to facilitate advance fee payment for patently illegal activities, would be unenforceable. The plaintiff would be branded as a party to a culpable crime by the Nigerian courts."

• **Economic and Financial Crimes Commission Act 2004**

The Economic and Financial Crimes Commission (EFCC) Act was adopted in June 2004. Under this act, the Commission has the power to investigate all financial crimes relating to terrorism, money laundering, drug trafficking, etc. Sections 14-18 stipulate offences within the remit of the Act. This includes offences in relation to financial malpractices, offences in relation to terrorism, offences relating to false information and offences in relation to economic and financial crimes [Chawki, 2009].

• **Money Laundering (Prohibition) Act 2004**

According to Chawki (2009) the Money Laundering (Prohibition) Act 2004, makes provisions for prohibition of the laundering of the proceeds of crime or an illegal act. Although advance fee fraud is not expressly mentioned in the Act, proceeds of the scam would appear covered under section 14(1) (a) which, prohibits the concealing or disguising of the illicit origin of resources or property which are the proceeds of illicit drugs, narcotics or any other crime. The Act also implicates any person corporate or individual who aids or abet illicit disguise of criminal proceeds. Section 10 makes life more difficult for money launderers. Subsection (1) places a duty on every financial institution to report within seven days to the Economic and Financial Crimes Commission and the National Drug Law Enforcement Agency any single transaction or transfer that is in excess of N1m (or US$7,143) in the case of an individual or N5m (US$35,714) in the case of a body corporate (Chukwuemerie, 2006). Any other person may under sub-section (2) also give information on any such transaction, or transfer. Under sub-section (6) even if a transaction is below US$5,000 or equivalent in value, but the financial institution suspects or has reasonable grounds to suspect that the amount involved in the transaction is the proceed of a crime or an illegal act it shall require identification of the customer. In the same way, if it appears that a customer may not be acting on his own account, the financial institution shall seek from him by all reasonable means information as to the true identity of the principal (Chukwuemerie, 2006). This enables authorities to monitor and detect suspicious cash transactions.

- **Computer Security and Critical Information Infrastructure Protection Bill 2005**

In 2005, the Nigerian government adopted the Computer Security and Critical Information Infrastructure Protection Bill (known as the Cybercrime Bill). The Bill aims to '*secure computer systems and networks and protect critical information infrastructure in Nigeria by prohibiting certain computer based activities*' and to impose liability for global crimes committed over the Internet. The Bill requires all service providers to record all traffic and subscriber information and to release this information to any law enforcement agency on the production of a warrant. Such information may only be used for legitimate purposes as determined by a court of competent jurisdiction, or other lawful authority. The Bill does not provide independent monitoring of the law enforcement agencies carrying out the provisions, nor does the Bill define 'law enforcement agency' or 'lawful authority.' Finally the Bill does not distinguish between serious offences and emergencies or minor misdemeanours. As a result it may conflict with Article 37 of Nigeria's Constitution, which guarantees the privacy of citizens including their homes and telephone conversations, absent a threat on national security, public health, morality, or the safety of others (Chawki, 2009),

- **AFF and other Fraud Related Offences Act 2006**

Another relevant legislative measure in the fight against advance fee fraud on the Internet is the Advance Fee Fraud and other Fraud Related Offences Act 2006. This is a replacement of an Act of the same title passed in 1995. The act prescribes, among others, ways to combat cybercrime and other related online frauds. The Act provides for a general offence of fraud with several ways of committing it, which are by obtaining property by false pretence, use of premises, fraudulent invitation, laundering of fund obtained through unlawful activity, conspiracy, aiding, etc. Section 2 makes it an offence to commit fraud by false representation. Subsection (2)(a) and (2)(b) makes clear that the representation must be made with intent to defraud. Section 3 makes it an offence if a person who is being the occupier or is concerned in the management of any premises, causes or knowingly permits the premises to be used for any purpose which constitutes an offence under this Act. This section provides that the sentence for this offence is the imprisonment for a term of not less more than 15 years and not less than five years without the option of a fine.

Section 4 refers to the case where a person who by false pretence, and with the intent to defraud any other person, invites or otherwise induces that person or any other person to visit Nigeria for any purpose connected with the commission of an offence under this Act. The sentence for this offence is the imprisonment for a term not more than 20 years and not less than seven years without the option of a fine.

According to section 7, a person who conducts or attempts to conduct a financial transaction which involves the proceeds of a specified unlawful activity with the intent to promote the carrying on of a specified unlawful activity; or where the transaction is designed to conceal or disguise the nature, the location, the source, the ownership or the control of the proceeds of a specified unlawful activity is liable on conviction to a fine of N 1 million and in the case of a director, secretary or other officer of the financial institution or corporate body or any other person, to imprisonment for a term, not more than 10 years and not less than five years (EFCC Act, 2004).

However, while in previous laws the onus was on the government to carry out surveillance on such crimes and alleged criminals, the new law vests this responsibility on industry players, including ISPs and cybercafé operators, among others. While the Economic and Financial Crimes Commission (EFCC) becomes the sub-sector regulator, the Act prescribes that henceforth, any user of Internet services shall no longer be accepted as anonymous. Through what has been prescribed as due care measure, cybercafés operators and ISPs will henceforth monitor the use of their systems and keep a record of transactions of users (Adomi, & Igun, 2008). These details include, but are not limited to, photographs of users, their home address, telephone, email address, etc. So far, over 20 cybercafés have been raided by the EFCC as of August 7, 2007. The operators appear set to comply with the law by notifying users of the relevant portion of the law, corporate user policy, firewall recommendation, protection procedure, indemnity and right of disclosure, and so forth (Adomi and Igun, 2008)

2. Administrative Measures

Administrative measures chiefly involve the setting-up of special bodies by the Nigerian government to combat advance fee fraud. Equally important, however, are the technical measures which these bodies then take to prevent and/ or prosecute this activity, and these will also be examined below. It must be emphasised that many European countries have established special computer units to take specific measures against cybercrime. The following are some examples of these bodies (Chawki, 2009):

- ### The Economic and Financial Crimes Commission (EFCC)

The Economic and Financial Crimes Commission (EFCC) is a Nigerian law enforcement agency that investigates financial crimes such as advance fee fraud (419 fraud) and money laundering. The EFCC was established in 2003, partially in response to pressure from the Financial Action Task Force on Money Laundering (FATF), which named Nigeria as one of 23 countries non-cooperative in the international community's efforts to fight money laundering. The Commission is empowered to investigate, prevent and prosecute offenders who engage in

"money laundering, embezzlement, bribery, looting and any form of corrupt practices, illegal arms deal, smuggling, human trafficking, and child labour, illegal oil bunkering, illegal mining, tax evasion, foreign exchange malpractices including counterfeiting of currency, theft of intellectual property and piracy, open market abuse, dumping of toxic wastes, and prohibited goods" (Ribadu, 2006, p.4).

The Commission is also responsible for identifying, tracing, freezing, confiscating, or seizing proceeds derived from terrorist activities. EFCC is also host to the Nigerian Financial Intelligence Unit (NFIU), vested with the responsibility of collecting suspicious transactions reports (STRs) from financial and designated non-financial institutions, analyzing and disseminating them to all relevant Government agencies and other FIUs all over the world (Ribadu, 2006). In addition to any other law relating to economic and financial crimes, including the criminal and penal codes, EFCC is empowered to enforce all the pre-1999 anti-corruption and anti-money laundering laws. Punishment prescribed in the EFCC Establishment Act range from combination of payment of fine, forfeiture of assets and up to five years imprisonment depending on the nature and gravity of the offence (Ribadu, 2006). Conviction for terrorist financing and terrorist activities attracts life imprisonment. It must be mentioned that EFCC has excellent working relationship with major Law Enforcement Agencies all over the world (Ribadu, 2006). These include the International Criminal Police Organization (INTERPOL), the United Kingdom (UK) Metropolitan Police, Federal Bureau of Investigations (FBI), Canadian Mounted Police, the Scorpions of South Africa, etc.

NIGERIAN FINANCIAL INTELLIGENCE UNIT (NFIU)

In 2005, the EFCC established the Nigerian Financial Intelligence Unit (NFIU). The NFIU draws its powers from the Money Laundering (Prohibition) Act of 2004 and the Economic and Financial Crimes Commission Act of 2004. It is the central agency for the collection, analysis and dissemination of information on money laundering and terrorism financing. All financial institutions and designated non-financial institutions are required by law to furnish the NFIU with details of their financial transactions. Provisions have been included to give the NFIU power to receive suspicious transaction reports made by financial institutions and non-designated financial institutions, as well as to receive reports involving the transfer to or from a foreign country of funds or securities exceeding \$10,000 in value (International Narcotics Control Strategy Report, 2006). The NFIU is a significant component of the EFCC. It complements the EFCC's directorate of investigations but does not carry out its own investigations. It is staffed with competent officials, many with degrees in accounting and law (International Narcotics Control Strategy Report, 2006; Chawki (2009)).

The Nigerian Financial Intelligence Unit (NFIU) is playing a pivotal role in receiving and analyzing Suspicious Transactions Reports (STRs). As a result, banks have improved their responsiveness to forwarding records to the NFIU (International Narcotics Control Strategy Report, 2006). Under the EFCC act, whistle-blowers are protected. Nigeria has no secrecy laws that prevent the disclosure of client and ownership information by domestic financial services companies to bank regulatory and law enforcement authorities (International Narcotics Control Strategy Report, 2006). The NFIU has access to records and databanks of all government and financial institutions, and it has entered into memorandums of understandings (MOUs) on information sharing with several other financial intelligence centres. The establishment of the NFIU is part of Nigeria's efforts toward removal from the NCCT list (International Narcotics Control Strategy Report, 2006).

- **Nigerian Cybercrime Working Group (NCWG)**

The NCWG is an inter-agency body comprising law enforcement, intelligence, security as well as ICT agencies of Government and key private sector ICT organizations (NCWG website, 2008). It was established by the Federal Executive Council (FEC) on the recommendation of his Excellency President of Nigeria on March 31 2004. The group was created to deliberate on and propose ways of tackling the malaise of Internet 419 in Nigeria. This includes (NCWG, 2008):

- Educating Nigerians on cybercrime and cybersecurity;
- Undertaking international awareness programs for the purpose of informing the World of Nigeria's strict Policy on Cybercrime and to draw global attention to the steps taken by the Government to rid the country of Internet 419 in particular and all forms of cybercrimes;
- Providing legal and technical assistance to the National Assembly on cybercrime and cybersecurity12 in order to promote gen-

eral understanding of the subject matters amongst the legislators;

- Carrying out institutional consensus building and conflict resolutions amongst law enforcement, intelligence and security Agencies in Nigeria for the purpose of easing any jurisdictional or territorial conflicts or concerns of duties overlap;

- Reviewing, in conjunction with the Office of the Attorney General of the Federation, all multilateral and bilateral treaties between Nigeria and the rest of the World on cross-border law enforcement known as Mutual Legal Assistance Treaties (MLAT), for the purpose of amending the operative legal framework to enable Nigeria secure from, as well as render, extra-jurisdictional assistance to its MLAT Partners in respect of cybercrime.

3. Technical Measures

Criminals are often quicker to exploit new technologies than law-enforcers who, to some extent, always seem 'behind the game'. In order to salvage Nigeria from the negative consequences of cybercrime, the government has been making frantic efforts to ensure that this malaise is nipped in the bud. These efforts are discussed below (Chawki, 2009):

- **Regulation of Cybercafés**

Cybercafés in Nigeria render overnight browsing, a special internet service is offered by cybercafés from 10.00 p.m. to 6.00 a.m. This service allows users who have a lot to obtain from the net to do so at a minimal cost (Adomi, 2007). Though overnight browsing is very important and useful to cybercafé users, it was banned by the EFCC and the Association of Cybercafé and Telecentres Owners (ATCON) in Nigeria (Adomi, 2008). The ban is coming on the heels of several attempts by the EFCC to arrest the ugly trend through raids,

arrests, and precautions of cybercafés and cyber-criminals as a result of the constant embarrassment posed to the Nigerian Federal Government by their nefarious activities. Some Nigerian fraudsters have perfected the act of using the internet via cybercafés as their criminal platform to dupe unsuspecting citizens across the globe (Adomi and Igun, 2008). This ban on night browsing is likely to negatively affect clients who use the cafes for academic and other useful and positive purposes in night browsing sessions. Other decisions of EFCC and ATCON reached to combat cybercrime include (Adomi and Igun, 2008):

- Undertaking international awareness programs for the purpose of informing the World of Nigeria's strict Policy on Cybercrime and to draw global attention to the steps taken by the Government to rid the country of Internet 419 in particular and all forms of cybercrimes;

- That each sector of the telecom industry, namely the global system for mobile communication operators, private telecomm operators and cybercafés should come up with a due care document that would be a standard guide and proffer measures for the effective policing of cybercrime in Nigeria;

- That all cyber cafes must be registered with the Corporate Affairs Commission, NCC and EFCC;

- That cybercafés will now be run on membership basis instead of pay-as-you-go;

- All cybercafés must install acceptable hardware surveillance;

- The architecture of cyber cafes must be done such that all computers are exposed;

- ATCON members must subscribe to registered and licensed ISP in the country;

- Each cybercafé is expected to be a watchdog to others, as they have been detailed to have direct access to EFCC (Adomi & Igun, 2008).

•   **Government Partnership with Microsoft**

The Government of Nigeria and Microsoft Corp signed a Memorandum of Understanding defining a framework for co-operation between Microsoft and the Economic and Financial Crimes Commission (EFCC) of Nigeria with the aim of identifying and prosecuting cybercriminals, creating a safe legal environment and restore hundreds of millions of dollars in cost investment (Chawki, 2009). This agreement is the first of its kind between Microsoft and an African government and will give the EFCC access to Microsoft technical expertise information for successful enforcement. The Memorandum combats issues such as spam, financial scam, phishing, spyware, viruses, worms, malicious code launches and counterfeiting. Microsoft is expected to instruct Nigerian investigators on techniques of extracting useful information from PCs compromised by botnet attacks, how to monitor computer network to detect such attacks, and how to identify the people behind them. Microsoft will also provide leads on spam emanating from Nigeria, enabling the authorities to pursue investigations more quickly and successfully (Adomi & Igun, 2008).

FUTURE TRENDS

The future of cyber crime is uncertain, however, cyber terrorism is a threat to society that will always exist, and law enforcement agencies must continue to expand into the "cyber" universe. A terrorist can do more with a computer than he can with a gun. People rely heavily on telephones, computers, and the Internet. Criminal information is stored, and often transmitted via Internet, as well as legal, medical, and other confidential information. People would be surprised if they knew how much information is stored on the Internet, and on computers, about them (Thomas, 2006).

Law enforcement agencies must expand their services, and most of all, reach out into the commu-

nity. Teenagers are one of the top cyber criminals. Providing community awareness workshops for both parents, and teenagers, are important. Community policing is just as important, there must be online leadership promoting "safe neighbour-hoods" and providing citizens with information on possible security breaches on their systems and networks (Thomas, 2006).

Most home users and businesses do not realize the vulnerabilities in their networks or systems until it is too late. Law enforcement agencies must cross the line into the information age and take an active part in identifying such vulnerabilities in a timely manner. Law enforcement must target those criminals who create spam, spyware, adware, viruses, and worms. They are the heart of corruption on the Internet. One must work in a pyramid, finding the creators of the viruses and worms will create a better Internet (Thomas, 2006)

Although there are already examples of codes of conduct, which have been introduced online to tackle the problem of cybercrime and advance fee fraud, a self-regulation system should also be introduced. Self-regulation is based on three key elements (Chawki, 2009): first, involvement of all interested parties (government, user associations, and access providers) in producing new strategies; secondly, implementation of these strategies by the party concerned; thirdly, evaluation of the measures taken. Self-regulation can be backed by clear legal regulation and this is what the term 'co-regulation' means. A co-regulatory system is one in which the public authorities accept that protection of the society can be left to self-regulatory schemes, but reserve the right to intervene, if self-regulation fails to work.

For the purpose of prosecuting advance fee fraud scammers, we should use local, regional and national co-ordination and information sharing mechanisms; national liaison officers posted overseas or links with liaison officer networks; Europol and its Liaison Bureaux; Interpol's National Contact Bureaux; Eurojust; and direct bi-lateral contacts. Channels that already exist for other purposes should be activated and adapted. Finally, we need to make the laws more effective by improving the quality of criminal codes and increasing the penalties to match the seriousness of loss. Laws will be effective against scammers who are deterred by criminal law and frightened by the prospect of incarceration; however, there will always be scammers who are motivated to engage in online fraud to overcome these laws and the efforts of the criminal justice community(Chawki, 2009):.

CONCLUSION

The embracement of the Internet culture in Nigeria has come with a lot of mixed feelings. The proliferation of ICTs and progressive development in digital transactions and communications have created new opportunities and opened up new windows which have resulted in the emergence of new forms of criminal behaviour and cybercrime. Advance fee fraud ranks amongst the most important and virulent forms of cybercrime; not only due to its adverse impact on the development of cyberspace but also due to the diversity of means and methods that could be utilized in committing this crime as well as the inherent risk of using advance fee fraud as a leeway and instrument to commit other crimes using the stolen identities of victims. Besides other domestic approach such as legislating against cybercrime menace, this chapter agrees with Chawki (2009), that cybercrime especially advance fee fraud should be subject to a global principle of public policy that aims at combating and preventing this form of organized crime through raising global awareness, increasing literacy rates, coordinating legislative efforts on national, regional and global levels, establishing a high level global network of cooperation between national, regional, and international enforcement agencies and police forces.

Just as it is with other forms of crime, it is likely that cybercrime and its perpetrators will continue developing and upgrading to stay ahead of the law; hence the Internet community must therefore engage in a collective effort to curb the Internet of the demeaning crimes it is helping to fuel. For Nigeria, the EFCC and of course ICT professionals will have to take decisive steps and formulate policies that will help in drastically reducing this menace at Internet access points. A combination of sound technical measures tailored to the origin of spam in conjunction with legal deterrents as currently being pursued by the EFCC will be a potent remedy (Longe and Chiemeke, 2008)

REFERENCES

Adomi, E. E. (2007), Overnight Internet browsing among cybercafé users in Abraka, Nigeria. *Journal of Community Informatics, 3*(2). Retrieved October 30, 2009 from http://ci-journal.net/index.php/ciej/article/viewPDFInterstitial/322/351

Adomi, E. E., & Igun, S. E. (2008). Combating cybercrime in Nigeria. *The Electronic Library, 26*(5), 16–725. doi:10.1108/02640470810910738

Aghatise, E. J. (2006). *Cybercrime Definition. Computer Crime Research Center*. June 28, 2006. Available online at www.crime-research.org

Awe, J. (n.d.). *Fighting cybercrime in Nigeria*. Retrieved October 30, 2009 from http://www.jidaw.com/itsolutions/security7.html

Chawki, M. (2005). A Critical Look at the Regulation of Cybercrime. *The ICFAI Journal of Cyberlaw, 4*(4).

Chawki, M. (2006). *Anonymity in Cyberspace: Finding the Balance between Privacy and Security*. Revista da Faculdade de Direito Milton Campos.

Chawki, M. (2006). *Le Droit Penal à l'Epreuve de la Cybercriminalité*. Lyon, France: University of Lyon III.

Chawki, M. (2008). *Combattre la Cybercriminalité*. Perpignan, France: Editions de Saint Amans.

Chawki, M. (2009) Nigeria Tackles Advance Fee Fraud. *Journal of Information, Law and Technology.* Retrieved May 21st, 2009. From: http://www2.warwick.ac.uk/fac /soc/law/elj/jilt/ 2009 _1 /chawki/chawki.pdf

Chawki, M. & Wahab, M. (2006). Identity Theft in Cyberspace: Issues and Solutions. LexElectronica, *11*(1).

Chukwuemerie, A. (2006). Nigeria's Money Laundering (Prohibition) Act 2004: A Tighter Noose. *Journal of Money Laundering Control, 9*(2). doi:10.1108/13685200610660989

Economic and Financial Crimes Commission (Establishment) Act. (2004). Retrieved 10th May, 2009. From http://efccnigeria.org/index.php?option=com_ docman&task=doc_download&gid=5

Money Laundering (Prohibition) Act, of the Federal Republic of Nigeria. *(2004).*

Longe, O.B. (2006). Web Journalism In Nigeria: New Paradigms, New Challenges. *Journal of Society and Social Policy.*

Longe, O. B., & Chiemeke, S. C. (2006). The Design and Implementation of An E-Mail Encryptor for Combating Internet Spam. In *Proceedings of the Ist International Conference of the International Institute of Mathematics and Computer Sciences,* (pp. 1 – 7). Ota, Nigeria: Covenant University.

Longe, O. B., & Chiemeke, S. C. (2008). *Cybercrime and Criminality in Nigeria –What Roles are Internet Access Points in Playing?* Retrieved May 21st, 2009. From: http://www.eurojournals.com/ejes_6_4_12.pdf

Longe, O. B., & Longe, F. A. (2005). The Nigerian Web Content: Combating the Pornographic Malaise Using Content Filters. *Journal of Information Technology Impact, 5*(2), 59–64.

Lyman, M., & Potter, G. (1998). *Organized Crime.* Upper Saddle River, NJ: Prentice Hall.

Mcconnell International. (2000). *Cybercrime and Punishment?* Archaic Laws Threaten Global Information.

Oriola, T. (2005). Advance Fee Fraud on the Internet: Nigeria's Regulatory Response. *Computer Law & Security Review, 21*(3).

Paper delivered at Financial Lecture Series in Lagos, August 10, 2006.

Parker, D. B. (1998). *Fighting Computer Crime: A New Framework for Protecting Information.* Chichester, UK: Wiley Computer Publishing.

Ribadu, N. (2006). *Money laundry in Emerging economies: Nigeria as a case study.* A

Smith, A. D., & Rupp, W. T. (2002). Issues in cybersecurity: understanding the potential risks associated with hackers/crackers. *Information Management & Computer Security, 10*(4), 178–183. doi:10.1108/09685220210436976

Sylvester, L. (2001). *The Importance of Victimology in Criminal Profiling.* Retrieved May 21st, 2009, from: http://isuisse.ifrance.com/ emmaf/ base/ impvic.html

Thomas, D., & Loader, B. D. (2000). Introduction - Cybercrime: Law Enforcement, Security and Surveillance in the Information Age. In *Cybercrime: Law Enforcement, Security and Surveillance in the Information Age.* New York: Taylor & Francis Group.

Thomas, J. (2006). *Cybercrime: A revolution in terrorism and criminal behaviour creates change in the criminal justice system*. Retrieved November 1, 2009 from http://www.associatedcontent.com/article/44605/cybercrime_a_revolution_in_terrorism_pg12_pg12.html?cat=37

Wilson, C. (2000). Holding Management Accountable: A New Policy for Protection Against Computer Crime. In *National Aerospace and Electronics Conference . Proceedings of the IEEE, 2000*, 272–281.

KEY TERMS AND DEFINITIONS

Crime: An illegal act or activity that can be punishable by law.

Cybercrime: Illegal activities carried out over the Internet, such as malicious hacking, virus distribution, data theft, extortion, fraud, forgery, child pornography, trafficking in illegal substances, digital copyright infringement, etc.

Cyberspace: Refer to the virtual world of digital communication, in which human beings interact with one another electronically via computer networks instead of face-to-face.

Cybercafé: A commercial place equipped with computers for the use of its customers.

Cybersecurity: Security issues relating to the cyberspace especially the Internet.

Fraud: Crime of deceiving somebody to get something.

Regulation: An official rule made by government and some other authority.

Scam: Dishonest plan for making money.

Legislation: Process of making and passing laws.

Chapter 8
Emerging Information and Communication Technology Policy Framework for Africa

Saul F.C. Zulu
University of Botswana, Botswana

ABSTRACT

While emerging information and communications technologies (ICTs) offer possible solutions to some of the problems of applying ICTs in Africa, there are many challenges that have to be addressed in order to create an environment that is conducive for harnessing these technologies. This chapter, therefore, reviews emerging ICTs and their potential for application in leveraging Africa's efforts towards meeting its development efforts. The Chapter highlights the digital divide barriers that may inhibit emergent ICTs in Africa. A review of current ICT policies of selected African countries indicates that the policies are geared towards application of ICTs other than their production. The review also reveals a lack of appreciation for emerging ICTs in Africa, both at the national as well as the sub-regional economic bloc levels. The chapter proposes policy frameworks for emerging ICTs for Africa that are necessary for creating an enabling environment for harnessing the emerging ICTs that will propel the continent into the 21st Century and beyond. The barriers to ICTs cut across many different issues. As such, they require multi-pronged policy approaches to address them. And that an emerging ICT environment must be anchored on a number of strategic policy frameworks including the legal, regulatory/administrative institutional framework, infrastructure, technology advocacy, financial, human resources, education and research frameworks. It concludes that Africa can prepare for its future by creating an appropriate environment for fostering the adoption and application of emerging technologies.

DOI: 10.4018/978-1-61692-012-8.ch008

INTRODUCTION

Information and communications technologies (ICTs) is a collective term used to describe the various technologies that are used in the processing of information including its coding, creation, storage, retrieval, manipulation, dissemination and transmission. ICT technologies include computers that are used for processing information, publishing that is used for coding and dissemination of information including broadcasting, and telecommunications, which are used for the transmission of information (Zulu, 1994; Moll, 1983). Since the invention of the digital computer and the birth of the micro-electronics industry, the ICT revolution has been marked by three major waves. The first wave, which ran from the time the computer was invented in the late 1940s up to 1970s, was characterized by centralized computing where many people were connected to one computer. The second wave of computing, which started in the 1980s with the introduction of the microcomputer, has run throughout the 1990s to the present, has been dominated by personal computing that has been marked by each person being tied to a personal computer. We are now on the threshold of the third wave, where computing is moving away from an individual to the environment. This is an era of widespread computing, which will be characterized by one individual-to-many computers, dominated by handheld, intelligent, and everyday devices with imbedded technology and connectivity. This is the era of pervasive computing, which Agoston, Ueda and Nishimura (2000) have described as being characterized by "Anytime/Anywhere, Any Device, Any Network, Any Data" total connected computing environment of the third wave (p.3).

Owing to a variety of factors, Africa and most of the developing world were largely by-passed by the first two waves of the ICT revolution, which were the building blocks for entry into the digital age. The consequence of this has been what has been termed as the 'digital divide' that currently exists between the developed countries of the North and developing countries of the South. The digital divide also exists within the countries of the South between the majority of the rural-based citizens who have no access to ICTs and a tiny urban-minority that has access to ICTs. Emerging ICTs can assist the digitally excluded regions of the world, including Africa to leapfrog the digital divide and catch up with the digitally connected world. Emerging ICTs, if well harnessed through the creation of an appropriate environment, offer opportunities for bridging the digital divide in Africa that can be used to leverage its development efforts.

The purpose of this chapter is to discuss issues of the digital divide in Africa and how emerging ICTs can be employed to address the issues of the digital divide towards achieving Africa's development goals. The chapter is presented in six main sections as follows: a discussion on the concept of the digital divide and its impacts on Africa; based on the global strategic vision of the Millennium Development Goals (MDGs), a discussion on how ICTs can contribute towards realizing issues of development in Africa; a review of developments in emerging ICTs and how they can be used to overcome the digital divide, a review of ICT policies of selected African countries and regional economic blocs in Africa vis-à-vis emerging ICTs, and; a presentation of a policy framework which Africa should put in place in order to create a conducive environment for harnessing the emerging ICTs to leverage Africa's development goals.

BACKGROUND

Emerging technology may mean different things to different people. As such, there is no standard, universally adopted definition of the term. The definitions of the term are therefore as diverse as the technologies to which they refer. According to Adomi (2009), emerging technologies are "those new and important technological developments in the field."(p.87). The SAE International Emerging Technologies Advisory Board (2009) (originally known as Stanford Applied Engineering) describes emerging technologies as consisting of "…new technological discoveries that possess the potential to significantly affect the mobility industries; existing technologies that have impending new or revised applications; and major issues and trends that may have profound, new technology …" (p.1). Other definitions and descriptions of emerging technologies are provided by Mutula & Wamukoya (2007).

Emerging ICTs are a result of the continuous development of technology. The technology life cycle is a continuum of processes that involve (Forrester, 2003):

- Development of new processes through research and development,
- Application of the new processes or technologies in the market for everyday use
- INNOVATION: (i.e. making improvements on the existing technologies to make them better and more efficient), and
- Creation of new technologies through research and development

The end product of this technology lifecycle is the production of qualitatively better technologies that are cheaper and more efficient. Since the invention of the first electronic computers in the 1940s, efforts in the ICT sector have been geared towards consolidating the technologies of the electronic age into better performing devices and systems to facilitate the processing and communication of information. This has resulted in the development of newer hardware, software and telecommunications systems that are based on emerging technology platforms. We briefly review these emerging ICT developments in hardware, software and telecommunication below.

Emerging Hardware Developments

There are three major emerging computing hardware technology platforms that can be discerned from the literature on computer hardware (Encyclopædia Britannica, 2009; Walker, 2002; Foster, 2005; The Foresight Nanotech Institute; 2008; Wang; 2008)). These are:

- Platform 1: making further improvements on the existing computing hardware technology by changing the architecture of the computer and increasing the density of downsizing or miniaturization of hardware components;
- Platform 2: grid computing, where computing power is maximized by linking several existing computers to operate as one supercomputing machine: the virtual supercomputer;
- Platform 3: creating computing technology based on a totally new technology paradigm known as nanotechnology.

Platform 1: Improving the Existing Hardware Technology

The major thrust in this platform has been to design computer hardware that is based on alternative architectures away from the Von Neumann serial processing architecture. This has seen the development of computer hardware that is based on:

- Parallel processor technology that makes it possible for more than one computer processor to share the same memory (i.e. the random access memory (RAM)), thereby allowing the processing of more than one set of instructions per time as opposed to von Neumann's serial processing architecture that allowed only one set of instructions to be processed per time. This technology has greatly increased the processing speed of computers;

- Multi-parallel processor (MPP) technology that allows many processors to process many different sets of instructions simultaneously on one machine, similar to the way a human body functions in absorbing and processing knowledge and information capable of simultaneously using the senses of sight, smell, hearing, touch and taste. Buyya (2001) provides a graphic tutorial on the operations of parallel computing hardware technology.

In tandem with the development of alternative computer architectures, has been the development of technology that makes it possible to produce physically smaller but very powerful computing devices. This has been achieved through further miniaturization of computer circuitry by the application of very large scale integration (VLSI) technology which makes it possible to pack many computer components on a very small circuit board. For additional information on VLSI computing technology see Encyclopædia Britannica (2009) and Wikipedia (2009).

The current ICT hardware devices are based on VLSI, parallel processor and MPP technologies. These technologies involve denser packing of computer processors on circuit boards. A combination of ultimate integration of computer circuitry such as VLSI and use of alternative computer architectures such as MPP has resulted in the production of the most powerful computing machines ever made before. These powerful computing machines, known as supercomputers, have processing speeds measured in teraflops per second, where 1 teraflop represents 1 trillion operations per second. IBM's Blue Gene/P, released in 2007 was at the time touted to be the fastest supercomputer in the world with a reported processing speed of 3000 trillion calculations per second (or 3 petaflops) (Knight, 2007).

Platform 2: Grid Computing

Also known as distributed or cluster computing, the emerging technology in grid computing involves linking existing computers in a grid to make them behave as one computer so that they are able to share their combined capabilities to produce computer processing power similar to a stand alone supercomputer. This approach enables an organization to increase its computing and processing power without having to spend a lot of money on buying a conventional stand-alone supercomputer whose prices are beyond the budgets of most organizations. Grid computing requires installation of special software and telecommunications hardware that makes it possible to link computers in the grid and make them work together as one machine in the same local area. Walker (2002) and Foster (2005) provide useful tutorials on grid computing.

Platform 3: Nanotechnology

The technologies that have been used so far in the construction of computer hardware have been based on serial architecture or alternative architectures such as parallel and multi-parallel architectures. All these technologies are based on miniaturization in the fabrication of computer hardware. As stated above, miniaturization technology has involved progressive massive packing of computer circuitry. The shorter the distance between computer circuits, the faster the processing speed of the computer. However, there is a physical limit to miniaturization of computer circuit boards. It is for this reason that a totally new technology platform is being proposed. The new technology paradigm shift is what has come to be known as nanotechnology. The nanotechnology technology platform is based on various fields of knowledge including bio-technology, physical sciences technology (atomic and nuclear technology), neuro-science technology, molecular physics, materials science, chemistry, biology, computer science, electrical engineering, mechanical engineering and cognitive science or what has been termed as the convergence of the sciences. The Foresight Nanotech Institute (2008) has defined Nanotechnology as:

...a group of emerging technologies in which the structure of matter is controlled at the nanometer scale, the scale of small numbers of atoms, to produce novel materials and devices that have useful and unique properties. Some of these technologies impose only limited control of structure at the nanometer scale, but they are already in use, producing useful products. They are also being further developed to produce even more sophisticated products in which the structure of matter is more precisely controlled.... The goal of nanotechnology is to improve our control over how we build things, so that our products can be of the highest quality... (p.1)

Nanotechnology will make it possible to achieve further miniaturization of the machines of the future by manipulating the internal constitution of matter at the atomic level. Just like the smallest element in living things is the gene, the smallest element in non-living things is an atom. It is the smallest because it cannot be divided into smaller particles, although it has positive and negative elements. Nanotechnology will make it possible to produce physically tiny ICT devices that are several times more powerful than machines that have been created using the existing technologies based chiefly on micro-electronics technology. Terms associated with nano-based computing platforms include DNA computing, liquid computing, wet computing, molecular computing (Farrier (2010), Hagiya (2000), Hsu (2010), Palmer (2010), Loygren (2003), Bonsor (2000), Gavaghan (2010), Ernest & Shetty (2005), Roco (2003). Nanotechnology will make it possible to produce different types of "intelligent" products including household appliances, motor vehicles, intelligent fabrics, tools and other products of

everyday use. It is these new machines that will lead us into a new ICT interconnected environment: what has been called the age of pervasive computing. Harper (2003), Wang (2008), Foresight Nanotech Institute (2008), The Centre for Responsible Nanotechnology (2008) and the National Nanotechnology initiative (2008) provide useful insights into nanotechnology and the new field of science behind it, known as nanoscience.

Emerging Software Developments

Many emerging software developments that drive the emerging ICT environment are taking place at both systems and applications level. They include developments in the following major areas:

- **Artificial intelligence** (AI): where efforts have been towards developing software that mimic human intelligence with powers to 'learn' 'reason' and 'draw conclusions'. AI software is already being applied in everyday use including banking (e.g. in Automatic Teller Machines); telecommunications (e.g. voice simulators when using prepaid recharge, credit balance information facilities, voice recognition dialing) and other areas;

- **Expert systems**: research efforts have been towards making software that is encoded with the expertise of a professional in any field of knowledge, which, given a problem it is able to 'reason' as a human professional in that area of expertise;

- **Web management and applications software** such as Web 2.0 that is providing platforms for creating new web-based applications such as social networking;

- **Telecommunications software** that is used to run emerging telecommunications systems such as wireless networks;

- **Open systems software**: this is a movement towards producing software that is non-proprietary which anyone can access, modify and use to create new applications without the need to seek permission of the creators of the original software. The goal of the open software movement is to make software accessible, shareable and promote innovation in software development. And that this initiative will ultimately entrench ICT use in society.

Emerging Telecommunications Technology

A lot of emerging developments are talking place in the area of telecommunications. Some of the major innovations have been in the following areas (Mutula & Wamukoya 2007; Afele, 2003; the National Council on Disability, 2008):

- The development of **broadband intelligent networks** powered by artificial intelligence communications software;
- The development of **wireless communications systems** that have rendered the expensive fixed line networks unnecessary. Wireless communications systems employ free space to transmit information. They are based on radio microwave. A number of wireless-based communication technologies are being developed and coming on the market. Some the significant wireless telecommunications which will impact on the communication of information and data in a networked environment include the following:
- Voice over Internet Protocol (VoIP) that makes it possible to make very cheap voice phone calls over the Internet;
- Radio-frequency identification (RFID) is an automatic identification method, relying on storing and remotely retrieving data using devices called RFID tags or transponders. This technology has many applications including tracking of the movement

files and documents in organization, tracking animals, people and good;
- Wireless Application Protocol (WAP) is a communications standard used for developing applications over wireless networks. The WAP standard makes it possible for people to send text-based information across wireless-based networks such as cell-phone systems. WAP also makes it possible to provide Internet services over cell-phones. Another significant communications standard is Wireless Fidelity (Wi-Fi) which has been designed to facilitate communication across different wireless-based networks;
- General Packet Radio Services (GPRS) is another mobile-based transmission standard that allows faster transmission rates of data than the speeds of ordinary modems. This technology has made it possible to send broadband messages (text, data, voice, picture) over cell-phones;
- Wireless Local Area Network (WLAN) is a new network technology that allows the establishment of local area networks based on free-space communication media. Using this technology, an organization can set up a Local Area Network (LAN) without the need for physical cabling as is the case with ordinary networks that employ physical cabling such twisted pair, coaxial cable, and fibre optics. The major advantage of this technology is that, it is cheap, neat and requires very little physical infrastructure;
- Bluetooth technology is another wireless technology that affords fast access between communications points. Unlike other wireless-based telecommunications technologies that are based on infra-red light waves that require a 'line of sight' between communications devices, Bluetooth-based devices can "sense" and recognize each other without the need for "seeing" each other to

effect a communication link. It facilitates, among others, roaming and remote mouse applications;

• Very Small Aperture terminal (VSAT) is a satellite telecommunications technology that allows individuals to directly send and receive Internet-based services without the need to go through an ISP gateway. It employs telecommunications devices linked to computers at the user's location that sends and receives signals, this is in turn connected to a space satellite that is linked to an earth-based satellite which is linked to a computer that serves as a hub and directs the flow of information in the network.

• The development of common communications standards such as the Integrated Services Digital Network (ISDN) technology is an international communications standard which makes it possible to send voice, video, and data over digital telephone lines or normal phone wires. This technology employs a transmission technique that allows ordinary telephone systems to handle large quantities of data. It turns a low frequency, narrowband transmission medium into a broadband, high frequency medium that is making it possible to communicate across different networks and applications.

Developments in telecommunications have also seen major innovations in the utilization of fixed-line communications media such copper-based telecommunications infrastructures by the introduction of new technologies that converts these 'inefficient' analog voice-based narrowband systems into digital broadband, high-speed communications networks such as the Asymmetric Digital Subscriber Line (ADSL) technology. This has been made possible by employing a technology that harnesses the portion of the line's bandwidth that is not used by voice to achieve a simultaneous voice and data transmission (Mutula & Wamukoya, 2007).

Another telecommunication innovation that utilizes existing infrastructure is Power Line Technology (PLT) that allows fast broadband transmissions using ordinary office and household electricity power wiring to carry radio signals that connect consumers in offices and homes, thereby making it unnecessary to install expensive network transmission media such as fibre optics to achieve connectivity (Intellon, 2007;)

Products and Services from Emerging ICTs

The combined current and future developments in the key ICTs areas of hardware, software and telecommunications are fostering the emergence of new hand-held, physically smaller devices and applications that are beginning to change the ways in which humans create, store, communicate, use and manage information and knowledge. Some of the leading hand-held ICTs include the National Council on Disability (2008):

• **iPods:** a data compression technology that has capabilities to store vast amounts of information that can be carried around to be used where it may be needed. This will make it possible to deliver information services to remote areas that have no access to ICT infrastructure. Other products that have emerged from this technology include iPhones, and other models that can capture video as well,

• **Personal Digital Assistants (PDAs):** also called palmtops, hand-held computers and pocket computers, are handheld devices that combine computing, telephone/fax, Internet and networking features. A PDA can function as a cellular phone, fax sender, Web browser and personal organizer. Other features of PDAs include handwriting recognition, voice input. PDAs extend access to computing facilities outside the homes and offices.

Other new hardware initiatives that will go long way in improving accessibility to ICTs in developing countries include the $100 Computer Initiative, spearheaded by MIT's Nicholas Negroponte whose goal is to produce a low cost, low power, networked laptop PC to improve accessibility by school children in poor countries: the prototype of the laptop known as the 'Green Machine' because of its distinctive colour, was unveiled at the UN Net Summit Tunis, Tunisia in 2005 (Twist, 2005); and the proposed development of a low cost, low power, integrated multifunction information appliance called PCtvt (personal computer television telephony) which includes the functions of a PC, TV, digital VCR, Video Phone and IP-Telephone as an all-in-one device designed for use by the poorest of the poor and even the illiterate (Reddy, Arunachalam, Tongia, Subrahmanian, & Balakrishnan, (2004))

Developments in wireless telecommunications will make it possible to link rural communities that currently have no access due to infrastructural difficulties. Cell-phone technology has continued to develop many capabilities that were unthinkable only a few years ago. The cell phone now has broadband capabilities that have made it possible to capture, receive, store and transmit information in different formats including: audio, video, text, data, computing and Internet access. These developments make the cell-phone to have the potential to be a multi-media information management and communications device of the near future. Further, the development of common communications standards will make it possible to link various networks that will connect individuals, communities, nations and across nations and continents in a truly networked digital environment

Developments in ICTs will continue and will result in the provision of better information services. If Africa is to benefit from this new ICT-driven future, the continent and its people will need to prepare for it. There is need, therefore, to create an environment that is conducive for emerging technologies. How can this be achieved? This question is addressed in the section on 'Emerging technology policy framework for Africa.'

THE DIGITAL DIVIDE AND AFRICA

Webopedia (2008) defines and characterises the digital divide as:

...the discrepancy between people who have access to and the resources to use new information and communication tools, such as the Internet, *and people who do not have the resources and access to the technology. The term also describes the discrepancy between those who have the skills, knowledge and abilities to use the technologies and those who do not. The digital divide can exist between those living in rural areas and those living in urban areas, between the educated and uneducated, between economic classes, and on a global scale between more and less industrially developed nations (p.1).*

Reddy, Arunachalam, Tongia, Subrahmanian, & Balakrishnan, (2004) have identified nine categories of the digital divide including infrastructure divide, access divide, literacy divide, language divide, information and knowledge access divide, jobs divide, health-care divide, entertainment divide and demographic divide. Mutula (2008) & Murelli (2002) have identified a number of factors that fuel the digital divide on the African continent and other developing countries including: the general poor economic conditions obtaining in most countries of sub-Saharan Africa; local content factors whereby most of the digital resources are irrelevant to the needs of Africans because they are dominated by foreign cultures and contexts; cultural and linguistic factors where digital resources are presented in foreign languages and depict foreign cultures and in addition, where the education systems have also adopted foreign languages such as French and English as mediums of instruction in schools and for conducting of research; HIV/Aids, where financial resources are being diverted towards meeting the challenges of the pandemic in terms of procuring medicines and putting in place other mitigation strategies instead of investing in ICT infrastructure; telecommunications constraints, which as summarised by Mutume (2005) and the ITU (2005), show that Africa has the most under-developed telecommunications infrastructure in the world characterised by having the lowest ICT indicators in terms of connectivity, access to radio, television, computers, and the Internet. Recent studies however show a positive growth in the telecommunications sector for the continent, as Africa has now become the world's fastest grow-ing telecoms market with an annual growth rate of 50%, an estimated 300 million people owning cell phones in 2008, and total subscribers estimated to reach the 400 million mark by the end of 2009 (*African Business*, 2009).

Other digital divide factors identified by Mu-tula (2008) include the brain drain, where Africa's intellectual resources are constantly migrating to Western countries in search of greener pastures; education budgetary factors, where all the levels of the education sector (from primary to tertiary) are under-funded and yet education is one of the pre-requisites to a sustainable adoption and application of ICTs; literacy challenges where a significant portion of the population, particularly in rural Africa, is still illiterate, a situation that is incompatible with sustainable application of ICTs; institutional level policies which have resulted in under-utilization of even the little available ICT resources; research and development factors, where the low ratio of scientists on the continent, relative to other regions of the world, continue to hamper research and development activities on the continent to drive the creation, absorption, adaptation and application of new knowledge and technologies such as ICTs; national ICT policies

and regulatory constraints that militate against creating a conducive environment for ICTs, and; political factors that result in inequitable distri-bution of national resources and development, leading to social and political instability. Jensen (2002) provides a detailed picture on the state of the digital divide in Africa.

ICTS AND THE ATTAINMENT OF MILLENNIUM DEVELOPMENT GOALS IN AFRICA

Millennium Development Goals (MDGs) are development targets that are designed to improve human living conditions by 2015. The MDGs were agreed upon by world leaders at a United Nations conference held in 2000 (Department for International Development 2008). They cover eight broad areas that are critical to human development including education; health (includ-ing reduction of child mortality and improving maternal health); gender equality; HIV/AIDS; the environment; eradication of extreme poverty and developing a global partnership for development. The international community agreed to dedicate its efforts towards attaining these goals as a way of accelerating development in under-developed regions of the world such as Africa.

ICTs are now recognized as general facilita-tors in different areas of human endeavour. They are important in fostering development at all levels of time and space: personal, community, organizational, national, and regional as well as global. Many ICT and development experts agree that the application of ICT can significantly con-tribute to the attainment of Africa's MDGs in the following areas (Okot-Uma, 2002; Murelli, 2002; Mbumwae, 2006, Jensen, 2002, Department for International Development, 2006; Chetty, 2005):

• **Education**: where ICTs can extend access to education through ICT-based platforms such e-learning and video conferencing.

The application of these education delivery platforms can go a long way in solving some of the problems of shortage of teachers and the physical infrastructure at all levels of the education spectrum in most African countries because ICTs can reach many people spread across vast geographical areas. This can be achieved very quickly and at much lower cost than the traditional approaches of building schools and other centers of education, training and retaining education sector personnel;

- **Health**: ICTs can leverage health delivery, for instance, through such platforms as telemedicine (for example via remote diagnosis of ailments and prescription of remedies). The application of such ICT-based health delivery systems can go a long way in mitigating the critical shortage of qualified health personnel in Africa. Public health campaigns such as vaccination of under-five children against diseases, HIV/AIDS, and basic hygiene, can be more effectively delivered by improved radio and TV reception and access by all citizens, including those in rural areas. Further, ICTs can result in better managed health information such as patient records and the provision of timely information for planning, monitoring and management of the health delivery systems;

- **Eradication of extreme poverty**: the introduction of ICT-based value added services in the telephony sector (both fixed and mobile) as well as Internet services, has created and continue to create new opportunities for employment for many households. For instance, since the introduction of the cell phone industry in the late 1990s, many new jobs have been created in the formal sector by cell phone service companies. In addition, many people are serving as self-employed agents who sell pre-paid phone recharge cards as well as other telecom-munications services. The mushrooming of Internet cafes and telecentres in urban centres of Africa is providing employment opportunities that never existed before.

- Another area in which ICTs can leverage the fight against poverty is by providing timely information to peasant farmers in such areas as weather patterns and forecasting, agricultural prices, markets for agricultural produce, sources of finance and agricultural inputs, information on agricultural services, etc. This information can assist farmers to plan effectively and make informed decisions in their agricultural activities. At the governance level, a better-informed government can better plan for resource mobilization and distribution to mitigate poverty through e-government platforms. For instance, ICTs can leverage tax revenue collection, through better tracking of tax payers as well as defaulters. A World Bank (2002) report ably summarises the potential of ICTs in fighting poverty thus:

Access to telecommunications and information services and to ICTs in general, provides crucial knowledge inputs into the productive activities of rural and poor households; makes large regional, national, and even global, markets accessible to small enterprises; and increases the reach and efficiency of the delivery of government and social services. Furthermore, access to ICTs gives the poor a voice, with which they can influence the decisions of policy-makers, and allows them to participate in the decision-making process"

- **Gender equity**: ICTs can be used as vehicles for disseminating information that can go a long way in educating society on gender issues that hinder human development. In addition, ICTs can be used to provide gender-focused information to empower women to participate fully in business, accessing of loans and land, and where to seek assistance on issues that affect them.

- **Environmental sustainability**: ICTs can leverage better management of the environment by providing accurate information and better monitoring of the environment so that preventive action is taken before environmental degradation sets in. For example, ICT-based environmental management systems such Geographic Information Systems (GIS) and remote sensing can go a long way in ensuring better mapping, management and monitoring of genetic resources (both flora and fauna) for sustainable use by present and future generations. In addition, ICTs can be used to deliver sensitization and education campaigns to the public on a wide range of environmental issues.

- **Developing a global partnership for development**: ICTs are already serving as the means for sharing global information, knowledge and experiences in all areas of human endeavour including science and technology, culture, monitoring and preservation of global biodiversity resources, politics, education and training, health, trade and commerce and mobilizing global responses to natural disasters (such as earthquakes, tsunamis, cyclones and hurricanes, droughts) as well as human made disasters (such as nuclear and industrial accidents, wars and genocides).

It is evident from the foregoing that ICTs can directly contribute to the realization of each of the MDGs in Africa.

While ICTs have the potential to significantly contribute to the attainment of MDGs in Africa, major challenges still remain however on what needs to be done to overcome factors which impede their application. One of the major challenges is

that the vast majority of the citizens of Africa are still excluded from these global experiences and resources because of the digital divide. Emerging ICTs can go a long way in mitigating some of the digital divide factors. In particular, emerging technologies can overcome most of the problems associated with poor infrastructures including such issues as accessibility, connectivity, low bandwidth, and the high costs that are associated with implementation of ICT projects.

POLICIES ON EMERGING ICTS IN AFRICA

This section of the chapter reviews national ICT policy documents of 10 selected African countries in order to determine the extent to which they cover emerging ICTs. The countries are Botswana, Burkina Faso, Egypt, Ethiopia, Mauritius, Mozambique, Nigeria, Rwanda, Tanzania, and Zambia. The review of the ICT policies was further done according to the four regional groupings to which the 10 countries are members. The regional groupings covered were the East African Community ((EAC): represented by Tanzania and Rwanda; the Common Market for Eastern and Southern Africa ((COMESA): represented by Zambia, Ethiopia and Egypt; the Economic community of West African States ((ECOWAS), represented by Nigeria and Burkina Faso; and the Southern Africa Development Community ((SADC): represented by Zambia, Tanzania, Botswana, Mauritius and Mozambique. It should be noted that some countries belong to more than one regional bloc (e.g. Zambia (SADC and COMESA) and Tanzania (EAC and SADC)

The findings on the extent to which national ICT policies in the surveyed countries embrace emerging ICTs are presented in summary form in Table 1.

The analysis generally shows that out of the 10 countries only two (Botswana and Mauritius) specifically mention emerging ICTs in their

Table 1. Content of ICT Policies of selected African countries (Sources: Indicated on the Table)

Country	Date of Policy	Key Strategies/objectives in the document	Role of Emerging ICTs
Ethiopia (The Communication Initiative Network, 2009)	January 2005	Human resource development, physical and ICT infrastructure development, ICT for governance/e-government, ICT industry and private sector development, electronic commerce, community access and service delivery, local content and applications development, ICT for research and development, ICT systems security and standards, the legal and regulatory environment, promotion of ICT in education, use of ICT in health, and ICT for agricultural modernisation.	None. Emphasis on adoption and application of ICTs
Mozambique (Sesan, 2004; Communication Initiative Network, 2008)).	September 2000	Education, human resources development, universal access, infrastructure and governance, agriculture, natural resources, the environment, tourism, electronic commerce, business protection, public protection, the academic institutions and research network, women and youth, culture and art, and social communication and the press	None. Emphasis on adoption and application of ICTs
Tanzania (Tanzania, United Republic of, 2002)	May 2002	Focus is on: service sectors, availability of universal access, peace, stability and unity: strategic ICT leadership, legal & regulatory framework, good governance: public service (e-government), ICT infrastructure, a well-educated and learning society: human capital and, local content	None. Emphasis on adoption and application of ICTs
Botswana (Botswana, Republic of, 2008)	August 2007	Community access and development, government, learning, health, economic development, infrastructure and security legislation and policy	One phrase in the whole document refers to emerging ICTs. Emphasis on adoption and application of ICTs
Zambia (International Institute for Communication (IICD), 2008)	March 2007	To promote the economy; to improve the provision of public sector services to rural communities and other disadvantaged groups; and to boost the performance of the public sector, education, health, telecommunications	None. Emphasis on adoption and application of ICTs
Egypt (PanAfrL10n, 2009)	1999	Objectives are to: create a vibrant and exportable CIT (communications and information technology) industry; support the development of a state-of-the-art national telecommunications network that provides an enabling environment for business and electronically links Egypt with the rest of the world; increase employment opportunities in the communications and information technology sectors; build an information society capable of absorbing and benefiting from expanding sources of information; develop and upgrade CIT systems	None. Although the policy goes beyond the mere adoption and application of ICTs to production and export of ICT products
Mauritius (Mauritius, Republic of, 2007)	September 2007	Objectives are to: provide a framework that will enable ICT to contribute towards achieving national development goals; develop the export markets for ICT Services and BPO/ITES; position Mauritius as a regional ICT centre of excellence and knowledge hub; ensure that ICT infrastructure and capacity are utilised effectively, are compliant with regional and international standards and are internationally competitive; establish a trusted and secure information infrastructure and a culture of cyber security at all levels of society; enhance the exploitation of ICT across the economy for increased productivity and efficiency; and transform Mauritius into an Information-based society where everyone has equitable and affordable access to ICTs	In addition to putting emphasis on production and export of ICT products, Mauritius has also established a special committee that addresses issues of emerging ICTs
Nigeria (Jidaw Systems Limited, 2009)	March 2001	Has 22 objectives that centre on creating an enabling ICT environment, legislative and regulatory environment, ICT production and export, socio-economic development, human capital development, education and training, access, infrastructure development	None. Although the policy goes beyond the mere adoption and application of ICTs to production and export of ICT products

continued on following page

Table 1. continued

Country	Date of Policy	Key Strategies/objectives in the document	Role of Emerging ICTs
Rwanda (United Nations Economic Commission for Africa (2008).	January 2000	Focus is on the following sectors: human resource development, education, civil and public services, service sector, private sector, infrastructure, legislative and governance	None. Emphasis on adoption and application of ICTs
Burkina Faso	October 2000	The focus of the policy is on: computerization of the state/administration, reinforcement of the national capacities and the quality of training and research, improvement of the economic potentials, development of community communication centers, and infrastructure development	None. Emphasis on adoption and application of ICTs

policies. Of the two countries that cite emerging ICTs, it is only Mauritius which has put in place an institutional and policy framework for dealing with issues of emerging ICTs (Seebaluck, (2006), Dabeesing (2006)). Botswana only devotes a phrase to emerging ICTs. The rest of the countries are silent on issues of the emerging ICTs.

Another finding from the review shows that out of the 10 countries surveyed, only ICT policies of three (Nigeria, Egypt and Mauritius) go beyond mere adoption and application of ICTs to the production and export of ICT products. This seems to suggest a consumption or user orientation in the ICT policies of most African countries.

On the regional economic blocs (EAC, COMESA, ECOWAS and SADC), the analysis reveals that none of these bodies are specifically addressing issues of emerging ICTs in their regional ICT policies. Current efforts in these bodies seem to centre on developing regional ICT policies in areas of coordination, harmonization and integration of national regulatory and legislative frameworks, and regional infrastructural issues among member states (Muchanga, (2004), ECOWAS (2009), EAC (2009, 2007), COMESA (2009, 2004)). The Africa economic bodies are yet to embrace emerging ICTs. This is in contrast to the European Union which is actively promoting the development of emerging ICTs through funding of research in universities and other research institutes (Community Research and Development Information Service (CORDIS, 2009)).

EMERGING ICT POLICY FRAMEWORK FOR AFRICA

While emerging technologies offer possible solutions to some of the problems of applying ICTs in Africa, there are many challenges that have to be addressed in order to create an environment that is conducive for harnessing these technologies. The barriers to ICTs cut across many different issues including legal, regulatory, infrastructure, technology advocacy, human and financial resources. As such, they require a multi-pronged approach to address them. The main objective of this section is to propose a policy framework and highlight the various strategies and actions that have to be taken in order to create an enabling environment for the adoption and sustenance of emerging technologies that will propel Africa into the 21st Century and beyond.

A policy framework provides the guidelines on the approved way of operating in relation to a particular matter. Although a number of countries in Africa have put in place national ICT policies to guide their legal and other ICT environment-setting frameworks, many are yet to do so. In the absence of policy guidelines, it becomes very difficult to implement ICTs because it is policies that guide activities.

Those countries that may have policies on some aspects of ICTs, (e.g. telecommunications), by and large, those policies tend to be protective of the existing ICT institutions that are owned by the government. Such protectionist policies tend to work against the introduction of emerging ICTs that are perceived to be threatening the status quo in those countries (Arnold, Guermazi & Mattoo, 2007). A case in point here is the decision by the Zambian government in 2007, to 'ban' VoIP technology for no other reason than that it was seen to threaten the market dominance of the state-owned telecommunications service provider (Malama, 2007). This decision to ban the much cheaper voice communication technology will no doubt, delay its entry onto the Zambian market for many years. Another policy issue common in most developing countries is the tendency to treat ICTs as a 'luxury' goods sector that attracts very heavy and prohibitive customs tariffs and other forms of taxation. High taxes on ICTs discourage investment and hampers growth of the ICT sector.

Creation of an ICT-friendly environment must be anchored on policies that clearly define the role of ICTs in national development efforts. It is only when ICTs are seen in the context of leveraging national development that their place in society will be appreciated. A policy framework, besides acknowledging the enabling role of ICTs in national development must act as a stimulus for promoting the growth of the ICT sector by removing all inhibiting barriers in the sector. Such an environment must be anchored on a number of strategic policy frameworks including the following:

- Legal framework
- Regulatory/administrative institutional framework
- Infrastructure framework
- Technology advocacy framework
- Human resources, education and research framework
- Financial resources framework

Legal Framework

Formulation of policies, per se, is not enough. Good policy formulation must be backed by enabling pieces of legislation. Legislation is very important in enabling any activity. The absence of ICT-friendly legislation or the presence ICT-unfriendly legislation militates against introduction of emerging technologies. Most of the developing countries, especially in Africa, still have outdated pieces of legislation on telecommunications, which still favour state control and makes it difficult for private sector investment to penetrate the telecommunications sector. ICT legislation must address, among other things supporting new policy initiatives on ICTs and reviewing and repealing the existing pieces of legislation that inhibit the growth of the ICT sector.

Regulatory/Administrative Institutional Framework

The administration of ICT-related industries, such as the telecommunications sector, is characterized by bureaucratic bottlenecks and red tape, where it takes a very long time to grant licenses to investors who wish to operate ICT-related businesses, particularly in the telecommunications sector. Further, ICT regulatory institutions tend to favour state-owned national telecommunications companies as noted by Frost & Sullivan (2008) the African Internet market is characterised by national incumbents which command the highest market share given that they control most of the infrastructure and most regulatory frameworks are geared in their favour. These bureaucratic bottlenecks have tended to discourage investors who would have brought new and innovative technology in those countries (Toure, 2006).

An effective institutional framework for the regulation and administration of the ICT sector is essential for the promotion of emerging technologies. Among other things, an effective regulatory institutional framework ensures fair play and accountability in the sector; quality in service delivery; fair pricing of ICT services and products to customers; fair competition among the players in the sector; and fair application of regulations to all the players in the sector. The existing institutional framework for regulating and administering the ICT sector must be overhauled to make them efficient, transparent and accountable in the way they conduct the business of their mandate. It is only when stakeholders have trust and confidence in the manner in which regulatory and licensing bodies conduct their business that investors can be attracted to the sector. As Pais (2006) correctly observes "The eradication of corruption at all levels is a prerequisite for creating a framework that is transparent, fair and amenable to economic growth" (p.2). In addition, the institutions that are responsible for regulating the ICT sector must be well supported in terms of legislation, policy, financial as well as provision of appropriately qualified human resources to enable them carry out their mandate effectively. These regulatory issues are essential in achieving an effective institutional regulatory framework.

Infrastructure Framework

Information technology can only thrive where the necessary infrastructure is in place. In most African countries, the basic physical infrastructure required to sustain or even apply ICTs, is not in place. For instance, electricity supply is a basic requirement for ICTs, and it is not adequately and reliably supplied in most of the developing countries, particularly in Africa. Telecommunications infrastructure (both fixed and space-based) is also not in place in most of Africa.

If national governments are not able to provide the essential infrastructures, then they should allow the private sector to invest in infrastructure either on their own or in partnership with government (Department for International Development, 2006). While governments will always play a major role in issues of infrastructure development,

private sector participation should be encouraged because the private sector has the financial resources as well as technical expertise to bring about a speedy availability of the requisite ICTs (Pais, 2006). As noted by Juraske (2007), Hewlett Packard's Public Sector Vice President for Europe, Middle East and Africa, "forming strong collaboration between African governments and the corporate sector is the smartest way to do meet the challenge of deploying ICTs in Africa…" (p.1) This is so because governments generally do not have adequate capacity (human, financial, technology) to cope with developments that require rapid changes in adopting new technologies in a fast changing sector such as the field of ICTs.

Technology Advocacy Framework

Technology can only be appreciated when it is understood in the environment where it is deployed or will be deployed. In some cases, people will be averse to a technology, not because it is a bad technology, but because they do not understand it. To this end, there is need to sensitise not only the policy makers (political leaders, government officials, members of parliament, institutional leaders, etc), but society as whole on the implications of the emerging technologies for the present and future.

Technology advocacy will go a long way in buying-in the support of the critical decision makers (Chetty, 2005). Once the decision makers begin to see the benefits of ICTs, most of the barriers that inhibit the sustainable promotion and application of ICTs in developing countries may be minimized. Governments by their very nature are conservative institutions that approach issues of technology very cautiously. Government conservatism on issues of technology is a barrier because in most developing countries, the private sector is usually very small to make a significant impact in the adoption and sustenance of a new technology. Without the participation of the government, a new technology cannot be sustained because it is governments that have the financial clout to spend on new technology.

Human Resources Framework

Appropriately trained human resources are a pre-requisite to the effective implementation of any technological innovation. As the Africa Policy information Centre (APC) Africa Policy Monitor (2009) has ably observed:

While policies and strategies must address the extension of the communications infrastructure through telecommunications reform to stimulate private sector growth and create job opportunities, this is a necessary but by no means sufficient condition for an effective ICT contribution to national development goals. ICT policy and strategies must also incorporate social goals by building human capacity and creating the conditions for the development of relevant applications and content.

The need for highly trained expertise is even more critical in a highly sophisticated field such as the ICT sector (Chetty, 2005). Highly trained ICT engineers in systems design and development, hardware, software, telecommunications, and experts in regulatory, policy and legal are required to drive and maintain a viable ICT sector. Africa and other Third World regions lack this critical ICT expertise.

All these areas of expertise will require to be put in place through an appropriate education, training and research policy framework that will drive the production of the requisite human resources and knowledge to drive ICT activities in Africa. Further, the critical ICT expertise must be retained within Africa by putting in place attractive financial rewards and other staff retention incentives.

Financial Resources Framework

The creation (through education, training, research and development), acquisition (for application), and maintenance of technology require a lot of financial resources. Most of the countries of Africa lack adequate financial resources to fund their development efforts. The little money that is at their disposal is allocated to what are seen to be priority areas of health and education and provision of other basic services. It is therefore a major challenge to advocate for investment in ICTs (even in terms of infrastructure) in the face of a government that has to meet immediate needs for the people.

However, introducing and sustaining emerging ICTs will require financial resources to support all the frameworks that have been highlighted above.

If African countries can buy-in the argument that ICTs are critical inputs into their national development efforts and also that emerging ICTs are an investment into the future, then policy makers can start prioritizing ICTs in national budgetary allocations that will go towards financing the frameworks that are being proposed.

FUTURE TRENDS

Developments in emerging ICTs will continue particularly in the areas of artificial intelligence software, wireless telecommunications technology and the development of molecular computing hardware based on nanotechnology. Future areas of research from the African policy perspectives would be to explore to what extent African governments, institutions of higher learning, research institutes and industry are addressing these emerging ICT developments in terms of their research agendas, financing of research and human resource development, training and its retention.

CONCLUSION

Technology has, throughout the history of humankind, provided the great quantum leaps that have propelled civilization and progress. ICTs and associated technologies are now at the centre of all areas of human endeavour: scientific and technological, social, economic, cultural, military, and security, etc. Emerging ICTs, if properly harnessed, can provide what Kuttan & Peters (2003) call the 'digital opportunities' to leapfrog and bridge the ever widening digital divide between Africa and the rest of the world and within and between African nations, communities and individuals. A review of the literature on ICT policies in Africa shows that, by and large, these policies are essentially designed to facilitate the adoption and application of ICTs in various sectors, with no emphasis on their production. The implication of this is that African countries will continue to source their ICTs from outside the continent. The study also seems to indicate a lack of appreciation of emerging ICTs both at the national as well as the regional economic bloc levels. It is therefore proposed that in order to prepare for the future, Africa must embrace emerging technologies by creating a conducive emerging ICT environment. Africa cannot afford to ignore ICTs again as has been the case in the first two waves of the ICT revolution. The development of nanotechnology and artificial intelligence software have dire consequences for sharpening not only the digital divide but the technology divide between those nations and regions that will possess and control this new industrial technology platform and those that will not have it.

Africa should not behave like the proverbial ostrich that buries its head in the sand in the face of a formidable foe, in the vain hope that by not seeing the enemy, the enemy will not see it! It is said that the future favours those who prepare for it. Africa can prepare for its future by creating an appropriate environment for fostering the adoption and application of emerging technologies.

REFERENCES

Adomi, E. E. (2009). *Library and information resources*. Benin City, Nigeria: Ethiope Publishing Corporation.

Afele, J. S. (2003). *Digital bridges: developing countries in the knowledge economy*. Hershey, PA: Idea Group.

Africa Policy Monitor, A. P. C. (2009). Retrieved on March 27 2009 from http://africa.rights.apc.org/?apc=he_1&w=s&t=21873

African Business. (2009, Thursday, January 1). Retrieved on March 27, 2009 from http://www.allbusiness.com/media-telecommunications/11745598-1.html retrieved on 27/03/2009

Arnold, J., Guermazi, B., & Mattoo, A. (2007). Telecommunications: the persistence of monopoly. In Mattoo, A., & Payton, L. (Eds.), *Services, trade and development: the experience of Zambia* (pp. 101–153). Washington, DC: The International Bank for Reconstruction and Development/World Bank.

BITPIPE. *COM*. (2008). Retrieved on August 15, 2007 from http:/www.bitpie.com.

Bonsor, K. (2000). How DNA Computers Will Work. Retrieved on July 6, 2010 from HowStuffWorks.com. <http://computer.howstuffworks.com/dna-computer.htm>

Botswana, Ministry of Communications, Science and Technology (2008). *Maitlamo: National Information and Communications Technology Policy*. Gaborone, Botswana: Government Printer.

Buyya, R. (2001). *Parallel processing: architecture and system overview*. Retrieved on March 27 2009 from http://www.cs.mu.oz.au/678/ParCom.ppt#352,1,Parallel Processing: Architecture and System Overview.

Chetty, M. (2005). Information and Communications Technologies (ICTs) and Africa's development. Retrieved on March 5, 2009 from http://www.nepad.org/2005/files/documents/124.pdf.

Common Market for Eastern and Southern Africa. (2009). *COMESA Regional e-Government Framework*. Retrieved on September 23, 2009 from http://programmes.comesa.int/attachments/144_COMESA%20e-GOV%20Framework%20PresentationKampala.pdf.

Common Market for Eastern and Southern Africa, Association of Regulators of Information And Communications for Eastern and Southern Africa. (2004). *Policy Guidelines on Universal Service/Access*. Retrieved on September 22, 2009 from http://programmes.comesa.int/attachments/118_Policy_Guidelines_Universal_Service.pdf.

Community Research and Development Information service. (CORDIS, 2009). *ICT - Future and emerging technologies: a continuing well established successful ICT scheme*. Retrieved on September 22, 2009 from http://cordis.europa.eu/fp7/ict/programme/fet_en.html

Dabeesing, T. (2006). *National ICT Strategic Plan Emerging Technologies and Standards Working Group*. Retrieved on September 22, 2009 from http://www.gov.mu/portal/goc/ncb/file/Presentation_NWGET.pdf.

Department for International Development (DFID). (2006). *Financing ICT for Development: the EU approach*. Retrieved on March 27, 2009 from http://www.dfid.gov.uk/pubs/files/eu-financ-wsis-english.pdf.

Department for International Development (DFID). (2008). *Millennium Development Goals*. Retrieved on March 27, 2009 from http://www.dfid.gov.uk/mdg/

Encyclopaedia Britannica. (2009). Very large-scale integration. Retrieved on March 27, 2009, from *Encyclopædia Britannica Online* from http://www.britannica.com/EBchecked/topic/626791/very-large-scale-integration

Ernest, H., & Shetty, R. (2005). Impact of Nanotechnology on Biomedical Sciences: Review of Current Concepts on Convergence of Nanotechnology With Biology. *Journal of Nanotechology online*. Retrieved on July 6, 2010 from http://www.azonano.com/details.asp?ArticleID=1242

Farrier, J. (2010, January 11). "Wet Computer" Will Mimic A Biological Brain. Retrieved on July 7, 2010 from http://www.neatorama.com/2010/01/11/wet-computer-will-mimic-a-biological-brain/

Foresight Nanotech Institute. (2008). Retrieved on October 10, 2007 from http://www.foresight.org/nano/

Forrester, E. C. (2003). *A Life-Cycle Approach to Technology Transition*. Retrieved on June 15 2008 from http://www.sei.cmu.edu/news-at-sei/features/2003/3q03/feature-4-3q03.htm

Foster, I. (2005). *Internet computing and the emerging grid*. Retrieved on October 9, 2007 from http://www.nature.com/nature/webmatters/grid/grid.html

Frost & Sullivan. (2007). Cited in *ZAMNET gains global recognition as market leader*. Retrieved on December 8, 2007 from http://www.zamnet.zm/newsys/news/viewnews.cgi?category=30&id=1197109256

Gavaghan, T. (2010). *DNA computing*. Retrieved on July 6, 2010 from http://EzineArticles.com/?expert=Thomas_Gavaghan

Hagiya, M. (2000). Theory and Construction of Molecular Computers. Retrieved on July 6, 2010 from http://hagi.is.s.u-tokyo.ac.jp/MCP/.

Harper, T. (2003). *What is Nanotechnology?* 2003 Nanotechnology 14. doi:10.1088/0957-4484/14/1/001. Retrieved on October 10, 2007 from http://www.iop.org/EJ/abstract/0957-4484/14/1/001

Hsu, J. (2010, January 11). Wet Computer Literally Simulates Brain Cells. Retrieved on July 7, 2010 from http://www.popsci.com/technology/article/2010-01/wet-computer-literally-simultes-brain-cells

Intellon. (2007). *What is powerline communications?* Retrieved on March 5, 2009 from http://www.intellon.com/technology/powerlinecommunications.php

International Institute for Communication and Development (IICD). (2008). *Introducing Zambia.* Retrieved on September 21, 2009 from http://www.iicd.org/countries/zambia.

International Telecommunications Union. (2005). *What's the state of ICT access around the world?* Retrieved on February 22, 2009 from http://www.itu.int/wsis/tunis/newsroom/stats/

Jensen, M. (2002). *Information and Communication Technology (ICTs) in Africa – a status report.* UN ICT Task Force "Bridging the Digital Divide in the 21st Century" Presented to the Third Task Force Meeting United Nations Headquarters 30 September – 1 Oct 2002. Retrieved on January 27, 2009 from http://www.unicttaskforce.org/thirdmeeting/documents/jensen%20v6.htm

Jidaw Systems Limited. (2007). *The Nigerian National Information Technology Policy.* Retrieved on August 22, 2009 from http://www.jidaw.com/itsolutions/ict4dreview2007.html

Juraske, I. (2007). *Some success tips for deploying ICTs in Africa.* Retrieved on January 10, 2008 from http://communications-online.blogspot.com/2007/06/some-success-tips-for-deploying-icts-in.html

Knight, W. (2007). IBM creates world's most powerful computer. *New scientist.com news service*, June 2007. Retrieved on January 15, 2008 from http://technology.newscientist.com/article/dn12145-ibm-creates-worlds-most-powerful-computer.html

Kuttan, A., & Peer, L. (2003). *From digital divide to digital opportunity.* Lanham, MD: Scarecrow.

Loygren, S. (2003). Computer Made from DNA and Enzymes. Retrievd on 6 July 2010 from http://news.nationalgeographic.com/news/2003/02/0224_030224_DNAcomputer_2.html

Malama, F. (2007, April 17). Let's take advantage of VoIP. *The Post.*

Mauritius, Ministry of Information Technology and Telecommunications. (2007). *National ICT policy 2007-11.* Retrieved on September 21, 2009 from http://www.gov.mu/portal/goc/telecomit/file/ICT%20Policy%202007-2011.pdf

Moll, P. (1983). Should the Third World have Information Technology? *IFLA Journal*, *9*(4), 296–308. doi:10.1177/034003528300900406

Muchanga, A. M. (2004). *Statement by Albert M. Muchanga SADC Deputy Executive Secretary Delivered at the Opening of The SADC Regional Seminar on Website Policy Development Held at President Hotel Gaborone.* Retrieved on September 23, 2009 from http://www.sadc.int/archives/read/news/56.

Mutula, S. M. (2008). Digital divide in Africa: its causes, and amelioration strategies . In Aina, L. O., Mutula, S. M., & Tiamiyu, M. A. (Eds.), *Information and knowledge management in the digital age: Concepts technologies and African perspectives.* Ibadan, Nigeria: Third World Information Services.

Mutula, S. M., & Wamukoya, J. M. (2007). *Web Information management: a cross-disciplinary textbook*. Oxford, UK: Chandos Publishing.

Mutume, G. (2005). *Africa takes on the digital divide: new information technologies change the lives of those in reach*. Retrieved on November 16, 2007 from http/www.africarecovery.org

National Council on Disability. (2007). *Over the horizon: potential impact of emerging trends in information and communication technology on disability policy and practice*. Retrieved on May 22, 2007from http://www.disabilityinfo.gov/

Okot-Uma, R. W. (2002). The challenge of the digital divide . In Murelli, E. (Ed.), *Breaking the digital divide: implications for developing countries* (pp. ix–xi). London: Commonwealth Secretariat.

Pais, A. (2006). *Bridging the digital divide by bringing connectivity to underserved areas of the world*. Retrieved October 2, 2009 from http://www.itu.int/osg/spu/youngminds/2006/essays/essay-adrian-pais.pdf

Palmer, J. (2010, January 11). Chemical computer that mimics neurons to be created. *BBC News*. Retrieved on July 6, 2010 from http://news.bbc.co.uk/2/hi/8452196.stm

PanAfrL10n. (2009). *Egypt*. Retrieved on September 21, 2009 from http://www.panafril10n.org/pmwiki.php/PanAfrLoc/Egypt

Reddy, R., Arunachalam, V. S., Tongia, R., Subrahmanian, E., & Balakrishnan, N. (2004). *Sustainable ICT for emerging economies: mythology and reality of the digital divide problem – a discussion note*. Retrieved on December 30, 2008 from http://www.rr.cs.cmu.edu/ITSD.doc

Roco, M. C. (2003). Nanotechnology: convergence with modern biology and medicine. *Current Opinion in Biotechnology*. Volume 14, Issue 3, June 2003. Retrieved on July 7, 2010 from http://www.sciencedirect.com/science/journal/09581669 346

SAE International Emerging Technologies Advisory Board. (2009). Retrieved October 2, 2009 from http://www.sae.org/about/board/committees/etab.htm

Seebaluck, R. (2006). *Improving Your Business Environment Through Public and Private Partnerships: Strategies that Work*. Retrieved on September 22, 2009 from http://idisc.infodev.org/proxy/Document.48.aspx

Sesan, G. (2004). *Africa, ICT policy and the millennium development goals*. Retrieved on August 22, 2009 from http://wsispapers.choike.org/africa_ict_mdg.pdf

Tanzania, United Republic of, Ministry of Communications and Transport (2002). *National ICT Policy of Tanzania*. Retrieved on August 22, 2009 from http://www.ethinktanktz.org/esecretariat/DocArchive/zerothorder.pdf.

The Communication Initiative Network. (2008). *Mozambique ICT4D National Policy*. Retrieved on September 22, 2009 from http://www.comminit.com/en/node/148306.

The Communication Initiative Network. (2009). *Ethiopia ICT4D National Policy* [Draft]. Retrieved on September 22, 2009 from http://www.comminit.com/en/node/148306.

The World Bank. (2002). *Telecommunications and information services for the poor: Towards a strategy for universal access*. Washington, DC: The World Bank.

Twist, J. (2005). *UN debut for $100 laptop for poor*. Retrieved on November 15, 2007 from http://news.bbc.co.uk/2/hi/technology/4445060.stm

United Nations Economic Commission for Africa. (2008). Burkina Faso: NICI Policy. Retrieved on August 25, 2009 from http://www.uneca. org/aisi/nici/country_profiles/Burkina%20Faso/ burkinapol.htm

United Nations Economic Commission for Africa. (2008). *Rwanda National ICT Policy.* Retrieved on August 17, 2009 from http://www.comminit. com/en/node/148446.

Walker, D. W. (2002). *Emerging distributed computing technologies.* Retrieved on August 10, 2007 from http://www.cs.cf.ac.uk/User/ David.W.Walker

Wang, Z. L. (2008). *What is Nanotechnology?* Retrieved on February 22, 2008 from http:// www.nanoscience.gatech.edu/zlwang/research/ nano.html

Waters, D. (2007). Africa waiting for net revolution. *BBC.* Retrieved on March 10, 2009 from http://news.bbc.co.uk/1/hi/technology/7063682. stm

Webopedia. (2008) Retrieved on January 15, 2008 from http://inews.webopedia.com/TERM/d/ digital_divide.html

Wikipedia. (2009). Very-large-scale integration. Retrieved on July 28, 2009 from http:// en.wikipedia.org/wiki/Very-large-scale_integration.

Zulu, S. F. C. (1994). Africa's survival plan for meeting the challenges of information technology in the 1990s and beyond. *Libri, 44*(1), 77–94. doi:10.1515/libr.1994.44.1.77

KEY TERMS AND DEFINITIONS

Artificial Intelligence: Software that has capability to 'learn' 'reason' and 'draw conclusions.'

Digital Divide: Disparities in accessing ICTs by people within and between in individuals, communities, urban and rural areas, regions, countries and continents.

Emerging ICTs: New and emerging ICTs that are based on the converged ICTs (comprising telecommunications, computing and publishing technologies).

Grid Computing: Also known as distributed or cluster computing, the emerging technology in grid computing involves linking existing computers in a grid to make them behave as one computer so that they are able to share their combined capabilities to produce computer processing power similar to a stand-alone supercomputer.

Information and Communications Technologies: Various technologies that are used in the in the creation, acquisition, storage, dissemination, retrieval, manipulation and transmission of information. In its broader context, ICTs include: Computers: used for processing of information; Telecommunications: used for the transmission of information; Publishing: used for coding/recording of knowledge and information. Publishing in its entirety including broadcasting and narrowcasting, the press and other media such as micrographics, audio-visuals, etc.

Millennium Development Goals: Millennium Development Goals (MDGs) are development targets that are designed to improve human living conditions by 2015. The MDGs were agreed upon by world leaders at a United Nations conference held in 2000). They cover eight broad areas that are critical to human development including education; health (including reduction of child mortality and improving maternal health); gender equality; HIV/AIDS; the environment; eradication of extreme poverty and developing a global partnership for development. The international community agreed to dedicate its efforts towards attaining these goals as a way of accelerating development in under-developed regions of the world such as Africa.

Nanotechnology: a new technology paradigm platform that is based on various fields of knowledge including bio-technology, physical sciences technology (atomic and nuclear technology), neuro-science technology, molecular physics, materials science, chemistry, biology, computer science, electrical engineering, mechanical engineering and cognitive science: technology development at the atomic, molecular, or macromolecular range.

Pervasive or Ubiquitous Computing Environment: Also called ubiquitous computing, refers to the Third Wave of the computing revolution dominated by computing devices that will control everyday life. This is an era of widespread computing which will be characterized by one-individual-to-many computers which are connected in a seamless web: the era which has been characterized as Anytime/Anywhere-->Any Device --> Any Network --> Any Data.

Policy: A concise, formal statement of principles which indicate how an organization country, region, etc will act in a particular area of its operation e.g. in relation to ICTs. A policy provides the guidelines on the approved way of operating in relation to a particular matter.

Chapter 9
Ethics and Social Issues Related to Information Communication Technology (ICT)

Nelson Edewor
Delta State Polytechnic, Nigeria

ABSTRACT

Information Communication Technology (ICT) has raised new ethical concerns about the protection of personal privacy, protection of intellectual property, user responsibility, acceptable access and use of information, software licenses and piracy. A good ICT policy must be able to adequately consider these, and many other associated issues. This chapter therefore describes these ethical issues and how to deal with them as an individual or an organization. It provides information on the concept of ethics and the technological advancements responsible for the ethical concern. It discusses privacy, information rights, and intellectual property rights and ethics policy. The Nigerian national intellectual property right laws were examined in line with World Trade Organization/Trade Related Aspects of Intellectual Property Rights (WTO/TRIP) compliance.

INTRODUCTION

In the rapidly changing technological environment in which we live; ethical issues are increasingly been raised, demanding attention and efforts towards resolution. Of particular interest for us and the information society are those related to information communication technologies (ICTs). The explosive growth of ICT and the use of its enabling technologies have had major impacts on society and thus raise serious ethical questions for individuals and organisations. These issues have been raised to a new and often perplexing level which has greatly affected the society in various ways. The pressing issues raised by ICT include the invasion of individual and corporate privacy, intellectual property rights, individual and societal rights, values preservation and accountability for the consequences arising from the use of ICT, etc.

These issues have thrown up important challenges in the area of employment; working conditions and individuality. However, not much

DOI: 10.4018/978-1-61692-012-8.ch009

progress has been made in addressing these issues and challenges associated with ICT. This is because of lack of clear understanding of the issues involved.

In this chapter, we will explore ethical and social issues/challenges that surround the use of ICT. Specifically, the chapter sets out to:

- Identify ethical issues/challenges and how the use of ICT has greatly invaded individual privacy and the protection of intellectual property.
- Analyze principles that can serve as the basis for ethical conduct by users of ICT.
- Identify ethics policy structure and framework for ICT use.
- Evaluate the impact of ICT on the protection of individual and collective privacy, information rights and intellectual property rights.
- Evaluate intellectual property rights protection laws in Nigeria with respect to WTO/TRIPS compliance.
- Research and apply a range of transferable knowledge required for ethical decision making and strategy.

BACKGROUND

Ethics is a reflection on morality. It refers to the principles of right and wrong in making choices by individuals. It has been described as the art and science that seeks to bring sensitivity and methods to the discernment of moral values (Carbo, 2006). Thus, ethics guide human and societal behavior. Capuro (2006) had no difficulty in asserting that ethics is an unending quest on explicit and implicit use of the moral code.

The subject of social and ethical implications of Information and Communication technology has been addressed in the literature. As noted by Carbo (2006) ethical considerations for ICT related issues first appeared under the topic "information

ethics" in the Annual Review of Information Science and Technology in 1992. This suggests that there is an ethical agenda associated with the use of ICT. Individuals and organisations therefore need to be ethically sensitive as they deploy ICT on their operations. The impact of ICT on human relationship has been tremendous. ICT has helped to enhance family relationship (e.g. mobile phones, palmtops, laptops, virtual conferencing and so on), as well help to separate family and friends from each other. ICT has enabled new friendship and relationships in virtual communities. How genuine are such relationship? What does it portend for individual satisfaction? In the workplace for instance, new kinds of jobs are being created such as data miners, web-counselors etc, but these opportunities are also endangered by problems of unemployment from computer replacing humans. A wide range of new laws, regulations, rules and practices are therefore needed if society is to manage these workplace and other changes and development brought about by ICT. Thus the society need to consider the following ethical and social challenges related to ICT use:

- Recognition for personal and corporate ethics associated with ICT.
- Striking a balance between ethical, economic and technological (Rogerson, 2008) as well as political considerations.
- Intellectual property rights issue (trademarks, patents, copyright and trade secrets).
- Non violation of privacy and associated rights amidst electronic information data mining.
- The opportunity to commit crime with ICT (computer crime).
- Legal issues and limitations.
- Consequence of using ICT.
- Professional responsibilities (Kallman and Grillo, 1996).

Every organisation, society or nation must recognize and address the aforementioned issues thrown up by the enormous and rapid change in ICT. The full impact of such changes needs to be considered, because it affects how individuals operate. To this end, ethics and social issues are closely linked. As observed by Bynum(1999) more and more of society's activities and opportunities enter cyberspace: business opportunities, educational opportunities, medical services, employment, leisure time activities etc. to this end it becomes imperative therefore to develop policies that will include all parties (ICT "have-nots" and "haves").

Globalization and advances in ICT has helped to fuel the trends responsible for these ethical issues. More so ICT is developing so rapidly that new possibilities emerge before the social consequences can be understood (Rogerson and Bynum, 1995) .Traditional barriers between regions, states countries and continents no longer exist, because of the interconnection brought about by the Internet technology. Gorniak (1995) pointed out that ICT has been responsible for genuinely global discussion about ethics and values, for the very first time. For example if an information is posted on the Web in a culture/setting, which does not consider such an information offensive but is accessed by someone in a culture/setting, where such information is considered offensive and outlawed; whose values apply? Should the values of the first culture/setting, be permitted or not? What about the values of the second culture/setting? Arising from this situations, what kind of conflicts might arise and how should they be tackled, and by whom? What policies would be fair to all concerned? Will this bring about better understanding between peoples and cultures, new shared values and goals, new national and international laws and policies? Will individual cultures become "diluted" homogenized, blurred? (Bynum and Rogerson, 1996). These are amongst the many social and ethical issues related to ICT.

The forces of globalization and advances in ICT have helped to heighten ethical and social concerns. The rapid development of ICT has a ripple effect on society, thereby challenging existing social structure that must be dealt with as an individual, organization and government. These issues throw up new situations hitherto not known. Reactions to these issues differ at all levels of the information society: individuals, organizations and government.

Little progress has been made in addressing the ethical issues associated with ICT, especially in developing countries, because of the level of ignorance pervading the society in relation to ethical issues associated with ICT. This is reflected in most laws not drafted with the benefit of the digital environment in mind. Furthermore social and ethical implications of ICT are enormous and mostly unknown (Bynum, 1999). However we need to find a balance between technology, shaping social events and vice versa. To this end some governments particularly in the Western World and most professional computing bodies have turned their attention to the problem and instigated measure to include consideration of social, legal and ethical issues associated with ICT (Sherratt, Rogerson and Fairweather, 2005). This is aimed at raising ethical awareness.

The Internet and digital firm technologies make it easier than ever to assemble integrate and distribute information, unleashing new concerns about the appropriate use of information, the protection of personal privacy and the protection of intellectual property (Laudon and Laudon, 2006). The issue of privacy is at the core of any ethical consideration.

Privacy has been interpreted differently by various persons. According to Froehlich (1992) privacy refers to the right to be left alone. This definition is strategically mirrored in the light of an electronically enabled environment. Its applicability is relatively narrow. Privacy is a situation of an individual to be free from any form of interference or surveillance. With the advent of ICT, there is little or no control over an individual's personal information. Every time an individual visits a site on the World Wide Web, personal information is been collected about that individual.

Information technology makes it possible for organization to collect, store, exchange and retrieve data and information about any individual. Millions of employees are subject to electronic and other forms of high-tech surveillance (Ball, 2001). With ICT personal claims to privacy is threatened. The unauthorized use of such information has seriously damaged the privacy of individuals. To worsen the case much of the Net and Web are easy targets by hackers.

Every ethical consideration must be addressed from socio cultural and individual standpoints. The vast majority of such challenges are largely unknown and still emerging as new technology of ICT generates them. Meanwhile every social and ethical consideration must take into account the following:

- Individual right to liberty: every action must be based on every individual's right to liberty without much restrictions and unnecessary inhibitions. Every society and culture must be recognized and respected.
- Equal rights and opportunities of every individual must be respected. Every action should take this into consideration.
- Action to be taken must be right and beneficial to the organization and society: take the action that benefits the generality of all concerned.
- Respect for human dignity: take the action that produces least harm
- Respect for human basic rights.
- The "Golden Rule;" do unto others as you would have them do unto others.

These guidelines are however not fool-proof, but can serve as the basics for ethical conduct by users of information technology. The non consideration of these ethical responsibilities by

individuals and organisations for their actions is primarily responsible for the damaging effects of ICT.

INFORMATION RIGHTS: CLAIM TO PRIVACY

It is obvious that the power of Information Technology to store and retrieve information has eroded the right to privacy of every individual. Privacy is the right to be left alone. This includes freedom from State interference in all matters. It involves the sanctity of confidentiality as a right of individuals, confidentiality in all matters, both private and non private. Privacy has been eroded by the widespread use of advanced Information Communication Technology. Confidential information on individuals contained in existing files and databases by financial institutions, government agencies and private organisations have been misused, resulting in the invasion of privacy. The unauthorized access and use of such confidential information has seriously hurt and damaged the privacy of individuals. Invasion of individual's privacy occurs when there is an intrusion into a person's private "information bank" with intent to expose, or encroach upon individual's privacy.

The claim to privacy is protected in existing statute books and laws of various countries, though differences exist from countries to countries. For example, in the United States, privacy laws attempt to enforce the privacy of computer based files and communications (O' Brien2003). The United States, electronic communications privacy Act of 1986 and the Computer security Act of 1987 prohibits the interception of data communication messages, stealing or destroying data or trespassing in Federal related computer systems. Also the US computer matching and privacy of 1988 regulates the matching of data held in federal agency files viz a viz eligibility verifications for federal programs. There are also protection laws about individuals in financial services, health information management, education and insurance.

An important law worthy of mention here is the Freedom of Information Act (FOIA) of the United States. This law provides public access to federal Government files. FOIA came alive on July 4, 1967. It is founded on the belief that government is accountable for its actions and that the public possesses a right to obtain information about those actions regardless of ideological origin or artistic content. FOIA requires government agency to publish descriptions of its operations. Any person or organisations is recognized by law under the FOIA Act to also obtain data from a government agency through a FOIA request.

In Europe it is the general belief that information privacy should be treated as human right. Europe has applied its data protection principles both to government and the private sector. This directive requires companies to inform individual when they collect information about them and how such information is to be stored and used. This suggests that the consent of the individual must be given before any information can be accessed. All EU member states are expected to translate these principles into their own laws. The agreement took effect in October 1998. A major concern of this agreement with the United States is its prohibition of trade with any nation that does not have adequate privacy laws. The United States has raised serious security implications to this, especially after the 9/11 attacks. The Data Protection Directive is in danger of adequate non-enforcement amongst member Nations vis a vis security implications.

In UK, the Data Protection Act of 1984 is aimed at enforcing good practices in the management of personal data/information by organization. Accordingly anyone "processing personal data must comply with eight enforceable principles of good practices, data must be;

- Fairly and lawfully processed.
- Processed for limited purposes.

- Adequate, relevant and not excessive.
- Accurate
- Not kept longer than necessary.
- Processed in accordance with the data subject's right.
- Secure
- Not transferred to countries without adequate protection

In 2000, the UK Government implemented new legislation to bring the Act in line with the Data Protection Directive of the European Union" (Davies,2002,p.297). There is no doubt that the spread and use of ICT has made the world a less private place. The Internet has been described as an "omnipresent network of surveillance". For example e-mail data/messages of individuals are monitored, existing companies' files and large database can be tracked, misused and altered. The Internet however, has gained wide acceptance, as an important forum and avenue for meaningful venture among individuals. Human society is at a crossroad.ICT that may be used as object and instrument of criminality is also beneficial to modern society in all ramifications. The choice is ours.

In Africa, privacy law, have not been given serious consideration. This is traceable to the absence of Information Policy amongst various countries in Africa. An information policy aims to provide legal and institutional frameworks within which formal information exchange can take place (Ifidon, 2006). An information policy regulates the use, storage and communication of information. In Nigeria for instance, there is absence of National Information Policy (NIP). This is further worsened by the un-conducive environment for the development and management of Information Communication Technology (ICT) and National Information Infrastructure.

THE NIGERIAN FREEDOM OF INFORMATION BILL

The freedom of information bill currently before the Nigerian Senate has had a chequered legislative history. It has been pending before the National Assembly for Nine years. The bill seeks to guarantee the right of the citizens to know. The freedom of Information Bill, when it becomes law, will give Nigerian citizens access to public records and documents, subject to certain exemptions specified in the Bill. The categories of information exempted from public access under the proposed law include:

- Information that may be injurious to the defense of Nigeria
- Information relating to the conduct of international affairs.
- Information that can interfere with ongoing law enforcement investigations.
- Information that can prevent a fair trial or undermine the security of penal institutions.
- Information that can contain trade secrets, financial commercial or technical information that belongs to the government and has economic value.
- Personal information.
- Information that is subject to solicitor/client privilege.
- Research information materials prepared by faculty members in an academic institution.

The freedom of information bill is based on the belief that every citizen have right of access to information held by the state or its agencies. It anchors that partnership between government and citizens is necessary in ensuring accountable government and development (Edewor, 2008). Given the general apathy of lawmakers towards the freedom of Information Bill which has hindered its successful consideration, the Bill is in danger of not been passed into law.

INTELLECTUAL PROPERTY

Intellectual property is creative works of individuals or organisations that have economic value. These works are usually intangibles; that is product of human creativity and imagination which are protected by law. This law prevents others from copying, duplicating and distributing such work of human ingenuity without authorization. This is aimed at securing economic rewards for their efforts. The rapid development of information technology has made it increasingly difficult to protect intellectual property. This is largely due to ICT capability to duplicate, share and copy information via several channels and networks. Patents, copyright, trademarks and trade secrets are all forms of intellectual property.

Patent law protects technological invasion. It requires that an inventor or creator of any given work is recognized by law and is given exclusive right and monopoly for the underlying concepts, ideas behind such work for a given period of time. This however differs from country to country. Copyright law recognizes and protects literary and artistic work such as paintings, sculpture, prose, poetry, musical composition, dances, photographs, motion pictures, radio and television programs, sound recordings and computer software programs. Copyright law protects the copying, distribution or duplicating, in whole or in parts of any intellectual property without permission. Trademarks protects brand words, symbols and slogans of goods and services. Trade secrets protect confidential information of any given business. Such confidential information pertaining to 'how, why, and essence' of the business or venture must be protected from the general public arena. Most intellectual property right expires after a specific period depending on the country.

However trade mark never expires, so long as it is in use to identify a product or service.

The growth of the Internet and associated technologies is a serious problem for the protection of intellectual property. The Internet facilitates the transmission of data/information freely around the world, even copyrighted information. This is particularly true in the sense that with the World Wide Web, virtually anything can be copied and distributed to millions of people around the world. Unauthorized persons can copy duplicate and share music and movies in the Net, through various devices. Digitized versions of books can be easily made available to millions of people to download through the Internet without permission; e-mail file attachment can also be used to share or copy copyrighted information across the globe. The entertainment industry is been threatened as copyrighted music and motion pictures continues to be traded for free without authorization. These are serious challenges to effective protection of intellectual property rights. The major ethical challenge is to determine the continued relevance of protecting copyrighted data/information when they can be copied and shared easily without authorization. There is therefore the need for the development of new intellectual property rights protection strategies in this era of rapid development of ICT, capable of protecting software, eBooks, music and motion picture. In the music and film industry broadcasting organization function as purveyors of copyright works. In addition, they contribute to the repertoire of works by their independent creations, which are also subject to copyright protection. In the same vein, broadcasting organizations can digitize texts, pictures, image sequences and sounds which are intellectual property. At the same time separate media can be combined to create multimedia and can be altered by the user in a number of ways. The questions are; do broadcasting organizations require fresh authorization of right owners? Does this violate any of the rights of the original copyright owners? What should the rules be and who should formulate and enforce them? A wide range of new laws are needed if society is to manage these issues associated with the digital environment.

In an attempt to protect intellectual property rights; most nations of the world signed the agreement on Trade Related Aspects of Intellectual Property Rights (TRIPS) in 1994. TRIPS is administered by World Trade Organization (WTO) to strengthen legal protection of intellectual property rights around the world. The TRIPS agreement is a harmonized standard in intellectual property rights (IPR) protection. Nigeria will be focused on in detail as a case study. This is to examine existing privacy laws and intellectual property rights laws compliance with the ICT environment, more so whether the provision of the existing laws are tailored to resolve ethical and social issues connected with ICT, directly or expressly.

NATIONAL INTELLECTUAL PROPERTY RIGHTS LAWS AND WTO/TRIPS COMPLIANCE IN NIGERIA

Nigeria is a signatory to TRIPS agreement and must therefore comply with all its stipulations. Nigeria was meant to comply by January 2000; however as at the time of writing, it is yet to overhaul its intellectual property law. The significance of TRIPS in terms of IPR is extensive.

There exist a body of law, enactment and decrees that regulate intellectual property rights issues in Nigeria. Many laws are well conceptualized; however the greatest problem is that they are rarely enforced. Moreover, there is some overlap and at times the law creates confusion as to which agencies are responsible for implementation and enforcement. At times, this leaves ambiguous the appropriate bodies to determine the violation of the law. Many of these laws are currently under review, but it is difficult to track exactly how much streamlining process is been debated and discussed (Obileye, 2001). The copyright law of

1998 (as amended) and the Patent and Design Act of 1990 are discussed hereunder;

The Copyright Law of 1998

Nigeria is signatory to the universal copyright convention in 1961. This led to the promulgation of the copyright decree of 1970, which established the Nigerian copyright council (Omoba and Omoba, 2009). This decree was replaced by the copyright Act of 1988. In 1989, the Nigerian copyright commission was established. The copyright Act of 1988 stipulates 50 years for creators or authors of copyrighted work. It is meant to protect the owner of a work from unauthorized copying, duplicating or distribution. It seeks to prohibit the circulation of such work without permission; however through a lack of implementation, this decree has not brought about effective organization of intellectual property rights protection. Also there is the problem of copyrighted work registration under this Act, which is cumbersome. This leads to manufacturers ignoring the registration process which ultimately creates enforcement obstacles. The Nigerian copyright commission is however grossly underfunded. Its infrastructure is dilapidated and it lacks personnel to carry out its mandate, while human and financial resources are not available to meet the scope of this law.

The present copyright law was not drafted with the benefits of the ICT environment in mind. In terms of enforcement, copyright law relies on the police: an institution also in decay, enmeshed in corruption for the detention of suspects. The copyright law is not adequately backed up with the logistics to establish adequate and competent regulatory functions/activities. The WIPO Act and the WIPO performance and phonogram treaty that Nigeria is a signatory, enjoin countries to make provisions in their domestic laws for sanctions against the circumvention of devices used by authors to protect works, but Nigeria is yet to align its domestic legislation to address this and other issues of concern (Adewopo, 2009) Copyright law

allows works to be used under conditions defined as "fair use". This means that limited copying of copyrighted work for academic purposes; as long as such work is acknowledged is allowed. This is not to be considered as an infringement on intellectual property right. Copyright law in Nigeria has little or no recognition for electronic resource, including the establishment of standards of competence, conduct and ethical practice in honor of property rights for computer professionals. Modern copyright devices have become so common that the duplication of copyright works cannot be effectively monitored. Moreover, the problem has been made difficult by the present unfavorable economic condition.

However there is need to raise the level of awareness of people about copyright law so that they can operate within the confines of the law. The existence of enforcement of a comprehensive copyright regulatory system supported by legislation is a prerequisite. While it is correct to state that copyright law is aimed at encouraging creative intellectual efforts, it should be noted that provision of access to existing knowledge and information is as important as the encouragement of intellectual property.

Patent and Design Act Of 1990

The existing patent and designs Act of 1990 requires 20 years span for any patented work. The Nigerian patent and designs Act provides for compulsory licensing of created works in unrestricted ways. However, the patent and designs Act is currently under review and a 1991 draft version is in circulation. This draft was written before the TRIPS agreement; thus work still needs to be done to make it compliant with TRIPS while creating an enabling environment framework for IPR protection. Under the TRIPS agreement, there are new requirements for compulsory licensing, which include obtaining permission of the patent holder if possible before usage.

There is little understanding of the TRIPS agreement among government workers in term of its relationship to IPR. Within several ministries, some senior officers did not even know about the patent and designs Act. This might be because of the nature of business practices in Nigeria. Moreover the notion of IPR is not intuitively linked to trade regime and protocol; and with the chronic problem of ministries not sharing information or networking. It is reasonably understood that the regulatory agencies may not realize its objectives when it comes to TRIPS and IPR protection. It is imperative that the federal ministry of information and communication, the ministry of science and technology and other governmental agencies, civil society, and others are considered "stakeholders" in the rewriting of the new patent law. For this to happen there needs to be thorough education (for government, professional organisations and NGOs) on the implications of TRIPS and intellectual property protection.

On the whole in Nigeria, violations and infringement of intellectual property rights is on the increase (Omoba and Omoba,2009).This is further exacerbated by the negative use of ICT to access, use, and copy, distribute and duplicate intellectually protected work. This is because of the problems of underdevelopment in every facet of the Nigerian landscape: politics, economy, education, commerce and trade, information technology etc. the Reproduction Rights Society of Nigeria (REPRONIG) is bedeviled with leadership problems and not active.

PROPOSAL FOR POLICY MAKERS: IPR AND TRIPS

Government should;

- Identify IP focal points within the different sectors and ministries in Nigeria as it relates to ICT.

- Establish contacts, perhaps a working group with ministries of science and technology and information and communication.
- Obtain reliable specialized legal advice.
- Develop a mechanism to monitor the ICT impact on IPR protection
- Promote standards for IPR that takes ICT into account.

ETHICAL POLICY FRAMEWORK

Organization and governments must develop policies that take into account related ethical and social issues. Organizations should develop ethical policy covering privacy, intellectual property and associated issues. Ethics policy must provide guidelines and operational procedures on issues such as privacy, software licenses and agreement, access to information, acceptable use of electronic information, user responsibility, permission rights, system quality, and acceptable ICT use, security restrictions, authorized use and users, etc. the effectiveness of any ethics policy is dependent on the administrator who is responsible for the development and enforcement of the policy.

Rogerson (1998, p.1) provides the following steps to help organisations to establish an ethics policy to address the ethical issues arising in the use of ICT.

1. Decide the organization's policy, in broad terms, in relation to ICT. This should:
   ◦ Take account of the overall objectives of the organization, drawing from such existing sources as the organizational plan or mission statement;
   ◦ Use the organization's established values, possibly set out in its code of practice, for guidance in determining how to resolve ethical issues;
   ◦ Get the scope of policy in terms of matters to be covered;

2. Form a statement of principles that would probably include

    ○ Respect for privacy and confidentiality;

    ○ Avoid ICT misuse;

    ○ Avoid ambiguity regarding ICT status, use and capability;

    ○ Be committed to transparency of actions and decisions related to ICT;

    ○ Adhere to relevant laws and observe the spirit of such laws;

    ○ Support and promote the definition of standards in, for example, development, documentation and training; and

    ○ Abide by relevant professional codes.

    ○ Identify the key areas where ethical issues may arise for the organization, such as:

      ▪ ownership of software and data;

      ▪ integrity of data;

      ▪ preservation of privacy;

      ▪ prevention of fraud and computer misuse;

      ▪ the creation and retention of documentation;

      ▪ the effect of change on people both employees and others; and

      ▪ global ICT.

    ○ Consider the application of policy and determine in detail the approach to each area of sensitivity that has been identified.

    ○ Communicate practical guidance to all employees, covering:

      ▪ The clear definition and assignment of responsibilities;

      ▪ Awareness training on ethical sensitivities;

      ▪ The legal position regarding intellectual property, data protection and privacy;

      ▪ The explicit consideration of social cost and benefit of ICT application;

      ▪ The testing of systems (including risk assessment where public health, safety and welfare, or environmental concerns arise);

      ▪ Documentation standards; and

      ▪ Security and data protection

    ○ Whilst organisations have a responsibility to act ethically in the use of ICT so to do individual employees. Those involved in providing ICT facilities should support the ethical agenda of the organization and in the course of their work should:

      ▪ Consider broadly who is affected by their work;

      ▪ Examine if others are being treated with respect;

      ▪ Consider how the public would view their decisions and actions;

      ▪ Analyze how the least empowered will be affected by their decisions and actions; and

      ▪ Consider if their decisions and acts are worthy of the model ICT professional.

The development of such ethical policy should require active participation of individuals, organizations, government, professional bodies and associations, NGOs and ultimately the world community.

FUTURE TRENDS

Violations and infringement of intellectual property rights occasioned by the use of Information Communication Technology (ICT) will continue to be on the increase. There would be continued copying, distribution and duplication of intellectually protected work on the World Wide Web.

Proprietary software and network piracy using electronic networks in digitized form is expected to increase. The world will continue to witness series of protest by companies, organizations and individuals in this regard. This would be further exacerbated by the near lack of global standards on acceptable practices. The case of developing countries is expected to grow worse as a result of the problems of underdevelopment.

Information Communication Technology (ICT) and the Internet will fundamentally, irreversibly and comprehensively affect our ethics and social systems. Clearly therefore, the critical role of government, organizations professional bodies and individuals at this time is heightened rather than diminished. The beginning of wisdom in the ICT world is realizing just how little of the future you can predict, therefore the role of stakeholders can only be best proposed, not predicted.

It is therefore proposed that organizations, government and professional bodies should put in place a robust information policy that takes into account ethical and social issues arising from the use of ICT. The policy is expected to address such issues such as privacy, software licenses and agreement, acceptable use of information in electronic environment, user responsibility, permission rights, acceptable ICT use, security and authorized use and users, among others. Training and skill development on the technical, legal, ethical and policy aspects of information communication technology (ICT) is expected to be addressed

CONCLUSION

The use of ICT has raised new and perplexing ethical issues and challenges. This is occasioned by the near absolute dependence on ICT as well as rapid advances in data/information handling and processing. These ethical issues have social impacts and therefore must be viewed and addressed frotum cultural, social, political standpoint. These issues should be considered before a full implementation of ICT. Ethics guide human behavior in all areas. When applied to ICT, it concerns the relationship of system with the people who use them. The major ethical issues thrown up by ICT involves protection of individual and collective privacy and confidentiality, ownership and use of software, data integrity, protection of intellectual property etc .Individuals and organisations must analyze their actions even as they use ICT for daily operations. This is necessary to determine who is to be affected; public perception of decision and actions, infringement on individual or collective privacy. We have a responsibility to act ethically in the use of ICT. This is because our actions have social consequences.

Advances in ICT have greatly affected individual and corporate information storage and dissemination practices. Information on the Net can be accessible, monitored, copied, duplicated and distributed at many points. Digitized books, motion pictures and music are easily distributed via electronic devices and the Internet. Software is easily pirated, although some database owners have been able to provide licensing agreements with their users, it is very costly and sometimes impossible to enforce the contracts (Zhu et al, 2002). The ease and convenience with which data can be copied, manipulated, duplicated and distributed without permission is increasing. Though there are several possible, legal mechanism for the protection of individual and collective rights; advances and use of ICT has made the protection of intellectual property difficult. There is therefore the need for ethical consideration by people who manufacture and use ICT. Organization should develop ethics policy statements to guide individuals in making appropriate use of ICT.

Intellectual property rights protection laws in Nigeria should be expanded and strengthen to incorporate sensitive areas in ICT usage. IPR laws should take into account ethical issues/challenges in relation to ICT.

REFERENCES

Adewopo, A. (2009, June 9). Pay subscription broadcasting: rights acquisition and infringement. *The Guardian*, (p.96).

Ball, K. S. (2001). Situating workplace surveillance: ethics and computer based performance monitoring. *Ethics and Information Technology*, *3*(3). doi:10.1023/A:1012291900363

Bynum, T. W. (2009). *Ethics in the information age, (Tech.Rep.)*. New Haven, CT: Southern Connecticut State University.

Bynum, T. W., & Rogerson, S. (Eds.). (1996). *Global information ethics*. USA: Opragon press.

Capuro, R. (2006). Towards an ontological foundation of information ethics. *Ethics and Information Technology*, *8*(2), 175–186. doi:10.1007/s10676-006-9108-0

Carbo, T. (2006). *Understanding information ethics and policy: integrating ethical reflection and critical thinking into policy development.* Retrieved February 28, 2009, from http:/www.ethicspolicy.pitt.edu

Davies, P. B. (2002). *Information systems: an introduction to Informatics in organizations*. New York: Palgrave.

Edewor, N. (2008). Freedom of Information Bill: issues, imperatives and implications for Nigerian libraries. *Ozoro Journal of General Studies*, *1*(1), 34–40.

Froehlich, T. J. (1992). Ethical consideration of information professionals. *Annual Review of Information Science & Technology*, *27*, 291–32.

Gorniack, K. (1996). The computer revolution and the problem of global ethics. In Bynum & Rogerson (Eds.), *Global information ethics*, (pp.177-190). USA: Opragon press.

Ifidon, S.E. (2006, March). *Planning without information: the bane of national development.* Inaugural lecture delivered at Ambrose Alli University, Ekpoma, Edo State, Nigeria.

Kallman, E. A., & Grillo, J. P. (1996). *Ethical decision making and information technology: an introduction with cases*. New York: McGrawHill.

Laudon, K., & Laudon, C. (2006). *Management Information Systems: managing the digital firm*, (9thed.). New Delhi, India: Pearson Education.

O'Brien, J.A. (2006). *Introduction to information systems: essentials for the e-business enterprises*, (9thed.). New York: McGraw-Hill.

Obileye, O. (2001). *Combating counterfeit drugs: the way forward.* Paper presented at the public hearing of the health and social services committee, Federal House of Representative, National Assembly, Abuja, Nigeria.

Omoba, O.R. & Omoba, F. A. (2009). Copyright law: influence on the use of information resources in Nigeria. *Library philosophy and practice*.

Rogerson, S. (1998). The ethics of information and communication Technologies: ICT in business. *IMIS journal*, *8*(2), 1-2.

Rogerson, S. & Bynum, T.W. (1995, June 9). Cyberspace: the ethical frontier. *London times*.

Sherratt, D., Rogerson, S., & Fair-weather, N. B. (2005). The challenge of raising ethical awareness: a case based aiding system for use by computing and ICT students. *Science and Engineering Ethics*, *11*(2), 299–31. doi:10.1007/s11948-005-0047-7

Zhu, H. (2002). *The interplay of web aggregation and regulations. (Tech.Rep.)*. Cambridge, MA: MIT Sloan School of Management.

KEY TERMS AND DEFINITIONS

Ethics: Value of right or wrong of any decision made or reached by an individual or organisation.

Privacy: A state of freedom from any form of damaging interference.

Intellectual Property: Intangible human work, protected by law.

Copyright: An authority granted an individual for his/her imaginative creation.

Freedom of Information: Ability to access information without any form of restriction or inhibition.

Information Right: The right of citizens to information as stipulated by law.

Ethics Policy: A set of activities, rules or strategy for determining whether an action or decision is wrong or right.

Unauthorized Access: Series of actions or activity performed without due permission.

Chapter 10
Framework for Effective Development of Information and Communication Technology (ICT) Policy in University Libraries in Nigeria

Okon E. Ani
University of Calabar, Nigeria

Margaret Edem
University of Calabar, Nigeria

ABSTRACT

ICT is transforming library practices and procedures the world over. The aim of this chapter is to address the access gap to ICT by highlighting the framework for effective development of relevant ICT policy in university libraries in Nigeria. The chapter explores variety of ICT infrastructure/services that are available in university libraries, sources of ICT funding, ICT policy priority areas, key ICT policy issues and strategies for formulating ICT policy. Questionnaire survey was used for the study. The findings of the study indicate that, there is a widespread use of Internet in the 14 surveyed university libraries; and only 5 of these libraries have computerized library services. NUC/ETF, library development fund (LDF) and university management are major ICT funding sources. ICT funding/budgeting, ICT infrastructure procurement/maintenance, ICT literacy/capacity building and ICT use are the highest ranked ICT policy priority areas in the surveyed libraries. And annual budgetary allocation to ICT in university libraries, training/capacity building for librarians, as well as, organization of ICT literacy programme for patrons are the highest ranked key ICT policy issues in university libraries. The chapter recommends that, each university should formulate relevant ICT policy for its library, besides the national ICT policy.

DOI: 10.4018/978-1-61692-012-8.ch010

INTRODUCTION

Information and communication technology (ICT) has a critical role to play in development efforts around the world (Sierra, 2006), particularly in developing countries (such as Nigeria). The widespread use of ICT in the provision of quality education and research towards national development has been advocated. Thus, developing countries have begun taking concrete actions to integrate ICT not only into their economic policies and development agendas, but in the formulation of ICT policy towards effective educational development of the citizenry, as education is the bedrock for nation building. This is so, because access to information and ultimately, knowledge, is essential to societal development, and ICTs are tools for effective and efficient information access, processing, storage and retrieval, management and dissemination in educational/research institutions (the universities), and their libraries in particular.

However, there exists a widespread disparity in access to ICTs both among and within countries and institutions (universities), a notion which is commonly referred to as "digital divide". The effort of narrowing these access gaps or barriers to information dissemination and access to knowledge is therefore a priority in promoting sustainable educational development in developing countries (Navas-Sabater, Dymond & Juntunen, 2002). And this could essentially be achieved through the provision of relevant ICT policy in the affected countries/institutions (universities). For instance, in Nigerian universities, especially in the libraries, poor/lack of access to relevant ICT infrastructural facilities such as robust Internet connectivity is widespread, and this is principally attributable to poor/lack of appropriate ICT policy in the university system, with the resultant effects of low ranking of Nigerian universities in global perspective in terms of quality education and research. While the rest of the world utilized the advantages provided by ICT, Nigerian universi-

ties have only started embracing it ((Editorial, ThisDay Newspaper, 2008)

Norris, Sullivan, Poirot & Soloway (2003) have also confirmed the low level of use of ICT in education, particularly limited access to computer technology and the Internet in developing countries. Thus there is dire need to tackle the problem of inequitable access to ICT or "digital divide" in Nigerian universities to enhance global competitiveness in our educational and research output. This chapter would therefore explore the state of information and communication technology (ICT) in Nigerian university libraries; and thereafter develop framework for formulation of ICT policy as a tool for effective applications of ICT in the libraries to support teaching, learning and research in the universities.

Timely access to information and knowledge in the academic environment such as the university is dependent on the degree of application and integration of modern ICTs in academic and research activities. The development of a stable, predictable and transparent policy, legal and regulatory framework is a prerequisite for any sustainable approach to improve access to information and communication services (Navas-Sabater, Dymond & Juntunen, 2002) in the university libraries in Nigeria. With ICT, faculty/students could access information beyond their immediate academic environments, nationally and internationally for their teaching, learning and research. The university libraries have been the major access points for information/knowledge by the staff/students towards quality teaching, learning and research. But the Nigerian university libraries are relatively less competitive in the application of ICTs, in the provision of information to their clientele.

The aim of this chapter is to address the access gap to ICT in Nigerian university libraries by highlighting the framework for effective development of appropriate ICT policy towards enhanced information-handling and management. Thus it is the objective of this chapter to explore different ICT infrastructural facilities that are available in the surveyed university libraries, their major sources of funding, and thereafter develop policy framework to sustain, regulate and control ICT funding/budgetting. The chapter would also investigate other key policy areas/issues in university libraries; these would include ICT literacy/capacity building, ICT use, ICT infrastructure procurement/maintenance, electronic collection development, resource sharing and copyright (Intellectual Property Right) to enhance effectiveness and efficiency in access to information and knowledge in the university libraries. This is in line with the postulation of Guislain, Qiang, Lanvin, Minges & Swanson (2006), that, there is a growing consensus that countries or institutions seeking to strengthen their education and research outcomes should make it a priority to improve ICT access and quality in their academic environments. And the universal approach towards equitable and sustainable access to and use of ICT in Nigerian universities to support quality teaching, learning and research is to develop relevant ICT policy particularly in the libraries. According to Information Development News (2006), this can principally be achieved by developing and implementing enabling policies that reflect national realities and promote a supportive international environment, as well as the mobilization of domestic resources nationally, regionally and institutionally. These policies should also be reflected in a transparent equitable regulatory framework to create competitive academic and research environment. Thus effective development of ICT policy in our universities or specifically in university libraries would enhance equitable and affordable access to ICT to both faculty/students to tap relevant information and knowledge with increasing efficiency and productivity. This is so, as the emergent of digital technologies and globalization has called for ICT policy which should be high on the list of priorities of academic libraries (Banou, Kostagiolas & Olenolou, 2008).

BACKGROUND

In the last decade, developing countries have experienced a revolution in information and communication technology (World Bank, 2008). ICT is crucial to educational development of any nation, as it improves the quality of teaching, learning and research in the university. ICT has wider applications in the practice of modern librarianship, and thus university librarians must work out modalities towards its effective adoption and integration in libraries for access to global information resources as well as national resource sharing (Ani, 2007). ICT is a term that refers to the various technologies that are used in the creation, acquisition, storage, dissemination, retrieval, manipulation and transmission of information (Zulu, 2008). Grace, Kenny & Qiang (2004) described ICTs as tools that facilitate the production, transmission, and processing of information. In a broader context, ICTs include, computers: which are essentially used for processing and storage of information; telecommunications: which are used for the transmission of information; and publishing technologies (e.g. the Internet), which are the technologies used for documenting, presenting and disseminating of information and knowledge (Zulu, 2008). Thus Ani & Biao (2005) have referred to ICTs as globalization tools that enhance access to research information universally.

Adeyeye & Iwela (2005, p202) defined policy as "the vision, goals, principles and plans that guide the activities" of governments, and organizations/institutions. Policy makers in education (universities) are therefore responsible for developing a vision and strategy for educational development, and mobilizing support and cooperation for implementing the vision and strategy from a wide range of constituencies (Mingat, Tan & Sosale, 2003) or stakeholders. These stakeholders in the university system in Nigeria include government ministries, lecturers, university administrators who do the work of delivering educational services;

students who are the immediate beneficiaries of the services provided, employers and public at large who look to the university system to supply the skilled labour and future leaders who can contribute meaningfully to the socio-cultural and technological development of the nation.

So, ICT policy is concerned with the provision of appropriate law, rule and regulation or guideline by a university library to direct, control and regulate its staff/patrons on the adoption and integration of ICT in information management, processing and dissemination (Uhegbu, 2007). Hence for optimal operation and implementation of ICT policy, appropriate legislation by relevant authority within the organization/institution (e.g. university management/senate) or outside the institution (university) by the supervisory body (e.g. National Universities Commission (NUC)) may be required. For instance, the national policy of Library Development Fund (LDF) by National Universities Commission (NUC) that 10% of government's subvention to each university in Nigeria should be set aside for funding of the university library has been discussed by Ifidon (2002). Banou, Kostagiolas & Olenoglou (2008) have therefore viewed ICT policy as a term that describes the ways of promoting and using ICT in the provision of library tasks/services. A well developed ICT policy is the key to effective and efficient information-handling and access to knowledge in university libraries.

Adeyeye & Iwela (2005) therefore considered ICT policy nationally and institutionally, as the vision for ICT and its link to national developmental goals through the provision of access to information and knowledge. Thus ICT policy may be national, institutional, regional or international in scope, and each level may have its own decision-making bodies. Institutional ICT policy would therefore be concerned with the principles and plans that guide the applications and integration of ICT in different institutions (such as the universities) toward the realization of the goals of these institutions. This chapter would examine ICT policy at both national and institutional levels as applicable to university libraries in Nigeria. The aim is to streamline ICT policy in the university libraries at the national level through the intervention of National Universities Commission (NUC), while at the same time encouraging the development of relevant ICT policy institutionally in each university library in order to allow for smooth and transparent applications of ICT in the provision of library tasks/services in line with technological development.

In view of the importance of ICT in socio-economic development of Nigeria, the Federal Executive Council (FEC) of Nigeria had approved the National Information Technology Policy in March 2001. And the National Information Technology Development was established in April 2001 to implement the national ICT policy (Adeyeye & Iwela, 2005). Nigerian national policy for information technology (IT) is the bedrock for national survival and development in a rapidly changing global environment, and challenges us to devise bold and courageous initiatives to address host of socio-economic issues, such as reliable infrastructure, skilled human resources and capacity building (Nigerian National Policy for Information Technology (IT), 2001). It is for this reason that every progressive country has a national IT policy and an implementation strategy to respond to the emerging global reality and thus avert becoming a victim of digital divide. The mission statement of the ICT policy in Nigeria is centered on the use of IT as the engine for sustainable development and global competitiveness. The objectives of the national ICT policy include the following (Adeyeye & Iwela, 2005; Nigerian National Policy for Information Technology (IT), 2001):

- To ensure that information technology resources are readily available to promote efficient national development
- To improve accessibility to public administration for all citizens, bring transparency to government processes

- To develop human capital with emphasis on creating and supporting knowledge-based society
- To empower the youth with IT skills and prepare them for global competitiveness
- To integrate IT into the mainstream of education and training
- To create IT awareness and ensure universal access in order to promote universal IT diffusion in all sectors of our national life
- To build a mass pool of IT literate manpower, and
- To establish appropriate institutional framework to achieve the goals stated above.

From the above national objectives of ICT policy, it is imperative to develop an institutional framework for relevant ICT policy in university libraries towards sustainable ICT access and use to support quality research and development (R & D) in the country.

The goal of a university the world over is to provide enabling environment with relevant infrastructural facilities for teaching, learning and research by the teaching staff/students. And no university would effectively and efficiently achieve this goal without a functional library equipped with relevant materials: current books/journals, modern information and communication technologies (ICTs) particularly the computers and the Internet (or related networks e.g. virtual libraries) to provide access to electronic resources (electronic books/journals, online data bases etc.) with information technology literate librarians. This informed the basic reason why a library is often described as the "heart" of an academic institution. Regrettably, most university libraries in Nigeria are far from meeting up this standard as is obtained by their counterparts in developed countries. The dearth of ICT facilities in university libraries in Nigeria has been widely reported in literature (Ani, 2005; Ani, Esin & Edem, 2005; Etim, 2006; Gbaje, 2007). Ani,

Esin & Edem (2005, p703), in their survey of 17 university libraries found that "only six of them are fully computerized", five university libraries have OPAC, and that "many university libraries are yet to join the internet and the information superhighway". In a survey, with 14 university librarians in Nigeria as respondents, Ani (2005, p69) found that "only four of the surveyed university libraries have websites on the Internet". Gbaje (2007) study had revealed that, the shortage of web technologies skilled librarians; poor information infrastructure and high cost of equipment have hindered university libraries from providing online information services. Though, Ani & Ahiauzu (2008) had reported an improvement in Internet connectivity in university libraries in Nigeria; the problems of low bandwidth and lack of sustainability of the available Internet services are major challenges that need to be overcome. Etim (2006) has confirmed that, in most African universities adequate Internet access presents a great challenge for university administrators. In Nigeria, the situation is worsened by the epileptic power supply in the country, as university librarians need additional funds to acquire and maintain generating plants for alternative power supply.

A recent report by Foster, Heppenstal, Lazarz & Broug (2008) to examine the problems faced by African authors in getting their work published in international journals and how these problems could be overcome, showed that access to electronic resources is patchy due to internet connectivity and electronic infrastructure issues. And that many institutions only have a limited number of PC terminals and downloading a full text PDF will usually take a long time due to bandwidth issues. Besides low bandwidth, the report also indicated that, the budget allocation for electronic resources and other ICT infrastructure is not provided for by the university managements to support ICT diffusion and application in university libraries. Foster, Heppenstal, Lazarz & Broug (2008) report also revealed a low publication output of researchers in Nigeria and other African countries. The poor quality of research from Nigerian universities is essentially attributed to lack of access to relevant electronic resources in the university libraries (Ani & Ahiauzu, 2008; Foster, Heppenstal, Lazarz & Broug, 2008). In the final analysis, Foster, Heppenstal, Lazarz & Broug (2008) opined that subscription to electronic collections by African (Nigerian) university libraries would increase the number of successful article submissions from Africa (Nigeria). This requires a reliable Internet connectivity in university libraries across the country, and huge financial funding to support subscription of relevant electronic resources. This is why there is need for the formulation, development and implementation of relevant ICT policy in university libraries in Nigeria to enhance effective adoption and integration of ICT in the provision of library tasks/services. Besides, ICT policy would ensure the provision of equitable, affordable and sustainable access to ICTs and electronic resources in libraries by the patrons, academic researchers/students. And these would correspondingly make Nigerian researchers/students to be active participants in the competitive global knowledge.

Therefore, the purpose of developing ICT policy in university libraries in Nigeria is for the creation of new users and the provision of satisfactory service to those already using them. The development of ICT policy in university libraries is of strategic importance for investing in ICT as a means to fully comprehending, managing and developing any kind of information and knowledge, which would shape erudite researchers and scholars and sensitize the students to improve their learning potential (Banou, Kostagiolas & Olenoglou, 2008). Thus ICT policy is necessary for efficiency and effectiveness in acquisition, management and maintenance of ICT resources (Adedibu, 2005/2006). For university libraries in developing countries (Nigeria) to catch up with their counterparts in developed nations in digital parity, proactive policies are needed to enable these universities to implement the latest low cost

technologies in the provision and dissemination of information. There is also a need to develop measures and ICT policies that would bridge digital divide in order to provide universal and equitable access to these vital technologies both within and outside the university environments (Mutula, 2008).

ICT POLICY ISSUES IN UNVERSITY LIBRARIES

There are variety of ICT policy issues that need to be tackled in university libraries in Nigeria, in order to accelerate the diffusion of ICT in information access and utilization among library patrons. Nigerian students, teaching/research staff need to be linked to global academic and research resources through equitable and sustainable access to ICT in university libraries. Thus we need to diagnose key issues that hinder the proliferation and diffusion of ICT in information management and dissemination in university libraries in Nigeria, and proffer appropriate solutions to these problems through the development of relevant ICT policy. In order to achieve this goal, empirical data were collected in this study from university librarians to facilitate the development of framework for ICT policy in Nigerian university libraries.

Research Methodology

A survey method was used in the study. Copies of questionnaire were mailed to university librarians in 20 federal universities in Nigeria as respondents in January 2009. A self-addressed and stamped envelope was enclosed with the mailed questionnaire to each of the respondents to return the completed questionnaire. The choice of the mailed questionnaire survey was principally to save cost of traveling to the 20 selected federal universities for data collection. And federal universities were selected based on their uniqueness on ownership, funding and generally policy on all the ICT issues raised in the designed questionnaire. Fourteen (14) duly completed and usable copies of questionnaires were retrieved within the stipulated period for data analysis representing a response rate of 70.0 percent.

Results and Discussion

ICT Infrastructure/Service Available in University Libraries

The emergence of modern ICTs have transformed the practice of librarianship the world over. With ICTs, newer library services are now offered to library patrons in university libraries across the country. The respondents were asked to indicate as in Table 1, different ICT infrastructural facilities/services that are available in their university libraries for their information-handling, processing, storage and retrieval, and dissemination. The results of the study have shown that, all the 14 surveyed university libraries (representing 10.0% of the total frequency response) have provided Internet services to their library patrons. The new Internet technology has the capacity to host a lot of academic and research resources to supplement inadequate library holdings in Nigerian universities. The Internet promotes and facilitates resource sharing in the country, besides enhanced access to global information and knowledge. With access to the Internet, Nigerian students and researchers can relatively compete with their counterparts in developed nations in knowledge acquisition, production and dissemination. Besides, the Internet, other major ICT facilities/services that are available in the surveyed university libraries are photocopiers (10.0%), scanners (9.4%), CD-ROM databases/search (7.9%), online databases/search (7.9%), electronic journals (7.2%), library LAN (7.2%) among others (Table 1).

From the findings of the survey, Library LAN is currently available in 10 university libraries,

Table 1. ICT infrastructure/service available in university libraries

ICT infrastructure/service	Response	%
Internet services	14	10.0
Photocopier	14	10.0
Scanner	13	9.4
CD-ROM databases/search	11	7.9
Online databases/search	11	7.9
Electronic journals	10	7.2
Library LAN	10	7.2
Electronic books	9	6.5
VSAT	8	5.8
Library website	7	5.1
Online Public Access Catalogue (OPAC)	7	5.1
Land line telephone	6	4.3
Intercom	5	3.6
Computerized library service	5	3.6
Radio link	4	2.9
Fax machine	2	1.4
Others	3	2.2

and "yet to be commissioned" in one other university. Computerized library service is provided by only 5 (3.6%) university libraries, while one other university library is "about to" be computerized. This shows that, most university libraries are still involved in the use of traditional/manual operations and procedures to provide key library and information services to their patrons. Similarly, only 4 (2.9%) university libraries have radio link, and "yet to be commissioned" in one other library. The need for the provision of wireless connectivity for sustainable educational development, particularly in university libraries to enhance resource sharing has been advocated by Livingston (2004). "Others" ICT facilities mentioned by the respondents are web camera, digital camera, multimedia projector, overhead projector, public address system (PAS) and printers.

The Need for Alternative Power Supply in University Libraries

Lack of constant public power supply has been a major impediment towards effective application of ICT in Nigeria, particularly in the public sector. This has hindered the integration of ICT in the provision of library and information services in Nigerian university libraries. Hence, the need for acquisition of generating plant by each university library as source of alternative power supply towards effective and reliable ICT application in the library cannot be overemphasized. The findings of the study indicate that, 13 (92.9%) of the 14 surveyed university libraries have generating plants to cater for the inadequacy of the public power supply. This is a major progress towards effective applications of ICTs in modern librarianship in Nigeria.

ICT Funding Source in University Libraries

The issue of ICT funding has been a major source of concern in the literature of ICT applications in Nigeria; for without adequate financial support, there can be no good resources for teaching and research, good staff (ICT personnel) cannot be recruited, and ICT facilities cannot be acquired (Ifidon, 2002). This is why according to Ifidon (2002), the words funds and funding are continuously belaboured by university library administrations towards integration of ICT in library procedures and operations. It was therefore considered as pertinent to explore different ICT funding sources in university libraries in Nigeria. The respondents were asked to select different sources of funding ICT in their libraries as shown in Table 2.

As shown in Table 2, the highest ranked ICT funding sources in Nigerian university libraries are NUC/ETF (24.1%), library development fund (LDF) (22.4%), university management (15.5%), and international organizations/agencies (13.8%) respectively. National Universities Commission

Table 2. ICT funding source in university libraries

ICT Funding Source	Response	%
National Universities Commission (NUC)/Education Trust Fund(ETF)	14	24.1
Library Development Fund (LDF)	13	22.4
University management	9	15.5
International organizations/agencies	8	13.8
Library internal revenue	4	6.9
Individuals/alumni	4	6.9
Special government intervention fund	3	5.2
Private/multinational companies/organizations	2	3.4
Loan	1	1.7
Others	0	0.0

(NUC) in partnership with Education Trust Fund (ETF) has made annual financial provision for each public university library in the country to support any project(s) of its choice. From the findings of the survey, it is apparent that, 22.4% university librarians prefer to use fund from the National Universities Commission (NUC)/Education Trust Fund (ETF) to support ICT projects in their university libraries. As indicated in Table 2, LDF is the second ranked source of funding ICT applications in university libraries in Nigeria. According to Ifidon (2002), the National Universities Commission (NUC) has directed that each year 10% of the total recurrent grant to each university should be set aside for the library's operation; this special fund is referred to as library development fund (LDF). The third major source of ICT funding is the fund provided by the university management, according to Bozimo (2005/2006), it is usually the take off funds for ICTs in the university libraries. Another important source of funding ICT in university libraries in Nigeria comes from international organizations/agencies. For instance, Bozimo (2005/2006) has reported that the bulk of funding the computerization of the ABU Library System came from a MacArthur Foundation grant, an international organization/agency. Therefore Foreign Direct Investment (FDI) on funding of ICT in university libraries should be encouraged. Individuals/ alumni and library internal revenue are also good sources of funding ICT in the surveyed university libraries. The results of the survey have also shown, occasionally, respective governments and private/multinational companies do provide special funds to support ICT applications in university libraries in order to make them globally competitive.

ICT Policy Priority Area in University Libraries

The respondents were asked to rank different areas of formulating ICT policy in order of priority in Nigerian university libraries as shown in Table 3. The findings of the survey have shown that, ICT funding/budgeting (16.3%), ICT infrastructure procurement/maintenance (16.3%), ICT literacy/ capacity building (15.3%), and ICT use (15.3%) are ranked as major ICT policy priority areas in the surveyed university libraries. According to Ifidon (2002), the quality and adequacy of the (ICT) resources and services available in university libraries are function of the level of financial support which the libraries receive from their governing authorities. He therefore calls on the librarians to take the trouble to examine in-depth the policy issues involved in funding. Therefore adequate policy on ICT funding, will discourage or minimize mismanagement of money meant for ICT applications in Nigerian university libraries. Besides funding, the need to formulate relevant policy that would guide the procurement of ICT infrastructure as well as their maintenance cannot be overemphasized. This would make respective university managements/university librarians to be conscious of the need to provide basic ICT facilities/services to their library patrons. One cardinal area that requires appropriate ICT policy according to the findings of this survey is ICT literacy/capacity building. Lack of/inadequate personnel has

Table 3. ICT policy priority area in university libraries

ICT Policy Priority Area	Response	%
ICT funding/budgeting	14	16.5
ICT infrastructure procurement/maintenance	14	16.5
ICT literacy/capacity building	13	15.3
ICT use	13	15.3
Electronic collection development	12	14.1
Copyright (Intellectual property right)	10	11.8
Resource sharing	9	10.6
Others	0	0.0

been one of the major impediments for effective application of ICT in Nigerian university libraries. Thus it is pertinent to develop appropriate ICT policy to improve/promote ICT literacy/capacity building among librarians in Nigerian universities. And this should be supported by providing a policy that would encourage effective use of ICT by library patrons in the universities. The findings of the study have also revealed the need for ICT policy on copyrights (intellectual property rights) and resource sharing respectively.

Key ICT Policy Issues in University Libraries

The survey put forward key ICT policy issues that need to be developed in university libraries for the respondents to examine and rank as shown in Table 4. Key ICT policy issue is concerned with the basic policy on ICT that is imperative for adoption and application of ICT in university libraries in Nigeria. In this, study, key ICT policy issues include annual budgetary allocation for ICT, ICT training/capacity building, ICT literacy programme for library users, implementation of ICT budget in university library, payment of annual ICT levy in university library by students, and budgetary allocation for electronic collection development.

Table 4. Key ICT policy issues in university libraries

Key ICT Policy Issue	Response	%
Annual budgetary allocation to ICT in university libraries	14	23.0
Regular ICT training/retraining and capacity building for all categories of library staff	13	21.3
Organization of ICT literacy programme for new library users annually	10	16.4
Monitoring of implementation of ICT budget in university library by library committee	9	14.8
Payment of ICT levy in university libraries annually by students	8	13.1
Allocation of 40% of budget for collection development to electronic resources (electronic collection development)	7	11.5

Annual Budgetary Allocation for ICT

Most scholars are generally concerned about the issue of inadequate funding of university libraries in Nigeria (Adedibu, 2005/2006; Foster, Heppenstal, Lazarz & Broug, 2008; Lawal, 2004; Mutula, 2008; Ojebode, 2007). Adedibu (2005/2006) had suggested that, due to inadequate funding of university libraries, and in order to use the little fund available in each library judiciously, there is need for librarians to have written ICT policy. Thus, responding university librarians in the survey are unanimous in their ranking of annual budgetary allocation for ICT in Nigerian university libraries (23.0%) as the most important ICT policy issue as in Table 4. It is obvious that, annual budgetary allocation for ICT will relatively increase the level of ICT diffusion in university libraries.

ICT Training/Capacity Building

Lack of/scarcity of skilled ICT personnel has been another major factor that impedes ICT applications in university libraries in Nigeria. ICT literacy is central to capacity building, which in turn impacts on IT development and use, particularly in university libraries (Adeyeye & Iwela, 2005).

As shown in Table 4, the high ranking of regular training/retraining and capacity building (21.3%) for all categories of library staff as a key ICT issue is targeted to address this problem. Therefore the development of appropriate ICT policy to enhance capacity building for librarians and other library staff will accelerate ICT diffusion and application in university libraries in the country.

ICT Literacy Programme for Library Users

Regular organization of ICT literacy programme for library users is ranked third by the respondents as a key ICT policy issue in Nigerian university libraries. Library users need to be trained on how to use and get familiar with the new technologies, ICTs – the computers, the Internet, OPAC, CD-ROM/online databases etc. for effective and efficient information access and retrieval in libraries.

Implementation of ICT Budget in University Library

According to Table 4, respondents are also of the view that, it is not enough to have annual budgetary allocation for ICT, but its implementation should also be given attention in ICT policy formulation. Hence there is a need for monitoring of implementation of ICT budget in university library by library committee or any other supervisory body in each university in Nigeria.

Payment of Annual ICT Levy in University Library by Students

The results of the study have also indicated that there is need for the development of policy framework for the payment of ICT levy in university libraries annually by students. Already, in a few Nigerian universities, students do pay certain amount of money for library fee to support library internal revenue (Ifidon, 2002), which is one of the common sources of funding ICT in Nigerian university libraries as shown in Table 2.

Budgetary Allocation for Electronic Collection Development

Collection development is the backbone of any library, especially academic library. Collection development policy is a vital instrument in making selections to support the curricula of any academic institution, and it should be brought to cater for all aspects of collection in the library (Adedibu, 2005/2006). Foster, Heppenstal, Lazarz & Broug (2008) have reported that, budgetary allocation for electronic resources is not provided for in African universities, particularly in Nigeria. However, the policy of allocating 40% of annual budget for collection development to electronic resources or electronic collection development in Nigerian university libraries is imperative, if library users must have access to relevant digital resources for their teaching, learning and research.

ICT Policy Strategy

The respondents were asked to rank different strategies for developing ICT policy in university libraries as shown in Table 5. The findings of the survey have shown that, formulation of relevant ICT policy would be a good strategy for enhanced ICT integration in the provision of library tasks/ services (14.1%). And the basic strategy to be adopted should involve the formulation of ICT policy both at the national as well as the institutional levels (13.1%). This is affirmed by the respondents as in Table 5 that: besides national ICT policy, each university should formulate relevant ICT policy for its library. The goal of national ICT policy according to the results of the study is to bring standardization in information management in university libraries in the emerging digital information environment. However, for effective national ICT policy in Nigerian university libraries, National Universities Commission (NUC)

Table 5. ICT Policy Strategy

ICT Policy Strategy	Response	%
Formulation of relevant ICT policy would enhance integration of ICT in the provision of library tasks/services	14	14.1
Besides national ICT policy, each university should formulate relevant ICT policy for its library	13	13.1
National ICT policy would bring standardization in information management in university libraries	12	12.1
There is need for national ICT policy in university libraries	12	12.1
National Universities Commission (NUC) should monitor the implementation of national ICT policy in university libraries especially during accreditation of programmes	11	11.1
University staff (teaching/non-teaching) should be involved in the formulation of ICT policy in the university libraries	10	10.1
National Universities Commission (NUC) should coordinate and streamline relevant ICT policy in university libraries	9	9.1
There is existing formal/official ICT policy in my university library	7	7.1
Students (undergraduates/postgraduates) should be involved in the formulation of ICT policy in university libraries	6	6.1
Only university administration/library management should be involved in the formulation of ICT policy in the university libraries	5	5.1

should monitor its implementation especially during accreditation of academic programmes in Nigerian universities. In order to underscore the importance of this strategy, one of the respondents has put forward the following comment:

"National Universities Commission (NUC) should not only monitor, it should enforce implementation of the ICT policy in the university libraries and make it one of the conditions for programmes accreditation"

The findings of the survey have also indicated that, major stakeholders such as university staff (teaching/non-teaching), and the students should be involved in the formulation of relevant ICT policy in university libraries. Thus the formulation of ICT policy in university libraries should not be the prerogative of the university managements only. For instance, there is need for the university management to involve students' representatives in the determination of how much each student should pay annually as ICT levy as earlier indicated in Table 4.

FUTURE TRENDS

According to Adeyeye & Iwela (2006), there is need for periodic review of ICT policy to reflect technological development and advances. Etim (2006) had posited that there would be progressive ICT applications in university libraries in Nigeria in line with the emerging global trend for equitable access to information. This study has confirmed this prediction in terms of Internet connection, as all the 14 surveyed university libraries are connected to the Internet compare to earlier reports (Ani, 2005; Ani, Esin & Edem, 2005). What need to be done to improve access is increase in bandwidth. Although, other ICT infrastructure/services as shown in Table 1 such as computerized library services, OPAC, library website, land line telephone, VSAT, radio link are available in a few university libraries; there are indications that efforts are being made to improve access to these facilities/services in the future. The results of the study have shown that Nigerian university libraries are making progressive efforts to fully integrate ICT in their library procedures/operations. For examples, one university library is "about to computerize" its library services, radio link and library LAN are "yet to be commissioned" in one other library. Apparently, with relevant ICT policy in place, access to these ICT infrastructure/services would be accelerated significantly within the next half decade. Presently, the issue of ICT policy seems to be new in university libraries, with

awareness and sensitization of major stakeholders: National Universities Commission (NUC), university managements, university librarians among others, in future appropriate ICT policy would be put in place to tackle the inequitable access to ICT in university libraries.

CONCLUSION

ICTs are modern tools for efficient information management, processing and dissemination in university libraries. But in Nigeria access to ICTs in university libraries is not equitable and sustainable. Many ICT infrastructure/services are still not available in most (of the surveyed) Nigerian university libraries. The solution to the problem of "digital divide" or inequitable access to ICTs in Nigerian university libraries lies in effective development of relevant ICT policy. Appropriate ICT policy would accelerate the rate of diffusion of ICTs, as it would remove the bottlenecks in the applications and use of ICT in Nigerian university libraries. Apparently, ICT has brought enormous transformations and innovations in librarianship in Nigeria. The development of relevant ICT policy would relatively increase access to ICT in university libraries and optimally enhances information management and utilization. Although the framework for effective development of ICT policy should be holistic in approach, the following recommendations would assuage the process:

1. Each university should develop an ICT policy for its library.
2. National Universities Commission (NUC) should coordinate the framework for the development of national ICT policy in university libraries.
3. National Universities Commission (NUC) should monitor the implementation of national ICT policy in university libraries e.g. through accreditation of academic programmes. That is, each university library

should meet and satisfy a certain minimum standard of ICT integration, adoption and application in the provision of library tasks/services before certain academic programmes could be accredited in such university.

4. Institutionally, university library committee (or any other supervisory body) should monitor the implementation of ICT policy in each university library.

5. The areas that need effective ICT policy should include but not limited to the following: ICT funding/budgeting, ICT procurement/maintenance, ICT literacy/capacity building, ICT use, electronic collection development, copyright (intellectual property right) and resource sharing.

6. Other stakeholders such as staff (teaching/ non-teaching) and the students should be involved in the formulation of relevant ICT policy in Nigerian university libraries.

7. Workshops/seminars should be organized to sensitize major stakeholders: National Universities Commission (NUC), university managements and university librarians on the imperative of developing and implementing relevant ICT policy in university libraries towards enhanced access to global electronic information resources to support effectiveness and efficiency in teaching, learning and research in Nigerian universities.

REFERENCES

Adedibu, L. A. (2005/2006). Collection Development Policy: The Case of University of Illorin Library. *Nigerian Libraries*, *39*, 79–91.

Adeyeye, M., & Iwela, C. C. (2005). Towards an Effective National Information and Communication Technologies in Nigeria. *Information Development*, *21*(3), 202–208. doi:10.1177/0266666905057337

Ani, O. E. (2005). Evolution of Virtual Libraries in Nigeria: Myth or Reality? *Journal of Information Science*, *31*(1), 67–70. doi:10.1177/0165551505049262

Ani, O. E. (2007). Information and Communication Technology (ICT) Revolution in African Librarianship: Problems and Prospects. *Gateway Library Journal*, *10*(2), 111–118.

Ani, O. E., & Ahiauzu, B. (2008). Towards Effective Development of Electronic Information Resources in Nigerian University Libraries. *Library Management*, *29*(6/7), 504–514. doi:10.1108/01435120810894527

Ani, O. E., & Biao, E. P. (2005). Globalization: Its Impact on Scientific Research in Nigeria. *Journal of Librarianship and Information Science*, *37*(3), 153–160. doi:10.1177/0961000605057482

Ani, O. E., Esin, J. E., & Edem, N. (2005). Adoption of Information and Communication Technology (ICT) in Academic Libraries: A Strategy for Library Networking in Nigeria. *The Electronic Library*, *23*(6), 701–708. doi:10.1108/02640470510635782

Banou, C., Kostagiolas, P., & Olenoglou, A. (2008). The Reading Behavioural Patterns of the Ionian University Graduate Students: Reading Policy of the Academic Libraries. *Library Management*, *29*(6/7), 489–503. doi:10.1108/01435120810894518

Bozimo, D. O. (2005/2006). ICT and the Ahmadu Bello University Libraries. *Nigerian Libraries*, *39*, 1–20.

Editorial, (2008). The ranking of our Universities. *ThisDay Newspapers* (Tuesday 9, December).

Etim, F. (2006). Resource Sharing in the Digital Age: Prospects and Problems in African Universities. *Library Philosophy and Practice*, *9*(1), 12–19.

Foster, K., Heppenstal, R., Lazarz, C., & Broug, E. (2008). *Emerald Academy 2008 Authorship in Africa.* Retrieved March 20, 2009 from http://info.emeraldinsight.com/pdf/report.pdf/

Gbaje, E. S. (2007). Provision of Online Information Services in Nigerian Academic Libraries. *Nigerian Libraries, 40,* 1–14.

Grace, J., Kenny, C., & Qiang, C. Z. (2004). *Information and Communication Technologies and Broad-Based Development: A Partial Review of the Evidence.* Washington, DC: The World Bank.

Guislain, P., Qiang, C., Lanvin, B., Minges, M., & Swanson, E. (2006). Overview. In *Information and Communications for Development: Global Trends and Policies.* Washington, DC: World Bank.

Ifidon, S. E. (2002). Policy Issues in the Funding of Nigerian University Libraries . In Lawal, O. O. (Ed.), *Modern Librarianship in Nigeria: A Festschrift to mark the Retirement of Chief Nduntuei Otu Ita, University Librarian, University of Calabar, 1977-1997* (pp. 49–54). Calabar, Nigeria: University of Calabar Press.

Information Development News. (2006). *Information Development, 22*(1), 7–21. doi:10.1177/0266666906062685

Lawal, O. O. (2004). *Libraries as Tools for Educational Development.* A Paper presented at NLA 42nd National Conference and AGM at Solton International Hotel & Resort, Akure.

Livingston, P. (2004). Laptops Unleashed: A Middle School Experience. *Learning and Leading with Technology, 31*(7), 12-15. Retrieved June 15, 2009 from www.usq.edu.au/material/edu5472

Mingat, A., Tan, J., & Sosale, S. (2003). *Tools for Education Policy.* Washington, DC: World Bank.

Mutula, S. M. (2008). Digital Divide in Africa: Its Causes, Amelioration and Strategies . In Aina, L. O., Mutula, S. M., & Tiamiyu, M. A. (Eds.), *Information and Knowledge Management in the Digital Age: Concepts, Technologies and African Perspectives* (pp. 205–228). Ibadan, Nigeria: Third World Information Services Limited.

Navas-Sabater, N., Dymond, A., & Juntunen, N. (2002). *Telecommunications and Information Services for the Poor: Toward a Strategy for Universal Access* (World Bank Discussion Paper No. 432). Washington, DC: The World Bank.

Nigerian National Policy for Information Technology (IT). (2001). Retrieved March 18, 2009 from http://nitda.gov.ng/document/nigeriaitpolicy.pdf/

Norris, C., Sullivan, T., Poirot, T., & Soloway, E. (2003). No Access, No Use, No Impact: Snapshot of Surveys of Educational Technology in K-12. *Journal of Research on Technology in Education, 36*(1), 15-27. Retrieved June 15, 2009 from http://search.epnet.com/direct.asp; http://ez/noxy.us.edu

Ojebode, F. I. (2007). Library Funding and Book Collection Development: A Case Study of Oyo State College of Education and Federal College of Education (SP), Oyo. *Gateway Library Journal, 10*(2), 119–126.

Sierra, K. (2006) Foreword. In The World Bank, (Ed.), *Information and Communications for Development: Global Trends and Policies.* Washington, DC: The World Bank.

Uhegbu, A. N. (2007). *The Information User: Issues and Themes.* Okigwe, Nigeria: Whytem Publishers.

World Bank. (2008). *The Little Data Book on Information and Communication Technology.* Washington, DC: The World Bank.

Zulu, S. F. C. (2008). Intellectual Property Rights in the digital Age . In Aina, L. O., Mutula, S. M., & Tiamiyu, M. A. (Eds.), *Information and Knowledge Management in the Digital Age: Concepts, Technologies and African Perspectives* (pp. 335–354). Ibadan, Nigeria: Third World Information Services Limited.

KEY TERMS AND DEFINITIONS

Computerized Library Service: The application of computers in the management, processing and dissemination of information in university libraries.

Digital Divide: A term use to describe inequitable access to ICT, in a specific term in university libraries.

Information and Communication Technology (ICT): Modern digital technology: computer, Internet and telecommunications that is used in acquisition, processing and dissemination of information in university libraries.

ICT Literacy: Ability to use ICT in processing, disseminating and accessing information in university libraries by librarians/patrons.

ICT Policy: Rules and regulations, guidelines etc. that allow for effective and efficient use of ICT in university libraries.

Internet: Global network of networks of millions of computers, that provides access to academic and research information in university libraries.

Library Development Fund (LDF): A financial policy by National Universities Commission (NUC) of setting aside 10% annual government's subvention to each public university in Nigeria in funding university library.

Library LAN: Local Area Network, a network of computers in a university library, which improves communication and access to information within the library.

OPAC: Online Public Access Catalogue, a computer network that provides access to bibliographic information in university library.

APPENDIX: LIST OF RESPONDING UNIVERSITY LIBRARIANS

1. University of Calabar, Calabar

2. Obafemi Awolowo University, Ile-Ife

3. University of Agriculture, Abeokuta

4. University of Port Harcourt, Port Harcourt

5. University of Nigeria, Nsukka

6. Federal University of Technology, Owerri

7. University of Benin, Benin

8. University of Maiduguri, Maiduguri

9. University of Ibadan, Ibadan

10. Michael Okpara University of Agriculture, Umudike

11. University of Ilorin, Kwara State

12. University of Abuja, Abuja

13. Federal University of Technology, Yola

14. University of Lagos, Lagos

Chapter 11
Gender and ICT Policy

Tracy Efe Rhima
Delta State University, Nigeria

ABSTRACT

This chapter is devoted to discussion of ICT and gender policy. It explores the need for gender consideration in ICT policy, gender issues in ICT policy, adoption of gender perspective in ICT policies, challenges for the adoption of a gender perspective in the formulation and implementation of ICT policies, case studies of gender and ICT policies in Asia, Africa, Europe, Latin America and Caribbean and Australia, gender approaches to ICT policies and programs, guidelines for policy-making and regulatory agencies. It was concluded that various national government have started addressing gender issues in their policies. Recommendation was given that policy makers should ensure that Gender considerations are truly included in national ICT policy.

INTRODUCTION

In the last decades, information and communication technology (ICT) has become a Powerful and widespread communications platform, particularly given the convergence of existing communications media with new communication technologies. ICTs can be used to increase access to employment, education or health services; strengthen political participation; improve transparency; provide a platform for diverse voices; and cross-cultural knowledge exchange. The social, political and economic changes wrought by ICTs have prompted certain shifts in development thinking. Development strategists now see, as recognized for example in the United Nation Millennium Declaration, the need to adapt ICTs as a way to avoid further marginalization, and also as a potential force for creating new economic growth opportunities and for pushing democratic boundaries (Genderit.org, n.d.)

DOI: 10.4018/978-1-61692-102-8.ch011

According to Marcelle (2000) ICTs have enormous potential to benefit girls and women in terms of enhanced income: generation opportunities, employment, and improved quality of life, but because technologies are not gender neutral, it is important to advocate for ICT strategies to reduce and manage the potential for ICTs to create economic and social exclusion and reinforce existing social disparities.

Bonder (2002) states that access to information, to knowledge and the interaction between cultures and social groups have never been so within the reach of humanity or as valued as in the last decades. The continuous innovation and global spreading of ICTs appears like a fundamental resources which has goals to attain which will inaugurate a change of era known as information society.

Subramaman & Saxena (2005) have reported that the development of ICT has been termed as ICT revolution due to its transforming potential affecting all dimensions of human civilization of our times which is unprecedented and the ultimate aim of the information society is the empowerment and development of all its citizens through equal access to and use of information. With the growth of infrastructure and access, ICTs are beginning to permeate even the most isolated regions. Access or lack of access to a medium that in some places has become a principal means of expression, economic survival, and decision making is vital for women. (ARC WNSP, 2005).

ICTs can be used to close gender gap by creating new jobs for impoverished women. Women, for instance, have been at the forefront of the village phone movement, selling airtime to rural people too poor to own their own phones. ICTs can also be used to enhance basic literacy and education for women and girls, provide job training and prepare women for careers in the ICT sector as well as to ensure health and safety (International Telecommunication Union, 2009).

ICTs are already being used by women's organizations to communicate their own agendas and

perspectives in order to effect women's empowerment and social change. However, women also need to be involved in the policy processes that define access to and use of these ICTs. (Radloff, 2005)

While there is recognition of the potential of ICT as a tool for the promotion of gender equality and the empowerment of women, a "gender divide" has also been identified, manifested in the lower numbers of women accessing and using ICT compared with men.

Unless this gender divide is specifically addressed, there is a risk that ICT may aggravate existing inequalities between women and men and create new forms of inequality. .If, however, the gender dimensions of ICT—in terms of access and use, capacity-building opportunities, employment and potential for empowerment—are explicitly identified and addressed, ICT can be a powerful catalyst for political and social empowerment of women, and the promotion of gender equality (United Nations, 2005)

Gender equality aspects need to be fully incorporated in all works which relate to ICT at national, regional and global levels, including in the development of policies and regulatory frameworks, projects and research and data collection. A basic starting point for incorporating gender perspectives in ICT initiatives is the use of gender analysis to ascertain the needs and priorities of both women and men and the manner in which policy-making, planning and other activities can really support equitable access, use and benefits, including employment opportunities (United Nations, 2005)

This chapter dwells on the need for gender consideration in ICT policies, gender issues in ICT policy, adoption of gender perspective in ICT policies, challenges for adopting gender perspective in formulation and implementation of ICT policies, case studies of gender issues in ICT policies of countries from different continents, gender approaches to ICT policies and programs and guidelines for policy making and regulation agencies.

BACKGROUND

ICTs are collectively defined as innovations in microelectronics, computing (hardware and software), telecommunications and optic-electronic micro-processors, semiconductors fiber optics that enable the processing and storage of enormous amounts of information along with rapid distribution of information through communication networks. Linking computing devices and allowing them to communicate with each other create networked information systems based on a common protocol. This has greatly altered access to information and the structure of communication extending the networked reach to many parts of the world. (Human development report, 2001)

The last decade of 20th century marked the shift to a global information society characterized by the rapid development of information and communication technologies that have blurred the boundaries between information, communication and the various types of media. ICTs refer to technologies and tools that people use to share, distribute, and gather information and to communicate with one another through the use of computers and interconnected computer networks. They are media that utilize both telecommunication and computer technologies to transmit information. (Ramilo & Villanueva, 2001).

According to Amuriat & Okello (2007) ICTs are technologies which facilitate communication, processing and transmission of information by electronic means, ICTs are tools that can enable the participation of poor women and men in economic and civil life and help them to move out of poverty. ICTs have change the face of the world we live in, They enable people to communicate with relatives, friends and colleagues around the world instantaneously, gain access to global libraries, information resources and numerous other opportunities. They are one of the driving forces of globalization The ICT sector is heterogeneous, extending beyond traditional classifications of industrial or services sectors and because production and diffusion of ICTs are of equal importance, the ICT sector intersects with a number of other areas of policy making. (Idowu, Ogunbodede & Idowu, 2003)

Amuriat & Okello (2007) see policy as law that draws upon a number of strategies to accomplish its vision, including mainstreaming gender into policy programs and implementation strategies; sensitization and awareness creation; ICT capacity development among rural peoples' building appropriate infrastructure; supporting favorable investment projects, stimulating production storage and dissemination of national information, and facilitate of access to public domain information. The government envisages that by implementing the strategies, major problems such as access, application and utilization will be appropriately addressed.

NEED FOR GENDER CONSIDERATION IN ICT POLICY

ICT is the driving force that is increasingly resulting in tremendous change in all aspects of our lives, including education, knowledge dissemination, social interaction, political engagement, health, economic and business practices. In the last decades ICT has become a powerful and widespread communications platform, particularly given the convergence of existing communication media with new communication technologies. ICT can be used to increase access to employment, education or health services; strengthen democracy; improve transparency; provide a platform for diverse voices; and cross-cultural knowledge exchange. The social, political and economic changes wrought by new information and communications technology have promoted certain shifts in development thinking. (genderit.org n.d)

Women's empowerment is central to human development, as a process of enlarging people's choices, this cannot be realized when half of the choices of the humanity are restricted. Targeted actions aimed at empowering women and righting gender inequities in the social and economic shape, as well as in terms of civil and political rights, must be taken alongside efforts to engender the development process. According to Subramanian & Saxena (2008) today's technological transformations as tools for human development requires shifts in national and global public policy. It puts forward "a global call for policy not charity to build technological capacity in developing countries. Jansen (1989) also states that unfortunately despite the potential ability of information to empower disadvantaged groups and despite the massive investments in information and communication technologies the information society has remained largely silent on gender issues. In our society today, there are evidences of a gender imbalance in the use of ICT that threaten to restrict women to be the equal partners. Beneficiaries of the emerging information society thus creating a gender based digital divide. It has been felt that unless concerted and corrective policy initiatives are taken, women well continue to be excluded from the information society. Engendering ICT policies and programs becomes relevant especially when there is much evidence to show that policy making in technological field often ignores the need, requirements and aspirations of women unless gender analysis is included (Marcelle, 2002). If gender issues are not articulated in ICT policy, it is unlikely that girls and women will reap the benefits of the information age Hafkin (2002) closing one's eyes to this fact can entrench inequality and even enlarge the gender gap, making ICT s "gender negative" technology. (UNDP, 2003).

Despite the views of many government policy makers that a well thought out general policy benefits all, there is no such thing as a gender blind or neutral ICT policy. The government has already gotten a gender equality policy that obviate the need to spell out gender issues in every policy sector, on the contrary, there is much evidence to show that policy making in technological fields often ignore the need, requirements and aspiration of women unless gender analysis is included (Marcelle, 2002). If gender issues are not articulated in ICT policy,it is unlikely that girls and women will reap the benefits of the information age (Hafkin, 2002).

Information and communication technology policies and regulation are developed, managed, and controlled in majority by men. One problem is that at both the global and national levels, decision-making in ICTs is generally treated as a purely technical area (typically for male experts), where little or no space is given to civil society viewpoints, rather than a political domain. Deregulation and privatization of the telecommunications industry is also making decision-making in this sector less and less accountable to citizens and local communities, which further compounds women's role in decision-making and control of resources.

Given the under-representation of women in ICT policy-making processes, women's needs and views are not reflected in ICT policy frameworks. If women are to benefit from ICT interventions, mainstreaming the perspectives and concerns of women is one of the important tasks that should be undertaken. Very few governments, however, involve women in processes of formulating national ICT strategies and policies, beginning with the nomination of gender-balanced teams, consulting gender and ICTs experts or supporting women's groups to provide inputs from a civil society perspective (Genderit.org, n.d.).

Another obstacle to drafting gender sensitive policies on ICTs, and mapping and analyzing their impacts on women's and men's lives, is the absence of comparable sex-disaggregated data on ICT access, use, education, employment, participation in decision-making and development, etc (Genderit.org, n.d.).

GENDER ISSUES IN ICT POLICY

During the past decade, women and other feminist movement has been very active in the use of electronic communications and internet tools. This has enable them to take advantage quite soon of ICTs for networking for their rights, raising awareness or the issues that concerned them, lobbing authorities and planning actions for women's empowerment and drastic social change. (Plou, 2005)

Gender issues in ICT are understood as adjunct to the larger issues on women's right and gender equality since the former also spring from differences and the continuing absence of recognition of these differences in the very production meanings, ideologies, languages, behavior, practices, policies and technologies. Yet these gender issues necessitate specific interventions as they arise within specific milieus and affect specific communities and group of women. Most of the women involve with ICTs are middle class young urban women, that are well educated, employed or with little economic problems. The digital gap, which also reflect other gaps that exist in our world today like social, economic and cultural barriers that marginalize large sectors of the population, is clear when it comes to considerate women's access to ICTs .(Plou, 2005)

According to Ramilo &Villanueva (2001) far greater numbers of women are now using new communication technologies and the internet in their work. The following issues remain critical for most women in the world (Ramilo &Villanueva, 2001)

- **Access and Know How:** Women's access to ICTs is dependent on several factors such as gender discrimination in jobs and education, social class, illiteracy, geographic location (North or South, urban or rural) influence the fact that the vast majority of the world's women have no access to ICTs or to any other sort of modern communi-

cation system. As information dynamics accelerate their migration towards the internet, people without access must suffer greater exclusion. In areas where access to new technologies is still not practical, more traditional media such as community radio, audio-visual media and popular media should continue to be used by and for women.

- **Education, Training and Skill Development:** Education, training and skill development are critical to ICT interventions. Illiteracy rate for women in developing countries are far higher than the percentage of men. Training methods are not customized to women's needs. Learning practices for women should be extended to girls and women, made gender sensitive (making training women-specific, ensuring ongoing user support and mentoring in the communities where women live). Other major concerns are illiteracy and languages as impediments to information access; the need to breakdown gender and cultural barriers to women's access to carriers in technology; and the design of software, which often does not respond to the need of women and girls.

- **Industry and Labour:** In the ICT sector, labour is highly sex segregated. Women are found in disproportionately high numbers in the lowest paid and least secure jobs. The concentration of women in clerical work in the information and communication industry does not translate further up the industry hierarchies. While women fill up the factories that produce computes components, in working conditions often damaging to their health very few are represented in computer systems administration and technical development. Women in low grade technical and service jobs also make up the largest group of computer users. Still many more women have been displaced due to increasing automation and computerization of workplace.

- **Content and Language:** The dominance of English language, often from countries in the North, on the internet is a major concern raised by women's organizations. Women's viewpoints, knowledge and interests are not adequately represented while gender stereotypes also pre-dominate the World Wide Web. Some of these concerns are an extension of those formulated previously in relation to sexism and portrayal of women in the media. But they also relate to a broader range of issues such as the need for women to systematize and develop their own knowledge and perspectives and make sure they are adequately reflected in these spaces.

Language barriers to information access require the development of applications like multilingual tools and databases, interfaces for non-Latin alphabets, graphic interfaces for illiterate women and automatic translation software.

- **Power and Decision-Making:** Women are under-represented in all ICT decision-making instructive including policy and regulatory institutions ministries that are responsible for ICTs, boards and senior management of private ICT companies whether at the global or national levels. One problem is that at both the global and national levels, decision making in ICTs is generally treated as purely technical area (typically for male experts) where civil society viewpoints are given little or no space, rather than a political domain. Representation is important in creating the conditions and regulations that will enable women to maximize their possibilities of benefiting from ICTs, and ensuring the ac-

countability of the institutions responsible for developing ICT policies.

- **Privacy and Security:** Privacy, security and internet rights are other vital thematic areas for women. They include having secure online spaces where women can feel safe from harassment, freedom of expression, privacy of communication and protection from "electronic snooping". They also include the passage of ICT legislation that can threaten human right. While many developing countries are grappling with the basic access and IT infrastructure issues, many countries in the Northern part of the globe are now defining the basic rights infrastructure for internet use and governance.

ADOPTION OF GENDER PERSPECTIVE IN ICT POLICIES

As stated earlier, the women and feminist movement has, during the past decade, been very active in the use of electronic communication and internet tool which they have adopted in their work as an important tool. Women have been able to take advantage quite soon of ICTs for networking for their rights, raising awareness for the issues that concerned them, lobbing authorities and planning actions for women's empowerment and social change. (Plou, 2005)

Most of the women that are involved with ICTs are middle class young urban women, that are well educated, employed or with little financial problems. The digital gap, which reflects other gaps that exist in our world today like social, economic and cultural barriers that marginalize large sectors of the population, is clear when it comes to considerate women's access to ICTs. Although the women's movement has adopted ICTs in their work as an important tool, it has paid little attention to ICT policies. Mostly, women have seen ICTs only as a tool and have left aside other issues, like their strategic use and the relevance of ICT policies that would facilitate more and better access to women, especially those in marginalized areas, unemployed, illiterate or who have not the chance to, even know how a computer works (Plou, 2005).

Government tends to be open to recognize gender as a cross cutting issue and even to address it within certain contexts (such as universal access programs). However, they are normally not able to make the link between gender and the need for gender analysis in many other aspects of ICT policy, including pricing and affordability issues, network development, universal service, fund selection criteria, licensing etc; even in those cases where gender has made it to the policy level, there is still need to ensure that implementation will reflect a gender perspective and work towards gender equality. (Selaimen, 2006).

As early as 1992 when the internet was fairly new in the developing world, a number of women organizations had already started to adopt the use of ICT to support their information, communication and networking initiatives. By the time of the Fourth World Conference on Women held in Beijing in 1995, various women organizations led by the Association for Progressive Communications (APC) women's networking support programme were not only training women in the use of electronic mail and the World Wide Web but were also raising awareness about the urgency of broadening media and communication concerns to include the new ICTs and addressing women's access to ICT and women's participation in the determination of how technologies are designed and deployed. Women have benefited less from, and been disadvantaged more by, technological advances; women need to be actively involved in this development of new technologies; otherwise the information revolution might bypass them or produce adverse effects on their lives (Ramilo & Villanueva, 2001)

CHALLENGES FOR ADOPTING GENDER PERSPECTIVE IN FORMULATION AND IMPLEMENTATION OF ICT POLICIES

The challenges for adopting gender perspective in ICT policies will be to ensure that the efforts of gender advocated do not simply live within the printed page of policy documents, rather they result in practical solutions that address gender equality in ICT development (Selaimen, 2006). The challenges are as follows:

Lack of Skills: The lack of appropriate skills and capacity to effectively operate equipment and utilize ICT tools was a major factor which prevented women's groups from fully using ICTs. This includes limited knowledge of computer and information management systems, hardware and software installation and maintenance and limited internet and non-internet based skills (Ramilo & Villanueva, 2001).

Given their limited access to schooling, women, especially those in rural areas, are also much less likely than men to have computer skills. Information literacy essentially involves using information contextually, a skill that women are less likely than men to have (Heeks, 1999).

Limited Finance: Many women, girls and their organizations find difficulty getting online due to limited finance. Cost includes ICT-related equipment, software purchase, maintenance, training and connectivity. In many countries, women face high access cost due to the presence of Internet service provider (ISP) monopolies, lack of reliable connections as well as limitation of facilities. Financial capacity which for Non Governmental Organizations (NGO) is often determined by donor resources also influences ICT access. Ramilo & Villanueva(2001)

Hafkin (2002) posits that when it involves paying for information access, such as at a rural information center or cybercafé, women are likely to have disposable income to do so (or hesitate to use family food, education and clothing resources for information).

Lack of capacity on the part of policy makers and implementation agencies to address gender considerations and conduct gender analysis and failure by most institution to integrate gender experts in all policy and regulatory teams (even when there is a generally accepted need to consider gender as an important area). And there is an emphasize that having women ICT professionals is important, but not a sufficient condition to ensure that gender analysis will be conducted there needs to be a services commitment to gender in ICT work and all capacity building programs need to reflect that (Jorge (2006) cited by Selaimen (2006).

Social and Cultural Barrier: Women tend to have less access than men to those ICT facilities which exist. Most rural information centers or cybercafés are located in places that women may not be able to visit frequently. Women also have problem of limited time because of their multiple responsibilities in running their homes which makes their leisure hours very few and their mobility is also more limited than men.

In some countries, especially Africa, traditional cultural attitudes discriminate against women having access to education and technology. Girls are not encouraged to take any jobs or get higher education. The alternative of doing two or three things at the same time is not realistically entertained (Common Wealth of Learning, 2001; cited by Hafkin, 2002).

CASE STUDIES ON GENDER AND ICT POLICY

In this section, case studies of gender and ICT policies of some countries are presented. All the countries covered in this chapter have existing IT policy frameworks and strategic ICT development plans. Several key result areas are common in these policy frameworks, namely, provision of networking and telecommunications infrastructure facilitating e-commerce and job opportunities, human resource development and promotion of good governance and citizen's participation.

Asia

Case studies on Korea, Japan and India.

Korea

The task of implementing the Korean government's gender policies rests on the Ministry on Gender Equality (MOGE) a cabinet-level agency which was created in 2001 to replace the Presidential Commission on Women's Affairs. The MOGE is also responsible for convening and coordinating the gender focal point for each government ministry .The ministry is a key player in the Korean government's drive to achieve the goals of cyber Korea 21 plan. It closely collaborates with the Ministry of Information and Communication on the implementation of the national ICT literacy campaign targeting some 2 million house wives as well as employees and staff of some 1,000 private institutes. On its own, the Ministry on Gender Equality (MOGE) has launched its own anchor projects aimed at advancing gender equality through ICT. These projects include the creation of a portal site for Korean women called women-net, and the building of the cyber IT education center which is targeting to train 1,000 female Information Technology (IT) experts within its first year. (Ramilo & Villanuella, 2001)

Japan

Before the closing of the last millennium, the Japan government passed the basic law for a gender-equal society which laid down the basic principles pertaining to the information of a gender-equal society, as well as set the direction these should take. The basic law for gender-equal society also clarified the responsibilities of the state and local government and citizens, set down provisions to form the basis of policies related to promotion of formation of a gender-equal society, and mandated the creation of a council for gender equality. Also highlighted in the plan are challenges such as eliminating all forms of violence against women, respecting the human rights of women in the media and supporting life-long health for women. (Ramilo & Villanuella, 2001)

India

In 1998, India set a goal to become an "information technology superpower" and one of the largest generators and exporters of software in the world within ten years. In May 2000, a working group called the "IT for the Masses" initiative embodies the Indian government's strategic vision of information technology as an enabler of new opportunities to bridge the gap between India's "have" and "have not". The working group tasked to set the priorities of this initiative identifies the poor, comprising 40% of the Indian population, as the immediate concern of the government in providing new opportunities through information technology. India has made tremendous headway in the knowledge-based industry and the computer software industry through human resource development. It is estimates that about 73,000 students are trained every year in the field of information technology and projects that there will be no shortfalls of personnel till 2008 even allowing for international migration. The government released the sum of Rs 2,800 crore ($633 million) for the training and re-training of IT professionals and teachers. India's IT for the Masses. (Ramilo & Villanueva, 2001).

Gender considerations do not figure in any of the schemes or programs being developed under the banner of "IT for the masses". The "IT the masses" initiative offer some opportunities for addressing gender concerns in the formulation and implementation of ICT policies (Ramilo & Villanueva, 2001)

Europe

Case studies on Albania Bosnia and Herzegovina.

Albania

Albania is situated in South Europe. Its national ICT strategy is one of its kinds in the region, with a marked effort to include women's need and views. Gender incorporation in ICTs is part of Albania's attempt to address growing disparities in income, gender and geographical location. Information and communication technologies were seen as powerful tools that can assist to bridge these disparities and support the socio-economic development of Albania. In order to take advantage of the potential of ICTs, countries need to elaborate a national vision or strategy that reflects the needs of various stakeholders, including traditionally magnetized groups. The government of Albania launched the participatory ICT strategy process at a national conference. Several expert working groups were established. The Albanian experience can inform other countries attempts to incorporate a gender perspective in ICT policy processes. Tinning and sustained presence of women's representation was a critical factor in the Albania case. Perhaps the Albanian national ICT policy could have fulfilled its aim of gender inclusiveness if the policy makes involved had gone through gender sensitization sessions in the beginning of the process (Gustainiene, 2005)

Bosnia and Herzegovina

Bosnia and Herzegovina is situated in the Middle East Europe, the current ICTs situation in Bosnia and Herzegovina is paying particular attention to the development of their national ICT policy strategy, and the responses towards the need to integrate gender, concerns both the women's national machinery and civil society organization .

For Bosnia and Herzegovina (BiH) analysis and understanding of the impact of ICTs upon women and men is far behind other in the Central and Eastern Europe and the Commonwealth of Independent States (CEE/CIS) region. This is caused by several factors. One, the country could not follow the initial development of ICT technologies because it was interrupted by the atrocious war in the early 1990s. Secondly, BiH is a whole decade behind the time when development of ICTs was at its peak. Throughout this time, the country has been struggling with post-conflict reconciliation and reconstruction as the two main factors shaping development processes. In addition, it seems that technology instructors and educational system operate on the same principle, which greatly diminishes the potential and interest of young Bosnian women in technology.

To change the trajectory and see an increase in women's participation in ICT networks, one of the core tasks is to institute and develop capacity of female technology experts within primarily male-dominated BiH ICT provider organization. Gender equality "train-the-trainer" project is currently taking place in Sarajevo at the E.Net centre and in the organization of the United Nations Development Programme (UNDP). BiH and the BiH gender mechanisms could serve as an example of successful contribution to the increase of the awareness of the relation between gender and technology. Governmental bodies such as the ministries of transport and communications, the ministries of education and culture, as well as non-governmental organizations and international organizations such as United Nations Development Program (UNDP), etc has defined strategies and develop programme objectives for realization of equality between women and men in the BiH information society and are also the main actors in defining strategies for reducing gender discrepancy in the ICT field. (Kosovic, 2006).

Africa

Case studies on Zambia and Uganda.

Zambia

According to (Zulu, 2005) Zambia, like many other African countries, has recognized information, knowledge and technology as major drivers of social and economic development. Zambia made a decision to adopt information and communication technologies (ICT) as part of national development, expecting to narrow the digital divide and leap-frog the development process. While the government is committed to pursue appropriate institutional, legal and regulatory measures to achieve broad policy goals, Zambia's national ICT policy framework still lacks gender integration. Zambia's Ministry of Communication and Transport produced a second ICT draft policy to undergo validation in November 2004 before being submitted to cabinet office. The policy validation process was away to involve multiple stakeholders in the policy development and the policy validation workshop was held and participants at the policy validation observed the absence of gender analysis in the draft document.

One of the challenges faced by Zambians in the validation workshop was to engender the draft ICT policy document. This is important as it can reduce the digital divide between men and women. Technology is still viewed as a male domain in Zambia and all over Africa. The document emphasized the need for civil society, Non Governmental Organization (NGOs), media and government to create awareness and sensitization on gender and ICT policy. The Zambian experience of policy formulation in relation to gender and ICT shows that there is a lack of awareness and sensitization in the country about the importance of gender and ICTs. (Zulu, 2005)

Uganda

The Uganda National ICT Policy framework, approved by Cabinet in December 2003 envisions a country where national development, good governance and human development, are sustainably enhanced, promoted and accelerated by the efficient application and use of ICTs including timely access to information.

The policy draws upon a number of strategies to accomplish its vision, which include mainstreaming gender into policy programs and implementation strategies; sensitization and awareness creation; ICT capacity development among rural people; building appropriate infrastructure; supporting favorable investment environment; supporting innovative ICT projects; stimulating production, storage, and dissemination of national information; and facilitation of access to public domain information. The government envisages that by implementing strategies, the Uganda Women Caucus on ICT (UWCI) was initiated; UWCI is comprised of women and gender practitioners working on issues of women and ICTs in Uganda. UWCI's mission is to engender ICT policy formulation, implementation, monitoring and evaluation. There are other groups like the Council for the Economic Empowerment of Women in Africa (CEEWA-Uganda), the Uganda National Council for Science and Technology (UNSCT), the Uganda Communication (UCC) and other NGOs have made effort to expand ICT centres in rural areas and to increase and expand infrastructure coverage. (Amuriat & Okello, 2007)

Latin America

Case studies on Latin America and the Caribbean.

The possibilities of Latin America countries to be integrated in the information society is to remember that this "global" tendency has taken place along with one of the most critical historical stages in the economic and social scenarios since the 1970s. The scandalous growth of poverty and

the levels of social inequity, together with the weakness of the national states and the lack of public investment in strategic sectors for human development, such as education or health, together with other alarming signs such as the lack of transparency of the state administration of budgets for social programs and purchase of technological infrastructure, A significant number of countries in the region show degrees of connectivity higher than expected according to the income level per inhabitant, and the gap that separates them from the leading countries in the field of information and communication technologies (ICTs) has, to some extent, been reduced, does not guarantee that in the next few years they will be automatically incorporated into the digital era. Unless additional efforts are made on the part of the state societal groups, it is highly probable that e-gaps will continue to or grow. The economic commission for Latin America and the Caribbean (ECLAC) recommends carrying out a systematic strategy that articulates promotion of technological capabilities in all the countries support transformation of the productive structures, development of national and regional production networks setting up a quality infrastructure. Bonder (2002) suggests that the general frame offers new "entry points" for the integration of women, not only as ICT users, but also as researchers, producers, workers, educators, project managers and in many other positions from which they can contribute through the new technologies to the economic growth with equity" as needed in the LAC region. The situation of women/gender and ICTs in the region shows that high appraisal of the opportunities that ICTs offer to women as a means of exchanging information and the use of ICTs has brought about a spectacular progress in terms of organization, articulation of demand legitimacy, current knowledge building and creation of alliances among women NGOs over the last decade. Countries from South America like Argentina, Bolivia, Brazil, Chile, Colombia Uruguay etc, and countries from Central America like Belize,

Costa Rica, El Salvador, Mexico etc and the Caribbean like Haiti, Jamaica, Cuba, Bahamas, and Puerto Rico etc. are making progress since the formation and implementation of ICTs in this region (Bonder, 2002)

Australia

The Commonwealth Office of the Status Women (henceforth, OSW) is the main government agency that provides policy advice to the Prime Minister and the minister assisting the prime minister for status of women on issues affecting women in the Australia. The OSW also develops strategies for addressing priority issues and concern in gender. These strategies are reviewed every three years. Currently the OSW's priority concerns are

- Economic self-sufficiency and security for women.
- A life time optimal status and position for women.
- The elimination of violence in the lives of women.
- The maintenance of optimal health and well being throughout women's live.

The Office of the Status Women (OSW), for its part is currently developing what it calls "women's data warehouse" and a web-based system that integrates statistical information about women's needs and concerns. This information system is being developed in consultation with organization engaged in gender analysis and policy development work (Ramillo & Villanueva, 2001)

GENDER APPROACHES TO ICT POLICIES AND PROGRAMS

Bonder (2002) has highlighted some fundamental conceptions, principles, objectives and actions that are being used and implemented for engendering ICT policies. The following table enshrines fun-

damental points of the theoretical and strategic discussion around this topic. Its aim is not to show an "evolutionary progress" from one conception to the next; neither does it imply a judgment of value of any of the strategies proposed. Its sole purpose is to enrich the elaboration of new proposals by taking advantage of what has been achieved so far in Latin America (Bonder, 2002)

GUIDELINES FOR POLICY-MAKING AND REGULATORY AGENCIES

The International Telecommunication Union Task Force on Gender Issues (2001) has set gender-aware guidelines for policy-making and regulatory agencies are intended to assist decision makers to conduct their work in such a way that both women and men are considered in the process i.e. both are involved in decision making.

The following are the set of guidelines for policy making and regulatory agencies, with discussions on each point, they are; General, Human resources, Training and licensing activities. The guidelines are to be used to ensure that gender analysis becomes an integral part of regulatory/policy activities Successful implementation of these guidelines requires the development and promotion of new policies within the institutions seeking transformation into a gender-aware environment. The process of implementing these guides should also be conducted with full participation of all parties and if possible, with participation of gender experts (for example, from gender units or consultants in the area) to ensure full understanding of the issues and avoid unproductive resistance to the process (International Telecommunication Union Task Force on Gender Issues, 2001)

- General
  - **Facilitate and promote the establishment of a gender unit within the regulatory agency, by ministry or**

Table1. Approaches to ICT Policies and programs (Source: Bonder, 2002)

Problem Definition	Perspective	Goals	Measures/Actions	Ethical/ Political Principles
• Unequal access and participation of women in ICTs as users, students, teachers, workers and professionals. *Explanations:* • Lack of economic resources, education and infrastructure • Cultural values and gender discrimination patterns in society and institutions.	"Deficit model" Women seen as a group in social and economic disadvantage. **Compensatory Strategy**	• Promotion of equal opportunities for all women in terms of access to ICTs as well as participation in educational programs and technology industries.	• Community based projects (telecenters or similar). • Educational and training programs, scholarship, public campaigns, provision of equipment and other incentives. • Networking	• Women's rights • Equal opportunities • Gender justices • Integration of women in the development and modernization of economy and culture: being part of the global society.
• Gender "nature" and characteristics of ICTs: focus on contents, formats, uses, impacts, regulations, etc. • ICTs as a field of power relations. • Devaluation/invisibility of women's need, knowledge, skills and technological culture. • Homogenization vs. diversity.	"Difference model" (Valorization of women's cultures, values and visions in/for ICTs). **Critical Strategy**	• Integration of women's needs, "ways of knowing" and relating with information and communication in educational, research and innovation project. • Generation of new contents, formats, tools, etc. • Deconstruction of technology discourses and dominant practices	• Emphasis on research and academic debates, cyber-feminist theories and innovative experiences. • Women-friendly training and educational projects. • Promotion of critical analysis of power/gender relations in contents, tools and ICT policies.	• Inclusivity • Diversity • Empowerment • KEYS for improving the quality and social uses of ICTs.
• How to change gender/power relations in and through ICTs. • Information/knowledge society: meaning, power and impacts on gender equality and human development.	**Transformative Strategy**	• Mainstreaming gender analysis, planning and evaluation in ICT policies, programs and projects at national regional and international levels. • Addressing all dimensions of ICTs (access, uses, appropriation, production, management, ownership, regulation, policies etc).	• Collection and dissemination of statistics and elaboration of gender indicators in ICTs. • (Interdisciplinary) research of gender relations in all dimensions of ICTs. • Lobbying and continuous dialogue among researchers, policy makers, women groups, corporate sector. • Networking and collaborative projects at regional and international levels. • Continuous evaluation of policies and programs. • Development of gender scientific science and technology education at all levels of the educational system. • Promotion of equal participation of women and men at all levels of the technology industry. • Assertive action and other measures to remove subtle obstacles preventing women professional development in S&T.	• Long term transformational strategies. • Building a new social paradigm: a gender fair knowledge society.

an inter-agency effort: Regulatory and policy making agencies should establish units or inter-agency units to promote gender awareness and perspective and further facilitate the process of mainstreaming gender in the institutions processes and work.

- **Review, revise or develop new regulations, circulars, issuances and procedures to remove any gender bias:** Regulatory and policy making bodies should review, revise or develop new regulations to remove any gender bias that adversely affect women (e.g. poor working conditions, lack of child-care facilities, lack of maternity leave, limited opportunities for training and advancement etc).

- **Promote Gender Analysis as Part of the Policy Process:** Gender analysis is needed to ensure that the policy process is based on a complete set of facts and a comprehensive analysis of the problem. Integrating gender analysis is a means to increase the quality and positive impact of the policy.

- **Develop and Establish Systems to Gather Gender Statistics:** Regulatory and policy making bodies should work in conjunction with national statistics bureaus or other statistics agencies (such as the census bureau) to develop sex disaggregated statistics and new gender-specific statistics (such as, access by gender or number of lines per women headed households).

- **Dialogue with other National Entities:** To assist in harmonizing national efforts, regulatory and policy-making agencies should promote contacts with other ministries and bodies that govern national policy on access and education issues.

- Human Resources
  - **Ensure Equal Hiring Opportunities for all Women and Men, Regardless of Race, Ethnicity, Class and Age:** Women and men should be afforded equal opportunity to all positions available in any institution. Institutions should ensure that all positions are advertised in public channels and in avenues available to a diversified group of candidates (i.e. women and men of different racial, ethnic, class and age groups).

In addition, where appropriate, establish "affirmative action" policies or quotas (based on the real levels of qualifies people for each position) to ensure equal opportunity and avoid any tendencies to increase labor segmentation and/or occupational segregation in the work place.

  - **Ensure that a Certain Percentage, Targeting 50% of all Supervisory and Management Positions are Occupied by Women:** Institutions should establish a policy to raise the number of women in all supervisory and management positions at all levels of work, from clerical to top management.

  - **Develop Campaigns to Attract Women Professional (Particularly for Technical and Decision making Positions:** Institutions should develop hiring campaigns to attract women professionals to work in technical, regulatory and policy-making positions. Such campaigns can be developed in coordination with other governmental bodies as part of an overall campaign to increase the number of women at all levels and areas of government work.

○ **Develop and ensure the Existence of Appropriate Support Systems for Professional Women and Men:** Institutions should provide (by themselves or in partnership with other government or non-government institutions) support systems for professional women and men, such as on-site support to deal with male/female tensions and other important issues, such as day care, access to training opportunities, and flexible work schedules. Such systems not only facilitate women's participation but also increase workers' productivity and dedication.

○ **Ensure that there are no Wage Disparities among the Gender and Establish a Policy to Eliminate any such Gaps:** Human resources divisions should ensure fair and non-discriminatory salary practices. Where disparities occur, these should be corrected immediately.

○ Training

○ **Ensure Equal Access to Training Opportunities:** Regulatory and policymaking institutions must ensure equal access to all training opportunities domestically and internationally. Women should not be discriminated against because of other responsibilities (e.g. motherhood) or their current professional level. Instead, institutions should attempt to provide the necessary conditions so that all employees have equal opportunity to attend training programs and consequently benefit from improved qualifications.

○ In order to ensure equal access, institutions should, among other things: *(1* invite women and men to attend programs, *2)* advertise training opportu-

nities at all departments, divisions, or work groups, particularly those with a greater number of women workers, and *3)* provide adequate conditions for women's participation (e.g. scholarships to cover training costs and a gender-aware environment, including women trainers for various specialties).

• **Promote Gender-Awareness Training Opportunities for Women and Men:** Institutions should promote gender-awareness and gender analysis training for all their employees.

• **Support Technical and Management Programs that Train Women Professional and Create Internship Programs with Educational Institutions:** Regulatory and policy making bodies should create partnerships with educational institutions to promote women's enrollment in educational programs and, where possible, develop an internship program to provide training opportunities for women in regulatory and policy making agencies. In addition, partnerships with educational programs (such as business, law or engineering) may also increase the potential for future business ventures headed by women in the telecommunications market place.

• Licensing Activities
Licensing activities are associated with the process of awarding authorization for provision of services in the telecommunications sector. This is usually the responsibility of the regulator, which, in addition to setting licensing criteria and rules, is responsible for awarding the licenses themselves. The following guidelines provide a basic checklist of issues that should be followed by regulators or those responsible for the licensing process. These guidelines assume that the implementing agency also follows the general gender aware guidelines for regulatory purposes.

- **A Certain Percentage of Licenses should be Awarded to Women-owned Companies and/or Companies with Women in top Management Positions:** Regulatory agencies should establish a policy to promote licensing criteria that gives preference to women-owned companies and/or companies with women in top management positions.

- **Develop and Market Licensing Procedures where Potential Women Owners can have Access to the Information:** Ensure that licensing procedures and advertisements are placed in public sources and in particular those that women have access to, such as newspapers, universities, local and regional commerce associations, women's organizations, the internet and specific web pages of interest to businesswomen.

- **Promote the Development of Business Assistance Programs and Partnerships with Expertise in Assisting Women Entrepreneurs:** Institutions should promote the development of business assistance programs or partnerships to ensure that interested women have access to all business related services, such as license application, development of successful business plans, access to financing and appropriate capital loans, and training programs, among others.

- **Develop License Award Criteria based on Social Responsibility of the Business as well as Universal Access Objectives of the Proposed Venture:** Institutions should develop licensing criteria that considers the companies' social responsibility record as well as their plans to contribute to universal access to communications, (such as willingness and plans to contribute to development projects, particularly those with a gender component, telecentre-type projects, funding of educational programs to promote disadvantaged youth to attend technical degrees, or projects specially targeting women).

- **Ensure that Licenses Awarded Contain Certain Conditions to Promote Gender Analysis and Mainstreaming for the Particular Company:** As part of the license awards, regulators should include provisions to ensure that licensees engage in programmes to mainstream gender in their organizations, by accepting the gender-aware guidelines for women technicians and managers, and by providing a gender-sensitive work environment.

The use of these guidelines themselves should be the rule for policy and decision makers. They can be used in two ways.

1. It is used as a checklist of issues to consider when making decision.
2. A consultative document to provide ideas on how to mainstream gender in regulatory and licensing agencies.

CONCLUSION

This chapter discussed the need for gender consideration in ICT policy, gender issues in ICT policies, adoption of gender perspective in ICT policies, challenges for adopting gender perspective in formulation and implementation of ICT policies, case studies of Asia, Africa, Latin America and Caribbean, Europe, and Australia. Gender approaches to ICT policies and programs, guidelines for policy making and regulatory agencies, future trends on gender and ICT policies, were treated. Various national governments have started addressing gender issues in their policies. However,

In the light of what has been presented in this chapter, the following are recommended:

- Policy makers should ensure that Gender considerations are truly included in national ICT policy, both in the development and implementation process advocacy is needed.
- More professional experts in both gender and ICT issues are involved at all aspect of ICT developed.
- There is urgent need for greater involvement of women in decision making especially on ICT policy related matters. .

REFERENCES

APC WNSP. (2005). *New gender and policy monitor help women make ICT policy a priority.* Retrieved 2nd November 2009. from: http://www. genderit.org/ en/index.shtml?w=9&x=91258

Bonder, G. (2002). *From access to appropriation: women and ICT policies in Latin America and the Caribbean. A Report from the United Nations Division for the Advancement of Women (DAW) Seoul Republic of Korea II to IV November 2002.* Retrieved December 23, 2009 from http://www. un.org/womenwatch/daw/egm/ict2002/reports/ Paper-GBonder.PDF

Burch, S. (1997). *Latin American women take on the Internet.* Retrieved 2nd November, 2009. Retrieved from http://www. apcwomen.org/net-support/articles/art.01.html

Genderit.org. (n.d). *Why gender in ICT policy?* Retrieved November 25, 2009 from http://www. genderit.org/en/beginners/whygender.htm

Gustaniene, A. (2005). *Gender focused ICT policy making.* Retrieved 2nd November 2009. From: http:// www. genderit.org/en/indexshtml?w=a&x=91489

Hafkin, N. (2002). *Gender issues in ICT policy in developing countries: An overview.* Retrieved 9th January 2009, from http://www.un.org/ womenwatch/daw/egm/ ict2002/reports/paper--NHafkin.pdf

Heeks, R. (1999). *Information and communication technologies, poverty and development.* Retrieved 9th January 2009, from http://idpm.man.ac.ik/ idpm/diwpf5.htm

Human Development Report. (2001). *Making New Technologies Work for Human Development.* New York: Oxford University Press.

Idowu, B., Ogunbodede, E., & Idowu, B. (2003). Information and communication in Nigeria. The health sector experience. *Journal of Information Technology Impact, 3*(2), 69–76.

International Telecommunication Union. (2009). *Special initiatives: Gender.* Retrieved December 23, 2009 from http://www.itu.int/ITU-D/ sis/Gender/

International Telecommunication Union Task Force on Gender Issues. (2001). *Gender-aware guidelines for policy-making and regulatory agencies.* Retrieved December 23, 2009 from http://www.itu.int/ITU-D/gender/projects/Final-GendAwrnGuidelns.pdf

Kosovic, L. (2006). *Gender and ICT policy in Bosnia and Herzegovina: Rethinking ICT development through gender.* Retrieved November 2, 2009, from: http://www.gednerit.org/en/index. shtml?w=a&x=95059

Marcelle, G. (1998). *Strategies for including a gender perspective in African information and communications technologies (ICTs).* Retrieved 9th January 2009, from http://www.un-instraw. org/des/martines.doc

Marcelle, G. (2000). *Gender and information revolution in Africa.* Retrieved November 2 2009, from http://www.genderit.org/en/index.shtml?w=98x=91489

Marcelle, G. (2002). *Gender equality and ICT policy.* Retrieved January 9 2009, from http://www.worldbankorg/gender/digitaldivide/world-bankspresentation.ppt

Plou, D. S. (2005). *Gender and ICT policies: how do we start this discussion?* Retrieved November 2, 2009. from http://www.gender.org/en/index.shtml?w=a&x=90862

Radloff, J. (2005). *What's gender got to do with IT?* Retrieved November 30, 2009 from http://www.apcwomen.org/node/273

Ramilo, G., & Villanueva, P. (2001). *Issues, policies and outcome: Are ICT policies addressing gender equality?* Retrieved December 27, 2009 from http://www.genderit.org/upload/ad6d215b74e2a8613f0cf5416c9f3865/UNESCAP_Gender_and_ICT_Policy_Research_Report_Feb02.doc

Selaimen, G. (2006) *Gender issue at all levels – From policy formulation to implementation.* Retrieved November 2, 2009. From: http://www.genderit.org/ en/index.shtml?w=a&x=93423.

Subramanian, M., & Saxena, A. (2008). *Gender mainstreaming of information and communication technologies (ICT) policies and programs.* A case study of Ghattisgarth State of India, Project Report Submitted by WSIS-2 Awarder.

UNDP. (2003). *Transforming the Mainstream: Gender Mainstreaming Past, Present and future.* Retrieved Dec. 29, 2009 from www.undp.org/gender/policy/htm

United Nations. (2005). *Gender equality and empowerment of women through ICT.* Retrieved December 23, 2009 from http://www.un.org/womenwatch/daw/public/w2000-09.05-ict-e.pdf

KEY TERMS AND DEFINITIONS

Development Strategies: Are the plans to achieve, or a potential force for creating new economic growth opportunities.

Formulation and Implementation of ICTs Policy: To create and carry out ICT policy that will address the needs of women and men, improve their well being and facilitate their participation in their development process.

Gender: The fact of being male or female

ICT Policy: Is an integrated set of guidelines, decisions, laws, regulations and other mechanisms to direct and enhance the production, acquisition and use of ICTs.

ICT: Refers to technologies and tools that people use to collect, process, organize and disseminate information.

Issues: Are the topics of discussion concerning gender and ICT policy globally.

Needs: Are necessities or importance of engendering ICT policy.

Chapter 12
Name Authority Control Paradigm Shift in the Network Environment

Mirna Willer
University of Zadar, Croatia

ABSTRACT

The purpose of this chapter is to give an international perspective and overview of the theory, standardization processes and following practices in the field of authority control, with particular view on the name authority control since the 1960s to the present. In the focus of interest of this chapter is paradigm shift in the field, and the possibilities of semantic web technologies in meeting library users' needs, as well as librarians' tasks to produce tools convenient to the user in the network environment.

INTRODUCTION

The theory and practice of Universal Bibliographic Control (UBC) of published intellectual and artistic production belongs to the field of organization of information within library and information science. The significance of the topic of name authority control, as part of the information organization which comprises bibliographic and authority data control can be viewed as the one which over-passes the boundaries of library, and, indeed, heritage institutions community. Today, practically every type of service depends, in one way or the other, on retrieving information provided through the

World Wide Web infrastructure. As we are faced with the explosion of information on the web, what we need at the end of a day is authoritative and authentic data. Libraries and other heritage institutions traditionally provide such data. Those institutions, however, are faced with the problems of how to process the ever growing quantities of traditional media as well as web resources in efficient and effective ways, and how to make them available globally to all for professional and/or private use, education or entertainment. Libraries in particular are challenged by the Internet information providers such as Google and Amazon which draw library users away from their services due to providing links to instantly available resources. However, it is exactly that those Internet

DOI: 10.4018/978-1-61692-012-8.ch012

information providers are ever more becoming aware of the need to retrieve, if not incorporate, vast amounts of organized information provided by heritage institutions into their services, while at the same time the heritage institutions seek the ways how to include their services into those of Internet information providers.

Without control of a name: whether personal, corporate body, work, place or topic, there would not be a successfully performed search nor retrieval of information from catalogues, bibliographies, lists, full text databases, etc. in any format or form of a service, as indeed of information provider services of various kinds. International agreements that started within the library community defined mechanisms to control the forms of names with the aim of recording national production of publications and exchange of authority data in line with economic and efficient management.

The theoretical foundation of the concept of Universal Bibliographic Control was straightforward: bibliographic description which comprises of bibliographic and authority data should be created in the source of its origin, such as a national bibliographic agency, and made publicly available. As part of that description, and represented by authority data, a bibliographic entity, whether a person, corporate body, topic or some other entity of interest to the user of a bibliographic service, should be represented by one form of a name, and that form of name should be establish by the bibliographic agency of the entity's origin. The aim of the UBC is to enable any user anywhere in the world in any of bibliographic sources (e.g., catalogue, bibliography, list) to find all the works (in whatever manifestation) by the chosen entity collocated under that one form. The efficiency of the concept for the user (not to mention the economy for libraries due to adopting already processed authoritative data into their services) is obvious: searching one or more distributed catalogues would get all the works by an entity (e.g., personal author) under one authorized form of name regardless of the form of name s/he started

her/his search with, as all forms would point to that one (nationally and internationally) agreed upon name. In other words, the user would not find her or himself in a situation where s/he would not know when to stop her/his search (S.M. Malinconico), i.e., under what other forms of name s/he would yet have to search!

The period between 1961 and 1991 in which the concept of UBC and appropriate bibliographic tools were developed by International Federation of Library Associations and Institutions (IFLA) is described in the first part of the chapter.

There are a number of reasons why such an ideal could not have been realized, yet the majority of them could be subsumed to one: the basic principle of information organization is the convenience of the user, so the form of the name should be adapted to user's expectations. And although it is a common understanding that the user of contemporary bibliographic services in a global information environment could be anyone (as indeed, it could have been in the UBC modelled environment, but, it must be admitted, under significantly more restricted circumstances), and that the (proclaimed) aim of these services is indeed to reach everyone, they are built to address the needs of those that are closest – the local community.

The second part of the chapter deals with development of ideas, concepts and tools as a reaction to these developments since the beginning of the 1990ies to the present. The divergence from the UBC concept was brought to the full attention of IFLA not primarily because of the lack of libraries meeting their users' needs, but because of the noticed lack of international exchange of authority data, the production of which was considered the most resource consuming library processing task. The reaction to the situation was prompt: two analytical studies sponsored by IFLA presented a blueprint of the problem and recommendations. The result of IFLA's answer to these recommendations was basically the recognition of the paradigm shift in practices in the field of name authority control. IFLA also recognized that certain problems inherent to bibliographic control could be solved, if not fully, at least to a considerate extent, in co-ordinating its work with other communities/sectors such as publishers and rights management communities, and also archives and museums. A new conceptual model which aimed at defining functional requirements for authority data was developed taking into consideration the complexity of the described context. Being a conceptual model, though, its aim is not to solve two issues that are at the core of the name authority control: the identification (numbering) of the bibliographic entity and/or its name that is of interest to the library user, and actual practices on which libraries' online catalogues (WebPAC: Web Public Access Catalogue) are based and which would have to be changed to meet newly appearing forms of users' needs. As to the latter, however, the conceptual model for authority data draws attention to certain issues that have been formulated into recommendations for amendments and/or changes of IFLA standards. The abstracting processes performed during the design of the conceptual model released certain common understandings that could bring further refinements into recording authority information, and also in expressing the complexities of relations that govern the present bibliographic universe.

Another major undertaking to bring national cataloguing theories and practices back under common denominator was a series of worldwide consultations through a series of meetings on cataloguing principles organized by IFLA. The purpose of these meetings was to review and replace the cataloguing principles of the 1961, principles on which majority of cataloguing rules have been based, with the new ones required in the contemporary theoretical and technological context.

To link the present with the possible future, the future trends part of the chapter deals with charting the territory for the name authority control as part of the broader context recognized in library literature, emerging practice and projects

within the semantic web phenomena. There are already certain initiatives within IFLA and national libraries that could provide a context for such an inquiry. The author finds that the solution to the "convenience of the user" problem could be searched in the semantic web technologies and their underlying concepts applied to the bibliographic entity/name with its respective metadata identified within a recognized namespace. In other words, the possible future trend is seen in the deconstruction of the present machine-readable record or in some other way formatted authority and bibliographic record into data and their relationships that could be understood (manipulated) by the machine in performing users' queries, and in designing new services.

The purpose of this chapter is therefore to give an overview of the theory, standardization processes and practices in the field of authority control, with particular view on endeavours which the International Federation of Library Associations and Institutions (IFLA) has invested into the field since the 1960s to the present. In the focus of interest of this chapter is paradigm shift in the field, and the possibilities of semantic web technologies in meeting library users' needs, as well as librarians' need to produce tools convenient to the user in the network environment.

HISTORICAL BACKGROUND: THE CONCEPT OF UNIVERSAL BIBLIOGRAPHIC CONTROL

Since the 1960s the theory and its practical implementation in the international environment of the Universal Bibliographic Control, and its constituent part authority control of names has been pursued within the International Federation of Library Associations and Institutions' (IFLA) bodies by numerous experts with basically one goal: to enable the library user to find a book in a library. However, such a requirement of an organizing system as is a library catalogue indeed comprises of two objectives: a crucial factor in organizing information in a system that would not be made just a finding list (a list of items, e.g., find a book) is to enable the user to find collocated information. Collocation means bringing together under given name editions, translations, adaptations, etc. of a work, i.e., expressions of a work, as well as their manifestations in a different physical format and specific copy characteristics. The aim of collocation is to inform the user of a context of the sought information, and even to educate him/her. The latter objective Svenonius emphatically considers as "what users have a right to expect from systems for organizing information", stating that "a final argument in defense of full-featured bibliographic systems [meeting both objectives, MW] is that they are required if knowledge is to advance. Progress depends on cumulative scholarship, which in turn depends on scholar's ability to access all that has been created by the human intellect" (Svenonius, 2000, p. 29). This argument brings us to two major issues that IFLA pursues through its activities: the Universal Bibliographic Control of all published documents through adhering to common standards, and the design of bibliographic information system based on those standards and their underlying theory. Collocation function of the catalogue is founded on name authority control.

IFLA's Cataloguing Principles and the Concept of Uniform Heading

International cataloguing theoreticians and practitioners met in 1961 in Paris for the *International Conference on Cataloguing Principles* (International Conference, 1963) with the aim to define international principles on which national cataloguing rules could be based. The Statement of Principles, so called Paris Principles was adopted by delegations from fifty-three countries representing national library associations and twelve international organizations. Such a major undertaking was the beginning of the international

work on standardization of cataloguing procedures and had fundamental impact not only on the future of cataloguing practice but also on the cataloguing theory.

The context for the adopted statement could be analysed through three major factors. The two have already been mentioned: cataloguing practice and theory, while the third one is the form of the catalogue of the time. The first factor, the one concerning the practice, was the objective to bring bibliographic and cataloguing traditions and procedures under one set of rules which would enable the production of the efficient and economically justified record for multiple uses and diverse services: library catalogues and, indeed, integrated library information systems, national/special bibliographies, centralized/union catalogues providing other libraries within the region and worldwide with a standard bibliographic description of a book. Building on the responsibility of each national bibliographic agency (i.e., national library) to provide such a description, IFLA developed its first core programme, the one of Universal Bibliographic Control (UBC). In *Universal Bibliographic Control: A Long Term Policy, a Plan for Action* the concept of UBC was defined as the one that "presupposes the creation of a network made up of component national parts, each of which covers a wide range of publishing and library activities, all integrated at the international level to form the total system" (Anderson, 1974, p. 11). The long term programme of UBC was defined in a series of resolutions confirming the responsibilities of national libraries for "making the definitive bibliographic record of its own publications in accordance with agreed international standards", and accepting "responsibility for establishing the authoritative form of names for its country's authors, both personal and corporate, and authoritative lists of its country's authors, personal and corporate" (Anderson, 1974, p. 47). That task was further confirmed by the thirteenth recommendation of the International Conference on National Bibliographies held in Paris in 1977

(International Conference, 1978) that a national bibliographic agency "should maintain an authority control system for national names, personal and corporate, and uniform titles, in accordance with international guidelines", and by IFLA's publications of international rules for structuring normalized forms of names. The rules and lists were intended to help cataloguers in creating such normalized forms for names of persons, corporate headings, anonymous classics, liturgical works of the Latin Rites of the Catholic Church, names of states, and higher legislative and ministerial bodies in European countries.

The theoretical background behind the above mentioned activities as adopted in Paris in 1961 was the concept of the already mentioned term "uniform heading". It was not only agreed that each national bibliographic agency would provide authoritative lists of names, i.e., its country's authors' names and titles of anonymous works for others to consult, but that forms of names as published in those lists and according to national cataloguing rules would be adopted by other national bibliographic agencies for recording those names in their catalogues. Such a form of name is expressed in a uniform heading. Putting into practice such an agreement the libraries would enable their users to uniquely identify the name of an author or a title in catalogues globally, and to find all works by that particular author and all editions, translations, adaptations of a work collocated under that one name.

According to the general principle of the Paris Principles in section 7 Choice of Uniform Heading, "the uniform heading should normally be the most frequently used name (or form of name) or title appearing in editions of works catalogued...", while the subsection 7.1 specifies the choice of that form in given bibliographic conditions. The given conditions being the appearance of editions in the original and in translations, thus subsection 7.1 requires that "preference should in general be given to a heading based on editions in the original language; but if this language is not normally used in the catalogue, the heading may be derived from

editions and references in one of the languages normally used there" (International Conference, 1963, pp. 91-96). Another bibliographic condition that should be referred to here is "when the author is known by more than one name or form of name", the catalogue should contain more than one entry (i.e., variant forms of name) to discharge the functions of the catalogue. The particular section in question is section 3.21, but the intention of quoting it here, as indeed in quoting 7.1, is to enable us to assess from the present situation conditioned by the change of technological environment, bibliographic conditions, functions of the catalogue and user requirements, what should have been taken into account while charting the rules for international co-operation in building catalogues and serving users of the time. Speaking of bibliographic conditions, comments on section 3.21 should be cited from the annotated edition of Statement of Principles prepared by Eva Verona and associates published in 1971 (Statement, 1971). The commentary and examples (here omitted) run as follows:

Variation in a personal author's name or form of name may arise from many causes, for example:

- Variant spellings of a name
- Different Romanization of a name originally not written in the roman script
- Different phonetic transcriptions, that is, different conversions of a name originally written in the roman script into a non-Roman script, for example, into the Cyrillic script
- Different linguistic forms
- Use of complete and incomplete forms
- Change of status
- Arbitrary or legal change of name or form of name
- Use of pseudonyms, nicknames, clandestine names assumed for political activities or other assumed names, generic appellations, etc.
- Use of the title of another work.

Variation in the name or form of name of a corporate body [...] may also arise from various causes, for example:

- Variant spellings of a name
- Different linguistic forms
- Use of shorter names or of official names
- Change of name or form of name (Statement, 1971, pp. 9-11).

The annotated edition was the result of decisions made during the first part of the IFLA's International Meeting of Cataloguing Experts, held in Copenhagen in 1969 (International Meeting, 1969).[1] The discussion prepared and led by Eva Verona analysed various interpretations of Paris Principles in national cataloguing rules designed since 1961. Other problematic areas in relation to the choice and form of a heading that were recognized as needing further comment in the annotated edition were the following: treatment of an author who wrote under different names for different types of material, change of name by a living author, works issued by dignitaries which, although appearing under their personal names, carried a collective authority, original forms of cities and states, nature of entries under states for laws, constitutions, etc., works produced by several authors, and transliteration according to a standard international system, with an exception of ancient Greek names which may be written in the Latin form. However, in order to advance international uniformity in relation to the choice of uniform heading, it was decided that the commentary should encourage the use "wherever possible of the original forms of names and titles, rather than the forms used in the language of the country in which the library is located" (Report, 1970, p. 110).

The third factor that had fundamental impact on Paris Principles was the technology, i.e., the prevailing use of the printed card catalogues. It should be noted, though, that the working paper presented as the last one on the conference was

The Impact of Electronics upon Cataloguing Rules by C. D. Gull (Gull, 1969). Gull draw an image of the electronic environment that at the time already started to influence the type of materials libraries were collecting, the nature of these new materials and the impact they should have on cataloguing procedures and rules. His primary concern in that context was to alert librarians to the new type of authorship: "non-human authorship, or to use a positive phrase, automatic authorship" (Gull, 1969, p. 281). That type of authorship, named "automaton" would be introduced into the new IFLA Statement of International Cataloguing Principles (Statement, 2009) almost forty years later.

The Principles prescribed the type (4. Kinds of Entry), use (5. Use of Multiple Entries), and function of catalogue entries (6. Function of Different Kinds of Entry): main entry, added entry and references which headings carry identification and collocation functions. Yet, the Paris Principles do not prescribe one specific type of entry: an authority entry which function in the catalogue is to gather all information about various forms of names of authors and titles in order to control their use in the catalogue and to function as a guide to users.

Being rooted in the technology of the time was realistic: even the structure of the machine readable cataloguing format: MARC developed at the time followed the order and elements of a manually produced printed catalogue card. As Michael Gorman described it n the article published in 1978 entitled *The Anglo-American Cataloguing Rules, Second Edition*, and then again thirty years later, the MARC record "remains an automated version of a manual catalogue entry" (Gorman, 1978, p. 210; Gorman, 1998). In continuation to the mentioned remark, Gorman warned of the lack of full understanding of the nature and function of the authority file (i.e., a set of authority entry records) in an automated catalogue which had considerable impact on the formulation of cataloguing rules. He stated: "The crucial questions of 'levels of information' in bibliographic records and of the nature of authority file records in machine systems have only been raised; they have come nowhere near to being resolved. In short, *AACR 2* could not take the effect of library automation fully into account because those effects have yet to be completely assessed and understood" (Gorman, 1978, p. 210). The lack of international guidelines that would enable national cataloguing rules to prescribe the content of such authority entry record was stated also by Eva Verona in a comment to the temporary rule on the authority entry in her *Code and Manual for the Compilation of Alphabetical Catalogues* (Verona, 1983, p. 636).[2]

IFLA Standards for the Creation of Authority Entry Records

Verona and Gorman were voicing the common problem. In 1978 during the IFLA General Conference in Štrbské Pleso in the then Republic of Checkoslovakia, Working Group on an International Authority System was set up as a result of a project established by IFLA International Office for UBC and endorsed by the IFLA Sections on Cataloguing and Mechanization. According to its chairman Tom Delsey, following the decisions of the UBC programme made in 1974, and "considerable activity, both at the national and international level, in developing guidelines for catalogue headings and in compiling authority lists... [w]e are now at a point where planning for the efficient exchange of authority data between and among national bibliographic agencies is critical" (Delsey, 1980, p. 10). The Working Group's goal was "to establish principles for the creation of authority files and procedures to facilitate international exchange of authority information" to be accomplished through the following terms of reference:

- To discuss and formulate the specifications for an international authority system to satisfy the bibliographic needs of the libraries.

- To develop the UNIMARC format for the exchange of authority data.
- To develop the methods for the efficient and effective exchange of authority data (Delsey, 1980, p. 10).

The Working Group developed a general framework based on objectives and requirements for the work needed to meet the given goals. Four levels at which an international authority systems might operate were recognized: at *level 1*, the exchange of authority data in print or microprint form in a standard format, at *level 2*, the exchange of authority data in machine-readable form, at *level 3*, a system established to provide for the unique identification (for bibliographic purposes) of persons, corporate bodies, etc. by means of a numeric identification, and at *level 4*, a system designed for creating and exchanging authority data in a network. The prerequisite for such a system would be "a record format of sufficient sophistication to support the kind of dynamic, multi-directional data transfer that would be involved, and flexible enough to carry simultaneously data originating from different sources, formulated according to different standards" (Delsey, 1980, p 11).

Guidelines for Authority and Reference Entries (Guidelines, 1984) answered to the first term of reference, or level 1 of the framework. The *Guidelines* defined a set of elements to be included in three types of entries: authority entry, reference entry, and general explanatory entry, assigned an order to the elements, and specified a system of punctuation for the entry in print and micro-print form. Three types of headings were defined: headings for personal names, headings for corporate bodies, including conferences and territorial authorities, and uniform titles for anonymous classics. The guidelines, however, were confined to defining only the broad structure of the entry, and did not "prescribe the actual form of headings, references, notes, etc., nor [did] it prescribe punctuation that is internal to those elements" (Guidelines, 1984, p. x). The specifications pertaining to the form of heading should be made according to IFLA rules for structuring normalized forms of names, and according to national cataloguing rules.

National cataloguing rules published before the *Guidelines* were issued, as two examples mentioned above show, do not either address the creation of authority entries or lack the necessary framework for specifying their content, as in Verona's case. What was missing, even after the *Guidelines* were published, was the common agreement about the content of the authority record, and the definition of the relationships between the authority file and the bibliographic file within a given catalogue. The practice confirmed these needs too. That could be found in the description of first automated bibliographic systems with integrated authority file (Malinconico & Rizzolo, 1973; Buchinski & Newman & Dunn, 1976) and theoretical considerations about the role of a machine based authority file in such a system (Malinconico, 1975; Durance, 1978; Gorman, [1979] 1982; Taylor 1984, 1989), as well as in voicing the needs for authority control in union or cooperative international authority files in Canada (Clement, 1980), the USA (Avram, 1984) and Great Britain (Oddy, 1986). Pat Oddy particularly clearly envisioned[3] the function and use of authority files as part of "full-featured bibliographic systems", which Svenonus would argue for yet fifteen years later:

The on-line catalogue of the future will contain two kinds of records – perhaps they correspond to the mechanical and the intellectual aspects of cataloguing. The first file will consist of standardized descriptions of bibliographic items. These individual records will be linked, often in more than one way, to the second file, consisting of discrete authority records for personal and corporate names. These authority records will be linked to each other by references. For the first time catalogues will escape from the linear structure found in card, fiche, or conventional MARC record files. The authority records will

fulfil their traditional function of recording the authoritative form of name for an entity, but they'll also provide access to the bibliographic records and aid their manipulation. It's with these last two functions that we can come to the kernel of the new prospects which authority files can now offer" (Oddy, 1986, pp. 3-4).

Only in 1991 IFLA published *UNIMARC/ Authorities: Universal Format for Authorities* (UNIMARC/Authorities, 1991) as a result from a work started in 1984, immediately after the *Guidelines* were finished. The second objective defined in terms of reference for the Working Group on an International Authority System was thus fulfilled. The machine-readable format for authorities was complementary to UNIMARC format for bibliographic records which first edition was published already in 1977. It had needed fourteen years of effort of international experts to produce an instrument for international exchange of authority data, and for implementation of authority control in those systems that were using UNIMARC formats as the internal format of their catalogues or integrated library information systems. That shows the complexity of the task voiced in the stated requirements for the design of such a format of the level 2 of the framework. *UNIMARC/Authorities* is based on the *Guidelines* in defining the type and structure of entries, and the content designators for the same entities with addition of those for family name, territorial or geographic name, title (uniform and collective uniform title) for works by individual author and topical subjects. The format, just as in the case of *Guidelines* does not specify the content of authority and reference records but refers this problem to the application of IFLA standards (rules and lists of uniform headings) and national cataloguing rules.

It is important to point out here the mention in the format's foreword of the third objective defined in the terms of reference, i.e., development of methods for efficient and effective exchange of authority data, or level 3 of the framework.

Following the *Guidelines*, the format contains a field for an International Standard Authority Data Number (ISADN), but it is not defined: it is "reserved for the ISADN", and therefore giving no direction as to its use and structure that could have been taken from specifications in the *Guidelines*, elementary as they are. Namely, the stipulation of the *Guidelines* is ambiguous. It specifies the function and the use of the number in relation to the authority entry (record) and to the heading in the same rule. The general rule in 1.7 specifies that the ISADN "serves to identify the number assigned to the authority entry for purposes of international exchange and control", while the specific rule, 1.7.1.2, makes a condition: "if an ISADN has been assigned to the heading given in area 1 [uniform heading], the ISADN must be given". Such a provision should be born in mind because of the future developments relating to the identification concept in this field.

However, the creators of the format had to present an answer to the task given by the third term of reference or level 4 of the framework. What they did was visualize requirements the future development of an international authority system would put on the extension of the format. In the preface to the format, its editor Christine Bossmeyer says: "When a model for an international authority system is worked out by IFLA, target for data element requirements may be set out so that records exchanged internationally will have more consistency. Such a model may also indicate the need to add data elements to UNIMARC/Authorities in order to accommodate and facilitate exchange in a worldwide environment" (UNIMARC/Authorities, 1991, p. 8).

Summing up: Achievements of IFLA's Work in the Field of Authority Control: 1961-1991

In the period of thirty years, between Paris Principles adopted in 1961 and publication of *UNIMARC/Authorities* in 1991, IFLA, together with international experts, national bibliographic agencies and services succeeded in consolidating the ground for the efficient and economically founded creation and exchange of authority data. Theoretically, bibliographically significant conditions were identified, while cataloguing principles and standards were based on the concept of uniform heading. To enable the theory to function, concession was made by principles and implemented in cataloguing rules to meet the needs of local users and traditions in allowing the use of forms of names in the language of the country in which the library is located.

Variations in abiding to the rule of uniform heading related to linguistic differences were not the only ones that were recognized. In the commentary to Paris Principles (Statement, 1971) Verona exposed other problematic areas in establishing uniform headings in different cataloguing rules published since 1961. She highlighted the problem of inconsistency in the definition and use of the term "author", which according to Principles includes "personal author" and "corporate body", and commented that "differences resulting from these definitions are particularly noticeable in the rules which differentiate between publications to be entered under the name of a personal author and those to be entered under the name of a corporate body, in the rules for publications containing a work which has been produced under editorial direction, and in the rules for collections" (Statement, 1971, p. 24). Definition of the "author" and the realization of the authority control over author's names has become again a point of discussion related to decision to bibliographically control web resources.[4]

What was not solved yet, and what IFLA recognized as needing to focus on in the following period, was to design a model for an international authority system, to define international standard number for authority data as part of it, and to develop methods of putting the system into function globally. The period of almost twenty years since then has been devoted to that goal.

APPLICABILITY OF PRINCIPLES OF UNIVERSAL BIBLIOGRAPHIC CONTROL IN INTERNATIONAL NETWORKED ENVIRONMENT

In the article entitled *Authority Control in an International Context* (Delsey, 1989) published towards the end of the above period in 1989, Delsey summarizes IFLA's efforts in building a model for an international authority system. Analysing implementation of UNIMARC Authorities format by national bibliographic agencies from the aspect of bilateral exchange of authority data he states that "when one looks more closely at the complexities of authority control on an international scale, one soon realizes that an infrastructure more sophisticated than that supporting simple bilateral exchange is required" (Delsey, 1989, p. 23). Although referring to the context of the design of infrastructure for international authority control, Delsey has opened up the issue of insufficiency of the concept of "bilateral exchange" on which IFLA founded development of its standards. He further assumes that

[w]ithout a central database to serve as a register of authority entries and their corresponding standard numbers it would be next to impossible for national agencies to established that vital link between the entries in their own authority files and the standard number that will identify the variant forms of heading that have been used by other agencies and which will be circulating as headings in bibliographic records emanating from perhaps dozens of different national agencies. ... [A]ll such processing would be dependent on the key function of registering standard numbers having been carried out in some sort of centralized fashion (Delsey, 1989, pp. 25-26).

Delsey would return to this topic fifteen years later with basically the same arguments yet which take into consideration the contemporary informa-

tion infrastructure (Delsey, 2004), showing that in 1989 he delineated problems that are present still today. He concludes his 1989 article with the statement that although the standard authority entry and the adequate machine readable format "might eventually be expected to function", there was still an enormous amount of work to be done on the technical aspects, administrative complexities and cost-benefits of various schemes for the implementation of an international authority system for authority control. It was likely, he states, "to be non-technical issues, in fact, that ultimately determine the extent to which the concept of Universal Bibliographic Control is realized as an international system for authority control" (Delsey, 1989, p 27).

Apart from different linguistic forms that Paris Principles explicitly made concession to, there were other bibliographic conditions that cause variations in personal and corporate body names, and which Verona enumerated in the commentary to Paris Principles (Statement, 1971, pp. 9-11). Another aspect of the problem were the mentioned theoretical and implementation issues related to the linked bibliographic and authority files within an online library system elaborated by Gorman, Malinconico, Oddy, Taylor and others.

There were still other specific questions that needed answers. Some of these were: Is there a change of the functions of the catalogue due to the change of technology?, i.e., the functions of the catalogue with integrated bibliographic and authority files; Is the UBC concept still relevant in the context of internationally exchange of authority files, and if not, which other concept can replace it?; What are the methods or procedures in establishing authority entry records, particularly as to their content; What is a nature of relationship between authority records within an automated authority file?; What are implications of reference structures to the authority control in an online catalogue?; Do online retrieval techniques influence the content of authority entry records?

IFLA's Studies on the International Exchange of Authority Data

Following the publication of the *UNIMARC/ Authorities format* in 1991, two important studies were issued under the auspices of IFLA UBC International MARC Programme. IFLA's concern was to pinpoint the reasons why the effective international exchange of authority data was hindered, while the exchange of bibliographic data was flourishing. The studies commissioned by IFLA were *Management and Use of Authority Files* by Marcelle Beaudiquez and Françoise Bourdon (Beaudiquez & Bourdon, 1991), and *International Cooperation in the Field of Authority Data* by Françoise Bourdon (Bourdon, 1993).

In the first study, Beaudiquez and Bourdon surveyed 15 automated authority files containing names of persons, corporate bodies and uniform titles for anonymous classics. The report suggested that the failure of international cooperation in the field of authority data was due to the fact that the division of labour defined by UBC had been tacitly ignored, and that there was a lack of precision in the terminology in the field of authority data. The aim of the second study, *International Cooperation in the Field of Authority Data: An Analytical Study with Recommendations*, was to "identify the current obstacles to the international exchange of authority data, whether in manual or automated form, and to submit recommendations to IFLA to be ratified and thus to contribute to international cooperation in this area" (Bourdon, 1991, p. 6).

In this seminal work, Bourdon first presents analyses of international standards and concludes that they showed gaps in regard to the definition of the typical content of authority records that were intended to be re-useable outside the context in which they were created. That failure resulted from a lack of clarity in the aims of international standardization, and the lack of identification of different functions of an authority file in an automated environment. In this respect, Bourdon identifies two types of authority files according

to the function they perform. *Management name authority files* are those that are designed simply to ensure the formal management of name headings to a given catalogue, and therefore can be used effectively with constant reference to the catalogue on which they are dependent. Such files are not suitable for the international exchange. The opposite is true for the second type of authority files, which Bourdon calls *identification name authority files*. They are characterized by unambiguous identification of the entity which is the content of the authority record through additions to headings and different types of notes (an addition to the management function), which makes this type of authority file independent from the catalogue in which it was created and therefore applicable for re-use in a new context (Bourdon, 1991, pp. 31-67). The majority of files of national libraries Bourdon analysed were of the first kind. The explanation for such a situation she finds in the development of shared cataloguing and the need for re-used authority data which "all created new functions for name authority files at the national and international level which were not at all among the objectives assigned to them by the national bibliographic agencies when they created them" (Bourdon, 1991, p. 67).

The means for unique and unambiguous identification of the content, or "subject" of the authority record, Bourdon emphatically finds in the design of the International Standards Authority Data Number (ISADN). She argues that the number should not be attributed to the authority form (uniform heading) but "to the whole of the identification authority record drawn up by the national bibliographic agency which is responsible for the author in question" (Bourdon, 1991, p. 80). She explicitly states that she thus takes further the suggestions of the IFLA Working Group on an International Authority System that the "number would serve to identify the *object* of the authority entry" (Delsey, 1989, p. 24), or indeed the follow up of the ambiguous treatment of the number's function as the identifier of *authority entry* (record) and a *heading* in the *Guidelines for Authority and Reference Entries* (stipulations 1.7 and 1.7.1.2).

Bourdon's profound understanding of the problem from theoretical and practical point of view resulted in a proposal which, however, got blurred along the experts' search for a solution since then, but could be seen today reaffirmed in the research of implementations of semantic web tools to (bibliographic) services. The idea to attribute a number (e.g., ISADN) to the identification authority record containing the authorized form of name of an entity with additional data elements to disambiguate it from other entities within the authority file of a national bibliographic agency, is a commonality today where a Uniform Resource Identifier (URI) identifies an entity and its attributes within a given namespace.

Towards a New Paradigm: Redefinition of the UBC Concept: From Bilateral Exchange to Global Sharing

Following the publication of *UNIMARC/Authorities* and the two studies by Beaudiquez and Bourdon, IFLA UBCIM organized a series of meetings, workshops and conferences with the strongly expressed goal to not only promote the programme of international exchange of authority data, but to see to its practical implementation (Bourne, 1992; Plassard & McLean Brooking, 1993; Willer, 1994). Related to the latter goal, one should also mention discussions held within IFLA Standing Committee on Cataloguing between 1991 and 1993 that were published in the notes form under the title *Problems and Prospects of Linking Various Single-language and/or Multi-language Name Authority Files* (Murtomaa & Greig, 1994).

This period of IFLA's intensive involvement in the international authority control is characterized by the work of two subsequently established working groups which findings and recommendations directly influenced the revision of *Guidelines* and

UNIMARC/Authorities, but, more importantly, promoted certain ideas and concepts which were considered the answer to the changed technological environment.

The Working Group on Minimal Level Authority Records and ISADN (WG on MLAR and ISADN) created under the auspices of IFLA UBCIM in 1996 analysed nine national authority formats and UNIMARC, and made a list of *Mandatory Data Elements for Internationally Shared Resource Authority Records*. In the introduction and recommendations section of its report the Group states that it "has come to realize that the IFLA goal of Universal Bibliographic Control by way of requiring everyone to use the same form for headings globally is not practical". It further recognizes "the importance of allowing the preservation of national or rule-based differences in authorized forms for headings to be used in national bibliographies and library catalogues that best meet the language and cultural needs of the particular institution's users" (Mandatory data elements, 1998, p. 1). The term "sharing" used in the title instead of "exchange" is of the importance. In order to "facilitate international sharing of authority data" the Group also proposes that IFLANET hosts libraries' authority files as a kind of virtual shared resource authority file. As to its second mandate, to resolve the issue of the identifier: ISADN, the Group refers this to some future time for review, recommending co-operation with publishing and archival[5] communities, and "waiting to see how the emerging international electronic environment and advances in developing technologies impact the linking of records" (Mandatory data elements, 1998, p. 2).

The first recommendation to relieve the UBC concept of the "uniform heading" was realized in the second edition of the *Guidelines for authority records and references* published in 2001 (Guidelines, 2001) by the introduction of the concept "authorized heading". This term, which now replaces "uniform heading", is defined as "the uniform controlled heading for an entity" (Guidelines,

2001, p. 2). The explanation of the reason for the change with reference to the WG on MLAR and ISADN's recommendations is described in detail in the introduction (Guidelines, 2001, pp. ix-x). The UBC concept of requiring the use of the same form for headings globally is considered "not practical and [...] no longer necessary", because "with computer capabilities developing more sophistication, we can link the authority records created in one country according to one set of cataloguing rules with those in another country to facilitate sharing of authority records and potentially to enable computer-assisted switching to display authorized forms" (Guidelines, 2001, p. ix).The term "access point" as a means under which a bibliographic or authority record can be searched in online catalogues, is also introduced but only in the definition chapter (Guidelines, 2001, p. 2), as it does not yet have a function within the *Guidelines*. Furthermore, a list of mandatory data elements which contained recommendations by the WG on MLAR and ISADN to the Permanent UNIMARC Committee for additions to the UNIMARC Authorities format was duly analysed and incorporated into the second edition of the format and its manual (UNIMARC Manual: Authorities format, 2001; Willer, 2004).

The Working Group on Functional Requirements And Numbering of Authority Records (FRANAR) was established in 1999 under the auspices of IFLA Division of Bibliographic Control and UBCIM as a continuation of the WG on MLAR and ISADN. The Group's first goal was to design a conceptual model for authority data that would correspond to the model for bibliographic data published in the study *Functional Requirements for Bibliographic Records* (Functional requirements, 1998) (FRBR). The FRBR is based on the entity relationship model, so the one for authority data should built on it in defining entities, attributes and relationships based on user tasks performed in relation to authority data. The goal was met in a publication that needed ten years to be finalized – *Functional Requirements for Authority Data:*

A Conceptual Model (Functional requirements, 2009) (FRAD). The time that took the Group to finalize the model shows the complexity of the problem they had to deal with. More recent description of the Group's background, its liaison activities and the working model at the time named FRANAR can be found in the article by its chairman Glenn E. Patton (Patton, 2004), while the model's impact on the future development of UNIMARC Authorities format is described in detail by Willer (Willer, 2007).

In the mentioned article Patton addresses also the Group's second term of reference, which was "to study the feasibility of an ... ISADN, to define possible use and users, to determine for what types of authority records such an ISADN is necessary, to examine the possible structure of the number and the type of management that would be necessary" (Patton, 2004, p. 93). Patton gives a short overview of the Group's dealing with the number, referring to Bourdon's proposal from 1991, recommendation of the WG on MLAR and ISADN, liaison with publishers (<in*d*ecs> and its successor project InterPARTY) and archives community, and with representatives of ISO/TC46/SC9 working groups that were developing standard numbers such as ISAN, ISWC, ISRC, etc. All this activity focused on one single question "'what exactly were we attempting to number?' Was it the entity regardless of the form of the heading used for that entity? Was it each different authorized heading for that entity? Was it the authority record itself to which the number implied?" (Patton, 2004, p. 95). The Group therefore decided to prepare a separate document which was eventually published in September 2008 under the title *A Review of the Feasibility of an International Standard Authority Data Number (ISADN)* (Tillett, 2008). The review prepared for the Group by Barbara B. Tillett was approved by the Standing Committee of the IFLA Cataloguing Section, thus giving the official IFLA's approval of its first recommendation which runs: "IFLA should not pursue the idea of an International Standard Authority Data

Number (ISADN) as it has been defined" (Tillett, 2008, p. 1). ISADN was thus put to death. Three following recommendations outline directions which could be taken at this time of technological development and state of the art in national and international endeavors in the field of name authority control. These are:

- IFLA should continue to monitor the progress of efforts of the ISO 27729 ISNI Working Group and the VIAF Project and any potential numbering that may result from those efforts. IFLA member institutions should also actively seek to influence the ISNI with a view to identifying common purposes with other communities.
- IFLA should continue to encourage the testing of various models to enable global sharing of authority information.
- IFLA should encourage the use of authority information in presenting improved catalog interfaces (Tillett, 2008, p. 1).

Parallel to the Group's work on the conceptual model for functional requirements for authority data, another group of IFLA's experts started the revision of cataloguing principles, the Paris Principles of 1961. In a series of international consultations and discussions within the IFLA Meetings of Experts on an International Cataloguing Code that started in 2003 and finished in 2007 (IFLA Meetings of Experts, 2009) the new *Statement of International Cataloguing Principles* (Statement, 2009) was adopted. The statement takes into account new technological environment: "online library catalogues and beyond", all types of materials and all aspects of bibliographic and authority data. It also includes "not only principles and objectives (i.e., functions of the catalogue), but also guiding rules that should be included in cataloguing codes internationally, as well as guidance on search and retrieval capabilities" (Statement, 2009, p.1). The statement builds on two corresponding conceptual models for functional

requirements for bibliographic and authority data: FRBR and FRAD, and user tasks defined in FRBR as critically analysed and extended by Svenonius (Svenonius, 2000, pp. 15-20). Namely, to meet the objectives of a "full-featured bibliographic system", Svenonius modified the original FRBR user tasks "to provide model independence, continuation with tradition, and navigation objective" (Svenonius, 2000, p. 20).[6] The Statement builds user tasks into the objectives and functions of the catalogue which run as follows:

The catalogue should be an effective and efficient instrument that enables a user:

4.1. to **find** bibliographic resources [...]:
    4.1.1.   to **find** a single resource,
    4.1.2.   to **find** sets of resources [...];
4.2. to **identify** a bibliographic resource or agent;
4.3. to **select** a bibliographic resource that is appropriate to the user's needs;
4.4. to **acquire** or **obtain** access to an item described [...] or to access, acquire, or obtain authority data or bibliographic date;
4.5. to **navigate** within a catalogue and beyond (Statement, 2009, p. 3-4).

Thus, Svenonius's plea for "continuation with tradition" has been met with in the first, finding objective being elaborated to fulfil identification (find a single resource), as well as collocation (find sets of resources) function of the catalogue as stated in the Paris Principles, and built into the cataloguing rules that were developed since.

In regard to the "uniform heading" concept and concession to different linguistic forms of the Paris Principles, the new statement takes the same stand. The name should be given in the original language and script (6.3.2.1), "6.3.2.1.1. but if the original language or script is not normally used in the catalogue.... [it] may be based on forms found on manifestations or in reference sources in one of the languages or scripts best suited to the users of the catalogue" (Statement, 2009, p. 5). The concession of the Paris Principles which principal aim was, indeed, the concept of uniform heading built into the UBC programme, has now been levelled up in the name of the "first principle" of the statement which is "to serve the convenience of catalogue users" (Statement, 2009, p. 1). The practice (or, perhaps, "best practice") has found its way into objectives (i.e., principles) of a (bibliographic) system. To describe this new situation, the creators of the statement had to establish a new suitable terminology. The partial abandoning of the term "uniform heading" for the term "authorized heading" was done already in the *Guidelines for Authority Records and References*. The term "access point", considered to be more appropriate in the online catalogue environment, gained eventually over "heading" of the card catalogue. The Statement having to deal with the fact that any bibliographic or authority data can be made an access point in an online environment,[7] defined two new terms: controlled and uncontrolled access point. The first named access points are recorded in the authority record, while the latter are defined in a negative sense: "not controlled by an authority record" (Statement, 2009, p. 13). According to this negative definition, anything that falls within the range of authority record is considered controlled. For example, a variant form of name, i.e. the form that was not chosen by a bibliographic agency as an authorized one, still makes a controlled access point. The name of a publisher, on the other hand, which one bibliographic agency does not have a need to control, can be retrieved in its uncontrolled form from a bibliographic record file, while in the case of another agency that controls publisher/printer's name for its antiquarian material, would be retrieved as a controlled access point from the authority data file.

To understand the complexity of the use of terms related to the name authority control, one has to revert to the FRAD conceptual model. The fundamental basis for the conceptual model constitutes of three types of entities and their relationships. That basis for the model was developed at quite a late stage of the Group's work

in a need to abstract even further from a fully developed model being a response to worldwide comments asking for clear concepts and definitions. The three basic entities are bibliographic entities, names and/or identifiers and controlled access points. The entities are described with their relationships as follows:

Entities in the bibliographic universe (such as those identified in the Functional Requirements for Bibliographic Records) are known by names and/or identifiers. In the cataloguing process (whether it happens in libraries, museums, or archives), those names and identifiers are used as the basis for constructing controlled access points (Functional requirements, 2009, p. 3).

The relationships–*known by* and [used as] *basis for*–are bidirectional and multiple. Such a definition is important as it is necessary to express all the complexities of the bibliographic universe, together with legalizing practices in "serving the convenience of catalogue users". For example, a bibliographic entity person is known by the name Lewis Carroll which is used as a basis for controlled access point following rules by one bibliographic agency. That same person is also known by the name Charles L. Dodgson which is used as basis for another controlled access point by the same rules. It could happen, though, that a certain bibliographic agency following different (national) cataloguing rules would recognize the second name as a separate name of the entity (i.e., a variant name form), but not as belonging to the separate bibliographic entity person, (i.e., other biographic identity). Such a situation would result in two different online catalogues as follows: in the first instance/catalogue, a user searching for works by Lewis Carroll would get (collocated) all the works under that name of the bibliographic entity person (in fact a pseudonym of Dodgson, and in the model considered as *persona* adopted by the individual); the same would be the case when searching under the name Dodgson. (There would be a *see also* reference or a link between these two access points for the real name and the pseudonym, however, this is not the point here.) In the second instance/catalogue, a user searching for works by the author named Lewis Carroll would get (collocated) all the works published under that name, but also the works written under the name of Dodgson, as they are considered to belong to the same bibliographic entity. That is, the rules that govern the latter access points in the catalogue, do not allow the concept of separate biographic identities of a person.

Thus, in the first instance, both forms of names would be *authorized forms of names* (by specific rules) and chosen for the *authorized access point*, while in the second instance only one form of name (of bibliographic entity person, i.e. biographic identity) would have been chosen. In the latter case, it would be necessary to decide which one is the *preferred name*: preferred by the user community which particular bibliographic agency serves. In the given example, the *preferred name* is Lewis Carroll, and that particular form will be *authorized form of name* (by the rules) and chosen for *authorized access point*. The second name, Dodgson, is considered a non preferred name, and is chosen as a *variant form of name* which can be used to access the authority record or as a link to authorized access point. The example uses terminology defined in the Statement of Principless, which when read sequentially is rather opaque. It should be also noted that a term *variant access point* is missing from the Glossary, and that it would be useful to define it in a positive sense. Namely, *variant form of name* is defined as "a form of name not chosen as the authorized access point" (Statement, 2009, p.13), thus blurring the entities that FRAD keeps specifically apart: the name and the controlled access point. Another argument for introducing the term would mean recognition of the function of a *variant access point* in an online catalogue (*see* reference structure).

Summing up: Further Achievements of IFLA's Work in the Field of Name Authority Control: Since 1991 to the Present

The period of the last twenty years saw intensive work in the field of authority control under the auspices of IFLA, equal to the one of the previous period. That work could be characterized by theory informed by practice, and practice informed by theory in its fullest sense.

The answer to the basic question "is there a change in the functions of the catalogue due to the change of technology?" can be found in IFLA's new *Statement of International Cataloguing Principles* which accepted Svenonius's emphatic intervention on formalizing the objectives (functions) of the catalogue. Svenonius insisted on expanding finding function of FRBR user task in order that the collocation function would be adequately dealt with. Thus, the answer to the question "is there a change in the functions of the catalogue?" is negative. What is different is the expanded and complex bibliographic universe, to which some of the tools should be adapted and some should evolve into the new ones.

IFLA has answered to these challenges by mobilizing a wide range of international experts, whose theoretical knowledge, research and practical experience built a consensus about common bibliographic tools. The milestone of these efforts was IFLA's study *Functional Requirements for Bibliographic Records* (FRBR) which impact crossed library community boarders.[8] The study is based on the entity-relationship model, in which the entities are defined as "the key objects of interest to users of bibliographic records" (Functional Requirements for Bibliographic Records, 1998, p. 4). Elements of a bibliographic record were thus put under scrutiny of their relevance for the user of the catalogue. In deciding to adopt the new type of modelling bibliographic data, IFLA's study has broken with the long tradition of linear representation and, indeed, linear structuring of data. The objections to such a structure and proposals for a disaggregated record structure were made already in the 1970ies (Durance, Buchinski, Gorman, Malinconico) and also later (Willer, 1999) in relation to designing MARC formats, and implementing authority control in automated library catalogues.

The tools in the field of name authority control which answered to the changed environment are *Guidelines for Authority Records and References, UNIMARC Manual: Authorities Format, Functional Requirements for Authority Data: A Conceptual Model* and *Statement of International Cataloguing Principles*. The FRAD conceptual model and the Statement of Principles have provided theoretical basis for harnessing the new bibliographic universe by reaffirming the concepts of authority control and fixing terminology.

What has been left yet unresolved is the task delineated in the late 1970ies – the international system for authority control. Summarizing discussions, reviews, proposals and practical achievements, three basic facts can inform future work: ISADN as an idea and key functioning element in an international authority system has been abandoned, i.e. there is no need to pursue it further; Bourdon's concept of numbering authority record, i.e. the need for the creation of *identification name authority record* by bibliographic agencies; and IFLA's monitoring research and practice, i.e., testing of various models to enable global sharing and re-use of authority data.

The new, third edition of *UNIMARC Manual: Authorities Format* (UNIMARC Manual, 2009) can be considered a prompt reaction to, as well as the implementation of these directions for future work. It incorporates both new concepts and the methodology developed by the FRAD conceptual model, as well as concepts and terminology of the Statement of Principles thus enabling the user of the format to build a comprehensive identification name authority record (Willer, 2009). As to the ISADN, the format makes field 015 International

Standard Authority Number obsolete, and introduces a new one": 003 Persistent Record Identifier. The field is defined as the one that "contains the persistent identifier of the record assigned by the agency which creates, uses or issues the record", explicating that it is "the persistent identifier for the authority record, not for the described entity itself". Referencing the record through a persistent identifier by "using the address which is displayed in the browser during an online session" (UNIMARC Manual, 2009), it is possible to put the record, but, more importantly, record's content to new use in new services.[9]

FUTURE TRENDS: SEMANTIC WEB AND "CONTINUATION WITH TRADITION"

At this point of time, one can state that IFLA, as the international body with the responsibility for the development and promotion of the programme of Universal Bibliographic Control globally, has made the major effort to meeting that goal. Based on the developed frameworks, conceptual models, standards and guidelines in the field of authority control, two main trends can be seen in their implementation.

The first one refers to the design of an international system for authority control, the one Delsey envisaged in 1989 as needing to be based on "an infrastructure more sophisticated than that supporting bilateral exchange", and which he revisited in 2004. Such a system could be identified with the development of the VIAF Project. VIAF: the Virtual International Authority File started in 2002 by the OCLC, the Library of Congress and the Deutsche Nationalbibliothek as "a proof of concept project to test the centralized union authority file model using OAI protocols" (Tillett, 2004, p. 38). The project was to test the viability of programmatically linking personal names in authority files from different national authority files regardless of format or cataloging rules used for the creation of such data. The project was joined subsequently by the Bibliothèque nationale de France, while in September of 2008, Barbara B. Tillett on behalf of the VIAF partners issued invitations to over 20 potential participants "signaling the official start of a full-scale production of the VIAF" (Lupe Cristán, 2008, p. 2).

The second trend can be seen in the visualization of "disaggregation" of what we now deal with as bibliographic and authority records in the semantic web environment, by using bibliographic infrastructure tools which are now in the development phase. Independently from IFLA work, but being fully aware of it, and working in "continuation with tradition", Gordon Dunsire recently published an article entitled *The Semantic Web and Expert Metadata: Pull Apart then Bring Together* (Dunsire, 2009). Dunsire researches the ways how to encode bibliographic and authority data – expert metadata to be understood by the machines, enabling their transformation into new uses and services.

Dunsire takes for an example a sentence "This presentation" - "has creator" - "Gordon Dunsire" expressed as an RDA triple (statement in the form of subject-predicate-object expressions), which, to be fully machine-processed requires three URIs embedded in an RDA/XML syntax. He further explains the process in the following way:

In RDF, the things requiring identification or URIs are the specific classes, properties, and instances associated with RDF subjects, predicates, and objects. In the example triple given above, the subject "This presentation" has an electronic location given by the URL http://cdlr.strath.ac.uk/ pubs/dunsireg/AKM2008.pps. The predicate "has creator" uses a property "creator" already defined in the Dublin Core metadata format with a URI http://purl.org/dc/terms/creator. And the object instance "Gordon Dunsire" has an entry in the Library of Congress Name Authority File which has been made available on the Web by OCLC

with the URI http://errol.oclc.org/laf/nb2001-72552.html.[10]

Both the predicate and object instances "creator" and "Gordon Dunsire" have URIs because they are entries in vocabularies which have been made available as "namespaces" in Semantic Web applications (Dunsire, 2009, p. 3).

Based on this technology, and having at disposal available bibliographic and authority records on the web, Dunsire's disaggregated bibliographic record of an FRBR model will have the following future:

So the original catalogue card, with explicit local content and implicit structure, has evolved into a multi-record aggregation with explicit structure and distributed global content shared amongst many such "records". If this is a truly different species, then the traditional library record based on the catalogue card has become extinct. ... The Semantic Web will allow machines to create a metadata record for a particular resource just-in-time- and on-the-fly, rather than have static records just-in-case. The benefits of metadata creation and maintenance by information professionals will be available to all (Dunsire, 2009, p. 11).

CONCLUSION

The turbulence of the past twenty years expressed in the need to chart unknown territories of the newly emerging bibliographic universe seems to be stabilized. What that period proved was that the breakthrough in theory, research and, indeed, practice owned its strength in building on the "continuation with tradition" (Svenonius). Therefore, it is expected that development in the field of authority control, and wider, in the field of Universal Bibliographic Control as such, will go in the direction of bringing newly developed conceptual models, bibliographic standards, and content of national and international catalogues into a semantic web environment.

The direction for the future work has been shown: obviously, it is not so direct and easily attained, but it is built on sound foundations. However, it could happen that real life forces which often cannot be controlled could impede (again) the realization of the design of an information system that aims at the ideal. It is now up to us, library professionals to visualize and realize new services for our users.

REFERENCES

Anderson, D. (1974). *Universal Bibliographic Control: A long term policy, a plan for action.* München, Germany: Verlag Dokumentation.

Avram, H. D. (1984). Authority control and its place. *Journal of Academic Librarianship, 6*(9), 331–335.

Beaudiquez, M., & Bourdon, F. (1991). *Management and use of name authority files (personal names, corporate bodies and uniform titles): evaluation and prospects.* München, Germany: Saur.

Bourdon, F. (1993). *International cooperation in the field of authority data: an analytical study with recommendations.* München, Germany: Saur.

Bourne, R. (Ed.). (1992). *Seminar on Bibliographic Records: proceedings of the Seminar held in Stockholm, 15-16 August 1990.* München, Germany: Saur.

Buchinski, E. J., Newman, W. L., & Dunn, M. J. (1976). The automated authority subsystem at the National Library of Canada. *Journal of Library Automation, 4*(9), 279–298.

Clement, E. A. H. (1980). The automated authority file at the National Library of Canada. *International Cataloguing, 4*(9), 45–48.

Delsey, T. (1980). IFLA Working Group on an International Authority System: A progress report. *International Cataloguing, 9*(January/March), 10–12.

Delsey, T. (1989). Authority control in an international context. *Cataloging & Classification Quarterly, 3*(9), 13–28. doi:10.1300/J104v09n03_02

Delsey, T. (2004). Authority records in a networked environment. *International Cataloguing and Bibliographic Control, 4*(33), 71–74.

Dunsire, G. (2009). *The Semantic Web and expert metadata: Pull apart then bring together: Presented at Archives, Libraries, Museums 12 (AKM12), Poreč, Croatia, 2008.* Retrieved March 25, 2009, from http://cdlr.strath.ac.uk/pubs/dunsireg/akm2008semanticweb.pdf

Durance, C. J. (1978). "What's in the name?" Summary of the workshop. In *What's in a name? Control of catalogue records through automated authority files* (pp. 225–227). Vancouver, Toronto: University of Toronto Library Automation System.

Functional requirements for authority data: A conceptual model. (2009). G. Patton, (Ed.). IFLA Working Group on Functional Requirements and Numbering of Authority Records (FRANAR). Final Report, December 2008. Approved by the Standing Committees of the IFLA Cataloguing Section and IFLA Classification and Indexing Section, March 2009. München, Germany: Saur.

Functional requirements for bibliographic records: Final report. (1998). IFLA Study Group on the Functional Requirements for Bibliographic Records. München, Germany: Saur. Retrieved March 25, 2009, from http://www.ifla.org/VII/s13/frbr/frbr.pdf

Gorman, M. (1978). The Anglo-American cataloguing rules, second edition. *Library Resources & Technical Services, 3*(22), 209–225.

Gorman, M. ([1979] 1982). Authority control in the prospective catalog. In M. W. Ghikas (Ed.), *Authority control: The key to tomorrow's catalog: Proceedings of the 1979 Library and Information Technology Institutes* (pp. 166-180). Phoenix, AZ: Oryx Press.

Gorman, M. (1998). The future of cataloguing and cataloguers. *International Cataloguing and Bibliographic Control, 27*(4), 68–71.

Guidelines for authority and reference entries (1984). Recommended by the Working Group on an International Authority System, approved by the Standing Committees of the IFLA Section on Cataloguing and the IFLA Section on Information Technology. London: IFLA International Programme for UBC.

Guidelines for authority records and references (2001). *Revised by the Working Group on GARE (Rev., 2nd ed.). München, Germany: Saur.*

Gull, C. D. (1969). The impact of electronics upon cataloguing rules. *International Conference on Cataloguing Principles, Paris, 9th-18th October, 1961 (1969). Report* (pp. 281-290). London: Clive Bingley.

IFLA Cataloguing Section FRBR Working Group. (n.d.). Retrieved March 25, 2009, from http://www.ifla.org/VII/s13/wgfrbr/

IFLA Meetings of Experts on an International Cataloguing Code. (2009). Retrieved March 25, 2009, from http://www.ifla.org/VII/s13/icc/

International Conference on Cataloguing Principles, Paris, 9th-18th October, 1961. (1963). *Report*. London: International Federation of Library Associations. Reprinted in: International Conference on Cataloguing Principles, Paris, 9th-18th October, 1961 (1969). *Report*. London: Clive Bingley.

International Conference on National Bibliographies, Paris, 1977 (1978). *Final report*. Paris: UNESCO.

International Meeting of Cataloguing Experts, Copenhagen, 1969: Report (1969). *IFLA Annual*. Also published in: Report of the International Meeting of Cataloguing Experts, Copenhagen, 1969 (1970). *Libri, 1*(20), 105-132.

International standard archival authority record for corporate bodies, persons and families: ISAAR(CPF). (1996). Ottawa, Canada: Secretariat of the ICA Commission on Descriptive Standards.

Lubetzky, S. (1979). The traditional ideals of cataloging and the new revision . In Freedman, M. J., & Malinconico, S. M. (Eds.), *The nature and future of the catalog: proceedings of the ALA's Information Science and Automation Division's 1975 and 1977 Institute on Cataloging* (pp. 153–169). Phoenix, AZ: Oryx Press.

Lupe Cristán, A. (2008). Virtual International Authority File: Update 2008. *SCATNews: Newsletter of the Standing Committee of the IFLA Cataloguing Section, 30*, 2–3.

Malinconico, S. M. (1975). The role of a machine based authority file in an automated bibliographic system. In *Automation in Libraries: Papers presented at the CACUL Workshop on Library Automation, Winnipeg, June 22-23, 1974*. Ottawa, Canada: Canadian Library Association. Cited according to: Carpenter, M. & Svenonius, E. (Eds). (1985). Foundations in cataloging: A sourcebook (pp. 211-233). Littleton, Co.: Libraries Unlimited.

Malinconico, S. M., & Rizzolo, J. A. (1973). The New York Public Library automated book catalog subsystem. *Journal of Library Automation, 1*(6), 3–36.

Mandatory data elements for internationally shared resource authority records: Report of the IFLA UBCIM Working Group on Minimal Level Authority Records and ISADN. *(1998). Frunkfurt/ Main, Germany: IFLA Universal Bibliographic Control and International MARC Programme. RetrievedMarch25, 2009, from*http://www.ifla. org/VI/3/p1996-2/mlar.htm

Manual, U. N. I. M. A. R. C. *Authorities Format.* (2001). (2ⁿᵈ rev. & enlarged ed.). München, Germany: Saur.

Manual, U. N. I. M. A. R. C. *Authorities Format.* (2009). (3ʳᵈ ed., ed. by Mirna Willer). München, Germany: Saur.

Murtomaa, E., & Greig, E. (Eds.). (1994). Problems and prospects of linking various single-language and/or multi-language name authority files: Notes. *International Cataloguing and Bibliographic Control, 3*(23), 55- 58.

Oddy, P. (1986). Name authority files. *Catalogue & Index, 82*(Autumn), 1 & 3-4.

Patton, G. (2004). FRANAR: A conceptual model for authority data. In Taylor, A.G. & Tillett, B. B. (Eds), *Authority control in organizing and accessing information: Definition and international experience* (pp. 91-104). Binghamton, NY: Haworth Information Press. DOI 10.1300/J104v38n03_09

Plassard, M.-F., & McLean Brooking, D. (Eds.). (1993). *UNIMARC/CCF: proceedings of the Workshop held in Florence, 5-7 June 1991*. München, Germany: Saur.

Report of the International Meeting of Cataloguing Experts, Copenhagen, 1969 (1970). *Libri, 1*(20), 105-132.

Statement of international cataloguing principles. (2009, February). Retrieved March 25, 2009, from http://www.ifla.org/VII/s13/icp/

Statement of Principles adopted at the International Conference on Cataloguing Principles Paris, October, 1961. (1971). Annotated ed. with commentary and examples by Eva Verona assisted by Franz Georg Kaltwasser, P.R. Lewis, Roger Pierrot. London: IFLA Committee on Cataloguing.

Svenonius, E. (2000). *The intellectual foundation of information organization.* Cambridge, MA: The MIT Press.

Taylor, A. G. (1984). Authority files in online catalogs: an investigation of their value. *Cataloging & Classification Quarterly, 3*(4), 1–17. doi:10.1300/J104v04n03_01

Taylor, A. G. (1989). Research and theoretical considerations in authority control. *Cataloging & Classification Quarterly, 3*(9), 29–56. doi:10.1300/J104v09n03_03

Tillett, B. B. (2004). Authority Control: State of the Art and New Perspectives. In A.G. Taylor & B. B. Tillett, (Eds.), *Authority control in organizing and accessing information: Definition and international experience* (pp. 23-57). Binghamton, NY: Haworth Information Press. DOI 10.1300/ J104v38n03_04

Tillett, B. B. (2008). *A Review of the Feasibility of an International Standard Authority Data Number (ISADN)*. Prepared for the IFLA Working Group on Functional Requirements and Numbering of Authority Records, edited by Glenn E. Patton, 1 July 2008. Approved by the Standing Committee of the IFLA Cataloguing Section, 15 September 2008. Retrieved March 25, 2009, from http:// www.ifla.org/VII/d4/franar-numbering-paper.pdf

UNIMARC/Authorities. *Universal format for authorities.* (1991). Recommended by the IFLA Steering Group on a UNIMARC Format for Authorities, approved by the Standing Committees of the IFLA Section on Cataloguing and Information Technology. München: Saur.

Verona, E. (1983). *Pravilnik i priručnik za izradbu abecednih kataloga. 2. dio: Katalozni opis.* Zagreb, Croatia: Hrvatsko bibliotekarsko društvo.

VIAF. *Virtual International Authority File.* (n.d.). Retrieved June 16, 2009, from http://www.oclc.org/research/projects/viaf/

Vrbanc, T. (2007). Choice of author headings for finite internet resources. In M. Willer & A. Barbarić (Eds.), *Međunarodni skup u čast 100-te godišnjice rođenja Eve Verona, Zagreb, 17.-18. studenoga 2005. = International Conference in Honour of the 100th Anniversary of Eva Verona's Birth, Zagreb, November 17-18, 2005* (pp. 459-475). Zagreb, Croatia: Hrvatsko knjižničarsko društvo = Croatian Library Association.

Willer, M. (1994). UNIMARC/Authorities: A new tool towards standardization. In M.-F. Plassard & M. Holdt, (Eds.), *UNIMARC and CDS/ISIS: Proceedings of the Workshops Held in Budapest, 21-22 June 1993 and Barcelona, 26 August 1993,* (pp. 19-36). München, Germany: Saur. The paper was also presented at the following seminars organized by IFLA UBCIMP Office: *UBC/UNIMARC: A seminar on Universal bibliographic Control and UNIMARC*, Lietuvos Nacionalnie Martyno Mazvydo Biblioteka, Vilnius, 2- 4 June 1994, Lithuania; *UNIMARC Workshop "Crimea '95",* Jevpatoria, Ukraine, 12-16 June 1995; *IFLA International Seminar Authority Files: Their Creation and Use in Cataloguing,* St. Petersburg, October 3-7, 1995. For reports see *International Cataloguing and Bibliographic Control.*

Willer, M. (1999). Formats and cataloguing rules: developments for cataloguing electronic resources. In *Program, 1*(33), 41-55.

Willer, M. (2004). UNIMARC format for authority records: Its scope and issues for authority control. In A.G. Taylor, & B. B. Tillett, (Eds.), *Authority control in organizing and accessing information: Definition and international experience,* (pp. 153-184). Binghamton, NY: Haworth Information Press. DOI 10.1300/J104v38n03_14

Willer, M. (2007). IFLA UBCIM Working Group on FRANAR recommendations for potential changes in the UNIMARC Authorities format. In Plassard, M.-F. *UNIMARC & friends: Charting the new landscape of library standards: Proceedings of the International Conference held in Lisbon, 20-21 March 2006* (pp. 61-68). München, Germany: Saur.

Willer, M. (2009, in press). Third edition of *UNIMARC Manual: Authorities Format*: Implementing concepts from the FRAD model and IME ICC Statement of International Cataloguing Principles. In *International Cataloguing and Bibliographic Control, 4*(38).

KEY TERMS AND DEFINITIONS

Access Point: A name (of a person, corporate body, work), term etc., under which a bibliographic or authority machine readable record may be searched. The word access point may also be used in terms such as "authorized access point" or "variant access point" to refer to the status or function of the access point in the automated, online catalogue. See also **Heading**.

Authority Control: A mechanism for creating consistency in a library catalogue, i.e. library information system.

Authority Entry/Record: An entry in a card catalogue for which the organizing element is a uniform heading for an entity such as person, corporate body or work. In an automated, online catalogue it is a record in an authority file for which the organizing element is the authorized access point. There is a change in terminology, but the function is the same in both types of technical implementations of the catalogue.

Cataloguing Principles: Objectives that govern design of a system, i.e., a library catalogue.

Heading: An initial element of an authority entry, used as the principal filing element in a (card) catalogue, bibliography or index. The word heading is also be used in terms such as "uniform heading" or "variant heading" to refer to the status or function of the heading in the catalogue. See also **Access point**.

FRAD: Functional Requirements for Authority Data, a conceptual entity-relational model in which each authority data element, its attributes and relationships are linked to a specified user task.

FRBR: Functional Requirements for Bibliographic Records, a conceptual entity-relational model in which each bibliographic data element, its attributes and relationships are linked to a specified user task.

IFLA: International Federation of Library Associations and Institutions, a body in charge of issuing international standards, guidelines and rules in the field of librarianship, accepted as a basis for national practices in the field of wide range of library functions, such as cataloguing.

Identification Authority Record: A record that contains authorized form of the entity with additional data elements such as qualifiers, notes and references which disambiguate it from other entities within the catalogue making it thus applicable for re-use in other contexts.

Name Authority Control: Authority control over names of persons, corporate bodies, families, titles of works. The term does not include the control over topical subjects, for which term subject authority control is used. It implies 100% precision and recall in searching an information system (catalogue), although not always attained.

Universal Bibliographic Control (UBC): International agreement that each national bibliographic agency create bibliographic and authority records for the "book" production within its territory and give it to other agencies for use. The aim is to record all publications worldwide for the advancement of human knowledge, understanding and wellbeing.

ENDNOTES

[1] The second part of the Meeting was dedicated to the definition of standard bibliographic description, ISBD.

[2] National Yugoslav cataloguing code at the time, now national Croatian code.

[3] Oddy builds her arguments on the example of the cooperative system Washington, later Western-Library Network "which provided *automated* authority control systems rather than card or fiche products" (Oddy, 1986, p. 4).

[4] For problems of the choice of author headings regarding the sources of information and different types of authorship in finite internet resources, see for example Vrbanc, 2007.

[5] The archival community is introduced here because the WG on MLAR and ISADN consulted with the representative from International Council on Archives/Commission on Descriptive Standards (ICA/CDA). Namely, the first edition of ICA/CDA standard parallel to the *Guidelines* had been published in 1996: *International standard archival authority record for corporate bodies, persons and families: ISAAR(CPF)*.

[6] The navigation objective could be traced to Seymour Lubetzky's concept of the syndetic structure of the catalogue (Lubetzky, 1979).

[7] See Svenonius's analysis of "open-ended objectives" (Svenonius, 2000, pp. 22-23).

[8] See FRBR Working Group website at: IFLA Cataloguing Section FRBR Working Group at http://www.ifla.org/VII/s13/wgfrbr/

[9] The proposal (2008/27) for this new field, 003, was prepared for the IFLA Permanent

UNIMARC Committee by the Comité français UNIMARC with the commentary that it will "allow one to point directly on the record in the catalogue from any quotation on the Web; it can be used in bibliographies and quotations, etc."

10 The identifier corresponds to the new field in the 3rd edition of *UNIMARC Manual: Authorities Format*, i.e. 003 Persistent Record Identifier: 003 http://errol.oclc.org/laf/nb2001-72552.html available from the OCLC LAF: Linked Authority File.

Chapter 13

Prevention and Regulation of Cyber–Crimes in the Age of Terrorism:
The Legal and Policy Model from India

S.R. Subramanian
Hidayatullah National Law University, India

ABSTRACT

India is the 12ᵗʰ nation in the world to have a special system of laws addressed to the information technology sector. Besides the general criminal law of the country, the Information Technology Act, 2000 incorporates a special legal framework relating to cyber-crimes. Looking differently, India is also a global hub of information technology and its allied services. Accordingly, the growth and development of the information technology sector and its contribution to national economy is phenomenal. It is in this context, the chapter examines and analyses the Indian ICT laws and policies in the backdrop of cyber-crime prevention and regulation, with the aim of offering a comprehensive model of ICT policy. It will discuss the extent of legal framework in the light of classification and criminalization of various cyber-crimes. Also, while examining the policy instruments, it will bring out the public and private initiatives on protection of information infrastructures, incident and emergency response and the innovative institutions and schemes involved.

INTRODUCTION

The unprecedented growth and development of information and communication technology (ICT) along with the open and hitherto unregulated nature of the internet and the anonymous feature of internet activities acted as the 'safe heaven' for criminal purposes. Besides a new range of technological offences, a number of traditional crimes such as theft, fraud and conspiracy can also be committed via the internet. In other words, internet can be the subjects of crime, it can be the site of a crime and it can also be a tool through which crimes can be committed (Kamath (2005). With the exponential increase in internet-related crimes, both in terms of number and sophistication, cyberspace present new challenges to the security and stability of the

DOI: 10.4018/978-1-61692-012-8.ch013

internet and raise serious concerns for policymakers and other stakeholders at all levels.

However, cyber-crime is a new discipline and hence the legal response to these rapidly growing illegal activities is still in the process of emerging. The dilemma of cyber-crime regulation is that it is caught between two diametrically opposite legal approaches (Gelbstein and Kurbalija (2005). The 'real law' approach, on the one hand, considers the internet as a natural evolution of existing technologies like telegraph and hence extends the application of prevalent legal rules to the internet. On the other hand, the 'cyber law' approach treats the internet as *sui generis* development and believes that it can only be regulated by special laws. Nonetheless, the practice of most government is that the existing law can be applied to the problems of internet, with varying levels of modifications.

However, the major challenge in regulating the cyber-crime is not the multiple categories or the magnitude of crimes, but that the law is inadequate to deter and prevent further violations. Hence, the investigation, prosecution and enforcement of cyber-crime is an enormous challenge for any criminal justice machinery. Moreover, the global phenomenon of the cyberspace also adds to the jurisdictional quagmire of the internet (Rao, (2004). This underlines the significance of the continual update of law and policies to keep pace with the latest technological developments to prevent it from being obsolete.

India is a global IT player and is a pioneer in the field of cyber-law, having brought the Information Technology Act in the year 2000. However, the Information Technology Act, 2000 was heavily criticized for improper treatment of cyber-crime. Even as the Statement of Objects and Reasons of the enactment claim, the law creates an enabling environment for electronic commerce and only incidentally addresses the issues of cyber-crime. Until the passage of Information Technology (Amendment) Act, 2008 (Act No. 10 of 2009),

the law was woefully inadequate to deal with the vital issues like 'cyber-terrorism', not to speak of the onerous challenges of use of internet for terrorist purposes.

It is in this context, the chapter examines and analyses the Indian ICT laws and policies in the backdrop of cyber-crime prevention and regulation, with the aim of offering a comprehensive model of ICT policy. It will discuss the extent of legal framework in the light of classification and criminalization of various cyber-crimes. Also, while examining the policy instruments, it will bring out the public and private initiatives on protection of information infrastructures, incident and emergency response and the innovative institutions and schemes involved.

CYBER-CRIMINAL LAW IN INDIA

India is the 12th nation in the world to have a special system of laws addressed to the information technology sector (Regulatory norms, 2006). Recognizing the potential contributions the information technology sector can make to the socio-economic development of the country, the legislation sought to create an environment for electronic commerce. It also incidentally criminalizes and punishes certain conduct prohibited under the law. These provisions are in addition to the general criminal law contained in the Indian Penal Code, 1860, itself amended by the Information Technology Act, 2000. Most importantly, a thorough overhaul of cyber-criminal law, inter alia, has taken place through the Information Technology Amendment Act, 2008[1].

Contraventions

The law vertically classifies the cyber-crimes into two types: contraventions and 'information technology offences'. While contraventions will attract financial sanctions in the form of compensation, 'information technology offences' are punishable with the sentence of imprisonment and/or fine. Chapter IX provides for Contraventions. Section 43 of the above Chapter stipulates that the following acts are contraventions when done without the permission of the owner or any other person who is in charge of a computer, computer system or computer network:

1. Accessing or securing access to the computer or computer resource
2. Downloading, copying or extracting any data or information from the computer network
3. Introducing or causing to be introduced any computer contaminant or computer virus into the computer or network
4. Damaging or causing to be damaged the computer or any other programmes residing in that computer
5. Disrupting or causing the disruption of the computer
6. Denying or causing the denial of access to any person authorized to access the computer/network by any means
7. Providing assistance to any person to facilitate access to the computer in contravention of law
8. Charging the services availed of by a person to the account of another person by tampering with or manipulating any computer etc.

Prior to the Amendment of 2008, hacking with computer system was punishable as an offence under Chapter XI, Section 66. However, taking into account the expert opinion that this provision may be used for unintended purposes, the Amendment has relegated the crime of hacking to the class of Contraventions and listed it as clause (i) to Section 43 (Duggal, (2005). Similarly, the Amendment brought another change with reference to the tampering of computer source code'. Though this crime was earlier punishable as an offence under Section 65, currently it merely constitutes a Contravention, especially, 'when there is no legal requirement to keep or maintain the computer source code'. It is submitted that the current scheme of classification is more in tune with the principle of proportionality of crime.

Data Protection

India is the global hub of information technology outsourcing. Hence, to maintain leadership, it is an imperative that it offers a system of robust data protection to the Indian corporate clients. The newly-inserted Section 43A provides that where a commercial organization possessing, dealing or handling any sensitive data or information in a computer resource which it owns, controls or operates, is negligent in implementing and maintaining 'reasonable security practices and procedures' and thereby causes wrongful loss or wrongful gain to any person, such organization shall be liable pay compensation to the person so affected.

It is also clarified in the legislation that for the purposes of data protection, 'reasonable security practices and procedures' (required to be adopted by the corporate organizations) would mean those security practices and procedures designed to protect such information from unauthorized access, damage, use, modification, disclosure or impairment, as may be specified in the agreement between the parties or any law for the time being in force or in their absence, any reasonable security practices and procedures prescribed by the Central Government in consultation with the professional bodies or associations. It is hoped that this new protection together with the provisions relating to confidentiality, privacy and other cyber-security provisions will mitigate the problems associated with the data protection and will enable India to compete with other leading destinations of outsourcing.

INFORMATION TECHNOLOGY OFFENCES

Chapter XI deals with the acts which are punishable as information technology offences. While some of these offences are substantive, others are procedural in nature.

1. **Source code attacks:** If any person with the knowledge or intention conceals, destroys or alters or causes another to conceal, destroy or alter any computer source code used for a computer, computer programme, computer system or computer network, when the computer source code is required to be kept or maintained by law for the time being in force, shall be punishable with imprisonment up to three years, or with fine which may extend up to two lakh Indian rupees, or with both (Section 65).

2. **Computer-related Offences:** The contraventions referred to in Section 43 committed with dishonest or fraudulent intention are punishable with imprisonment for a term which may extend to three years or fine up to five lakhs or both (Section 66). The pre-amendment version of the computer hacking is currently categorized as 'contraventions'.

3. **Sending offensive messages:** Any person who sends by means of a computer resource or a communication device a) any information that is grossly offensive or has menacing character or b) any information which he knows to be false, but for the purpose of causing annoyance, inconvenience, danger, obstruction, insult, injury, criminal intimidation, enmity, hatred, or ill-will persistently by making use of such computer resource or a communication device c) any electronic mail or electronic mail message for the purpose of causing annoyance or inconvenience or to deceive or to mislead the addressee or recipient about the origin of such messages shall be punishable with imprisonment for

a term which may extent to three years and with fine (Section 66A). The criticism that in India the cyber-crime is a crime without penalty with reference to a host of crimes such as cyber-defamation, cyber-stalking, cyber-nuisance, cyber-fraud and cyber-harassment has been addressed through this new provision.

4. **Dishonestly receiving stolen computer resource:** Whoever dishonestly receives or retains any stolen computer resource or communication device knowing or having reason to believe the same to be stolen computer resource or communication device, shall be punished with imprisonment of either description for a term which may extend to three years or with fine which may extend to rupees one lakh or with both (Section 66B).

5. **Identity theft:** Any person who with fraudulent or dishonest intention makes use of the electronic signature, password or any other unique identification feature of any other person, shall be punished with imprisonment for a term which may extend to three years and shall also be liable to fine which may extend to rupees one lakh (Section 66C).

6. **Phising:** The stealing of sensitive information of the net users by posing as trustworthy websites is popularly known as phising. The Act stipulate that whoever by means of any communication device or computer resource cheats by personation, shall be punished with imprisonment of either description for a term which may extend to three years and shall also be liable to fine which may extend to one lakh rupees (Section 66D).

7. **Invasion of privacy:** Whoever, intentionally or knowingly captures, publishes or transmits the image of a private area of any person without his or her consent, under circumstances violating the privacy of that person, shall be punished with imprisonment which may extend to three years or with fine not exceeding two lakh rupees or with both.

The striking feature of the provision is that the crime is committed not only when the image is captured in the form of photograph but it would include videotaping, filming or recording through any other means. The expression 'circumstances violating the privacy' is also expansively defined to include not only the circumstances in which a person can have a reasonable expectation that he or she could disrobe in privacy, but also when he or she would reasonably expect that his or her private area would not be visible to the public, regardless of whether that person is in a public or private place (Section 66E).

8. **Cyber-terrorism:** It is alleged that the immediate trigger for the passage of Amendment Act was the 26/11 Mumbai attacks (Duggal, (2005)[2]. The Act criminalized cyber-terrorism with punishment up to life imprisonment. According to Section 66F, cyber-terrorism can be committed in three ways. It may be due to 1) denial of access or 2) unauthorized penetration or access of computer resource or 3) through introduction of any computer contaminant. The level of intention that is required to commit any of the above acts is such as to threaten the unity, integrity, security or sovereignty of India or to strike terror in the people or any section of them. However, the offence of cyber-terrorism would not be made unless the conduct actually or in any likelihood causes the death or injuries to persons or damage to or destruction of property or disrupts or knowing that it is likely to cause damage or disruption of supplies or services essential to the life of the community or adversely affect the critical information infrastructure. Alternatively, where unauthorized or excess of authority in the penetration or access of information or data is concerned with the security of State or friendly or foreign relations or likely to cause injury to the interests of the sovereignty and integrity of India,

the security of the State, friendly relations with foreign States, public order, decency or morality, or in relation to contempt of court, defamation or incitement to an offence, or to the advantage of any foreign nation, group of individuals may also committing the offense of cyber-terrorism. It is provided that both the offenses of cyber-terrorism and the conspiracy to commit cyber-terrorism are punishable with imprisonment which may extend to imprisonment for life. The quantum of sentence for the offence of cyber-criticism was criticized as being disproportionate to the severity of the crime. However, the drafters might have considered that the prescription of death penalty for the crime will be a hindrance to the international cooperation such as mutual legal assistance and extradition. It is submitted that death sentence is a ground for refusal of cooperation in major international frameworks.

9. **Cyber-obscenity:** The post-Amendment law distinguishes between cyber-obscenity and cyber-pornography. According to Section 67, the publication or transmission of any material which is lascivious or appeals to the prurient interest or if its effect is such as to tend to deprave and corrupt persons who are likely, having regard to all relevant circumstances, to read, see or hear the matter contained or embodied in it. The crime attracts the punishment of imprisonment of three years and fine which may be up to five lakhs on first conviction and in the event of a second or subsequent conviction with imprisonment of five years and fine which may extend to ten lakhs rupees.

10. **Cyber-pornography:** The publication or transmission of any material which contains sexually explicit act or conduct in electronic form constitutes the offence of cyber-pornography. It shall be punishable on first conviction with imprisonment of either description which may extend to five

years and with fine which may extend to ten lakhs rupees and in the event of second or subsequent conviction with imprisonment of either description for a term which may extend to 7 years and also with fine which may extend to ten lakh rupees. However, as is the case with the obscenity in the physical world, cyber-pornography excepts certain material, act or conduct from its purview. They are ranging from literary, to artistic to religious purposes (Section 67A).

11. **Cyber-pornography of children:** Whoever, a) publishes or transmits or causes to be published or transmitted material in any electronic form which depicts children engaged in sexually explicit act or conduct or b) creates text or digital images, collects, seeks, browses, downloads, advertises, promotes, exchanges or distributes material in any electronic form depicting children in obscene or indecent or sexually explicit manner or c) cultivates, entices or induces children to online relationship with one or more children for and on sexually explicit act or in a manner that may offend a reasonable adult on the computer resource or d) facilitates abusing children online or e) records in any electronic form own abuse or that of others pertaining to sexually explicit act with children commits the offence of cyber-pornography of children.

12. **Breach of confidentiality and privacy:** The law provides that any person who has secured access to any electronic record etc., with the powers conferred under the law but without the consent of person concerned, discloses such material to any other person shall be punished with imprisonment for a term which may extend to two years, or with fine which may extend to one lakh rupees, or with both.

Powers for Enforcement of Cyber-Security Measures

Besides criminalization of these conducts, the Indian law also empowers the appropriate officials with the powers to prevent, investigate, detect and prosecute the cyber-criminals. It is the conferment of these powers without proper safeguards, though with enumeration of grounds of exercise of power, which have proved to be very controversial.

1. **Powers to intercept, monitor and decrypt (Section 69):** The Indian law provides that the appropriate government or its officers have the powers to direct, any agency of the government, to intercept, monitor, or decrypt any information transmitted, received or stored through any computer resource. However, having regard to the nature of serious violations of the civil liberties that it may cause, this power can be exercised only when it is necessary in the interest of the sovereignty and security of India, defense of India, security of the State, friendly relations with foreign states, or public order or for investigation of any offense. Moreover, the subscriber, or intermediary, or any person in charge of any computer resource legally bound to provide access or secure access to the computer resource generating, transmitting, receiving or storing such information or to intercept, monitor or decrypt the information or provide information stored in the computer resource.

2. **Powers to block public access of information (Section 69A):** The Central government or any of its officers by order direct any agency of the Central government or intermediary to block access by the public or cause to be blocked for access by public, any information generated, transmitted,

received, stored, hosted in any computer resource. This power can be invoked only when it is necessary or expedient in the interest of sovereignty and integrity of India, defense of India, security of the State, friendly relations with foreign states, or public order or for preventing incitement of commission of any cognizable offense relating to them.

3. **Powers to monitor and collect traffic data (Section 69B):** The Central government in order to enhance cyber-security and for identification, analysis and prevention of any intrusion or spread of computer contaminant in the country, by notification in the Official Gazette, authorize any agency of the government to monitor and collect traffic data or information generated, transmitted, received or stored in any computer resource. Moreover, the intermediary or the person in charge of the computer resource when called upon by the agency is under the legal duty to provide technical assistance to enable online access, or to secure and provide online access, to the computer resource generating, transmitting, receiving or storing such traffic data or information.

CYBER-SECURITY POLICIES IN INDIA

Though criminalization and prescription of deterrent punishment is an imperative for an effective system of cyber-crime free world, security in cyberspace can not be brought out by legal measures alone. The laws and regulations need to be supplemented by the policies of government and other stakeholders in the areas of the protection of critical information infrastructure, best practices and procedures, incident and emergency responses, data protection etc. While law comes into play only after the irreversible is done, policy has a huge potential for prevention. This emphasizes the right mix of both legal and non-

legal measures (Sadowsky, Dempsey, Greenberg, Mack, and Schwartz, (2003).

Governmental Policies

The government formulates policy not only for the protection its own assets, but it has the obligation to design the policy for protecting the national information infrastructure (Sadowsky et al. (2003):

- **Protection of Critical Information Infrastructure:** The governmental response to the problems of cyber-security led to the classification of critical information infrastructure. There are certain computers and networks, whether in government or with the private enterprises, which are of critical importance to the nation and those systems deserve the top-most attention. However, there is no uniform concept of critical information infrastructure throughout the world. The original Information Technology Act, 2000 does not define what a critical information infrastructure is. The newly-added Explanation to Section 70 of the Information Technology Act, 2000 meant them to be any computer resource, the incapacitation or destruction of which shall have debilitating impact on national security, economy, public health or safety.

- **Protected Systems:** Section 70 provides that the appropriate government, whether Central or State, by notification in the Official Gazette, declare any computer resource which directly or indirectly affects the facility of critical information infrastructure to be a protected system. It will also specify the persons who can access the protected system. If anyone accesses those systems in contravention of the above legal provision, the violator shall be punished with imprisonment for a term which may extend to ten years and shall also be liable to fine. The Central government will also prescribe the information security practices and procedures for such protected system.

- **National nodal agency:** The Central government will also designate an existing organization of the government as the national nodal agency for the protection of critical information infrastructure. It will be performing all the functions relating to the protection of such infrastructure including research and development.

- **Indian Computer Emergency Response Team (CERT-IN):** The Central government had notified the Indian Computer Emergency Response Team as the national agency for incident response. It was officially inaugurated on 19th January, 2004. It operates under the auspices of, and with the authority delegated by the Department of Information Technology, Ministry of Communications and Information Technology, Government of India (CERT-In, 2009). To ensure cyber-security of communication and information infrastructure, it performs the functions of a) collection, analysis and dissemination of cyber-incidents b) forecasts and alerts of cyber-security incidents c) emergency measures for handling cyber-security incidents d) coordination of cyber-incidents response activities and e) issue guidelines, advisories, vulnerability notes, white papers relating to information security practices, procedures, prevention and response and reporting of cyber-incidents.

Corporate Policies

However, effective cyber-security can not be ensured through legislative and governmental policies alone[3]. True to this, a distinguishing feature of the Indian environment is that the corporates, especially the IT and IT-related companies voluntarily consider it as the imperative of the success of their businesses. To this end, the Indian

corporates have taken a number of significant initiatives. NASSCOM, the national association of software and services companies, headquartered in Mumbai, is the premier trade body and the chamber of commerce of the IT-Business Process Outsourcing industries in India (About NASSCOM, 2009). It comprises of not only the Indian companies but also the multi-national corporations having a commercial presence in India (About NASSCOM, 2009). NASSCOM is instrumental in ensuring that the Indian information security environment benchmarks with the best across the globe (Information Security Environment, n.d). Besides creating an information security culture within the industry, it has also been interacting with the government on the creation of appropriate regulatory environment which will strengthen the initiatives from the industry (Information Security Environment, n.d).

- **NASSCOM's Security Initiatives:** NASSCOM, in its quest to integrate the robust information security environment within the industrial layers, initiated a diverse range of measures (Overview, 2006). They are: a) **Creating awareness with relevant stakeholders** i.e., it is popularizing the need, the legal requirement and the benefits of security practices among the information technology and related companies b) **Setting guidelines:** i.e., it advises member-companies in complying with the norms of leading security regimes including the Gramm-Leach-Bliley Act (GLBA), Health Insurance Portability and Accountability Act (HIPAA) and Sarbanes-Oxley Act c) **Defining standards:** it seeks to establish high standards of security required by the customers throughout the world d) **Security education:** i.e., it creates the model curriculum sensitized to the requirements of security and will train the entire range of personnel including IT professionals, support staff, security staff and audit and certification officials e) **Certifications:** it tries to introduce certification procedures f) **Shared services:** it also offers a set of services for the benefit of all the members such as, verification of employee's background and profiles etc (Overview, 2006).

According to NASSCOM, the following four initiatives and programs merit special attention:

4E Framework: The 4 E's are, namely, Engagement, Education, Enactment and Enforcement.

1. **Engagement:** NASSCOM strives to engage a rage of actors including government organizations, other industrial organizations, think tanks and universities. For instance, it partners with the Department of Homeland Security, Federal Reserve Board and Carnegie Mellon University etc (4E Initiative, 2008).

2. **Education:** As mentioned above, it seeks to educate all the relevant stakeholders to achieve a higher level of security.

3. **Enactment:** It also engages the law-makers and policymakers so that they attach due significance to the cyber-law making and implementation process (4E Initiative, 2008).

4. **Enforcement:** NASSCOM regularly in touch with the entire criminal justice machinery so that the laws and policies made are strictly enforced (4E Initiative, 2008).

*   **India Cyber Lab[4]:** The second major initiative of the NASSCOM was the India Cyber Lab, which is hailed as the innovative arrangement between the police and the information technology industry (NASSCOM, 2008). It is conceived out of the Cyber Safety Week organized jointly by the Mumbai Police and the NASSCOM in 2003. These labs provide the opportunity for the police officials to receive a week-long training at the hands of member companies (NASSCOM, 2008). The areas of training, according to the Association's report, computer operations, digital storage, forensic techniques, legal provisions, crime modus operandi, case studies and procedural practices. Besides the expansion of the concept to major cities in India, through outreach and outstation programmes the training was also offered to other officials of the banking sector, students, military officers and in other locations (NASSCOM, 2008)

*   **National Skills Registry for IT and IT Enabled Services (ITES) professionals[5]:** This third major initiative of the Association ensures that the unscrupulous persons are not appointed by the member companies across the industry. This program sees that the individuals employed by the organizations have their background and antecedents verified to avoid the chances of prevention of fake job resumes (National Skills Registry, 2007). The major advantages of this unique scheme are: a) theft of data by the employees b) enhances the trust and confidence of the clients d) when the employee attrition rate is very high due to the size of the industry etc it helps to collect the data of who is coming in and who is going out e) reduction of the cost of the profile verification in view of centralization f) standardization of employee verification process to deliver consistent results across the industry (National Skills Registry, 2007).

*   **National Skills Registry (NSR) process:** The scheme allows the employees to access the publicly-available national skills regis-

try website and upload the relevant data of personal identification, educational background, employment profile and biometric piece of evidence. This information is subjected to verification by the Empanelled Background Checking Companies (EBC), which is independent from the NASSCOM and its member companies. These details are accessible only to the employee and the subscriber-employer when authorized by the employee. The NSR website is hosted and maintained by the NSDL Database Management Limited (NDML), a subsidiary of the National Securities Depository Limited (NSDL). NSDL is a high-profile organization which hosts the securities of investors (National Skills Registry, 2007).

Data Security Council of India[6]

According to NASSCOM, the Data Security Council of India born out of the recognition that the rigid security parameters fixed by the customers can not be met by an individual company and it requires industry-wide solution (Data Security Council, 2008). Moreover, the problem can not solved by technology alone and the final solution to the issue of security is in the synergy of the three key resources: technology, people and business processes (Data Security Council, 2008). Also, the level of security to be achieved is the highest standards as even a single instance of breach will have the potential to tarnish the reputation of the entire industry both inside the country and overseas (Data Security Council, 2008). However, cost is also a problem as many smaller companies are looking for cost-effective solutions (Data Security Council, 2008). This demonstrated the urgent need for an effective Self Regulatory Organization (SRO) to establish, popularize, monitor and enforce data protection standards for the IT enabled industry ((Data Security Council, 2008).

The Data Security Council of India became a reality in April 2007. Though it is welcomed by the partners and clients of NASSCOM, the major drawback of the initiative is that it is just registered as a not-for-profit institution and lacks the legal powers to enforce the security standards. However, with the adoption of progressive enforcement standard as the guiding principle of the Data Security Council of India, it is hoped that a binding mechanism will be in place in future[7].

CONCLUSION

India has taken a number of initiatives to promote electronic commerce and to fight various criminal misbehaviors on the internet. The recent Amendment has also undertaken a thorough overhaul of the legal provisions dealing with cyber-crimes. Significantly, it has focused on strengthening the existing legal regime with additions of several new offences. It is hoped that the law and policies and their strict enforcement will ensure that the cyber-crimes can be effectively tackled by the law enforcement officials. At the same time, the policy instruments of the business organizations and their apex bodies will allow for the good measure of self-regulation by the private business houses and its employees. Though the first case of cyber-crime conviction was reported in India way back in 2003, a substantial revision of the penal provisions has happened only in 2009. Moreover, the major chunk of criminal prosecutions of cyber-crime still rely on the out-dated general criminal law of the country, namely, the Indian Penal Code (IPC), 1860, rather than the Information Technology Act, 2000.

However, the strengthening of the hands of criminal justice officials is not wholly without criticism. The biggest criticism of the Amendment is the erosion of privacy and other civil liberties at the altar of national security. The free hand to monitor and intercept any information not only on the grounds of national security but to carry out investigation of any offence raises serious concerns of human rights violation. This is an issue of enormous challenge and it is expected that appropriate safeguards will be incorporated to the mechanism in one way or the other.

Despite the limitations and shortcomings of the law of cyber-crimes, the model of combination of public and private initiatives, especially the pioneering initiatives of the NASSCOM deserve special attention. These developments highlight the huge potential of self-regulation and prevention to combat the menace of cyber-crime. It is hoped that the changes brought out by the Amendment Act will provide the way for the effective prosecution of cyber-criminals. Having placed the legal provisions in place, India can aspire to evolve the model cyber-criminal jurisprudence in the country.

REFERENCES

About, N. A. S. S. C. O. M. (2009). Retrieved March 18, 2009, from http://www.nasscom.in/Nasscom/templates/NormalPage.aspx?id=5365

Bantekas, I., & Nash, S. (2007). *International Criminal Law*. Abingdon, UK: Routledge-Cavendish.

Cert-in is the official website of the Indian Computer Emergency Response Team. (n.d.). Retrieved from http://www.cert-in.org.in/

Data Security Council of India. (2008). Retrieved March 18, 2009, from http://www.nasscom.in/Nasscom/templates/NormalPage.aspx?id=51973

Duggal, P. (2005). Cyber-crime in India: The Legal Approach . In *R.G. Broadhurst & P.N. Grabosky, (n.d.), Cyber-crime: The Challenge in Asia* (pp. 183–196). Hong Kong: Hong Kong University Press.

Gelbstein, E., & Kurbalija, J. (2005). *Internet Governance: Issues, Actors and Divide*. Malta: Diplo Foundation.

Indian Computer Emergency Response Team. (2006). *Information Security Policy for Protection of Critical Information Infrastructure*. New Delhi, India: Department of Information Technology, Government of India.

Information Security Environment in India. *NASSCOM Analysis*. (n.d.). Retrieved March 18, 2009, from http://www.nasscom.org/download/Indian_Security_Envomt_05_06_Factsheet_Final1.pdf

Information Technology Act. (2000). As amended by the Act 10 of 2009.

4E Initiative. (2008). Retrieved March 18, 2009, from http://www.nasscom.in/Nasscom/templates/NormalPage.aspx?id=5954

Kamath, A. (2005). *Law Relating to Computers, Internet and E-Commerce: A Guide to Cyberlaws, & the Information Technology Act, 2000 with Rules and Notifications*. Delhi, India: Universal Law Publishing.

NASSCOM. (2008). *NASSCOM 2007-08 Annual Report*. New Delhi: NASSCOM.

National Skills Registry for IT/ITES Professional. (2007). Retrieved March 18, 2009, from http://www.nasscom.in/Nasscom/templates/NormalPage.aspx?id=51441

National Skills Registry for Knowledge Professionals. (n.d). Retrieved March 18, 2009, from http://www.nasscom.in/upload/29264/NSR_Initiative_Overview.pdf

Overview. (2006). Retrieved Mach 18, 2009, from http://www.nasscom.in/Nasscom/templates/NormalPage.aspx?id=5957.

Rao, J.S.V. (2004). *Law of Cyber-crimes and Information Technology Law*. New Delhi: Lexis-Nexis.

Regulatory Norms in Indian cyber-space. (2006). Retrieved March 18, 2009, from http://www. nasscom.in/Nasscom/templates/NormalPage. aspx?id=6161

Sadowsky, G., Dempsey, J., Greenberg, A., Mack, B., & Schwartz, A. (2003). *Information Technology Security Handbook*. Washington, DC: The World Bank.

KEY TERMS AND DEFINITIONS

Contraventions: The type of crime which are punishable only with financial sanctions under the Indian Information Technology Act, as against the other type of crime punishable with imprisonment.

Cyber-Security: The protection of computer and its resources against unauthorized access.

Data Protection: The protection of data, personal or business, through legal, administrative, technical and physical means to ensure that the relevant information is not misused or abused.

ENDNOTES

[1.] The official version of the Information Technology Amendment Act, 2008 is available at the website of www.dsci.in. The Act is yet to be notified by the Central Government.

[2.] The Information Technology (Amendment) Bill, 2006 was pending for a long time with the Parliament.

[3.] "Undoubtedly, self-regulation by the high-tech industry can play an important role in preventing the proliferation of computer and internet-related crime". (Bantekas, I & Nash S. (2007). P. 73.

[4.] The official website of the India Cyber Lab is http://www.indiacyberlab.in. Also, http://www.nasscom.in/Nasscom/templates/NormalPage.aspx?id=5952.

[5.] The official website of the National Skills Registry can be accessed at: https://nationalskillsregistry.com/.

[6.] www.dsci.in is the official website of the Data Security Council of India.

[7.] "While self-regulatory mechanisms to combat the misuse of the new technologies have many advantages over the external regulation, to be effective self-regulation needs to be supported by appropriate national legislation and international agreements". (Bantekas, I & Nash S. (2007). P. 73.

Chapter 14

National Information and Communication Technology Policy Process in Developing Countries

Edwin I. Achugbue
Delta State University, Nigeria

C.E. Akporido
Delta State University, Nigeria

ABSTRACT

This chapter discusses national information and communication technology policy process in developing countries. It describes the need for information and communication technology policy, ICT policy development process, national ICT policy in developing countries, the role of an ICT policy in the developing country, factors affecting the formulation of national ICT policies and the future of national ICT policy was also discussed.

INTRODUCTION

Information and communication technologies (ICTs) have critical roles to play in development efforts around the world. There was a time when the benefits of applying ICTs in fighting poverty and promoting economic growth were not widely understood. Many in the development community questioned how high-tech (and often expensive) communication technologies could be used to alleviate such dire challenges as starvation, homelessness, and lack of basic education and health services. Lately, however, this view has given way to an understanding of ICTs as essential components of broader efforts to harness the free flow of information to increase voice, accountability, and economic development. In recent years, developing countries have started taking concrete actions to incorporate ICT into their national economic policies and development agendas. Many countries are preparing and implementing national ICT policies that emphasize the ubiquity of connectivity as well as new applications in areas such as e-government and e-business (World Bank, 2006), Policy makes a great difference

DOI: 10.4018/978-1-61692-012-8.ch014

regarding how countries are able to take advantage of the technological opportunities available to them and exploit them for good. Countries with progressive policies are seeing these technologies spread quickly. However, countries that are yet to formulate and integrate ICT policy have been plagued by slow growth of technology and the consequent lessening of support for economic and social development (Sarkar De, 2005, cited Adomi & Igun, 2008).

ICTs cannot substitute for a country's own good governance, economic reform or social policies. However, ICTs can be applied to support democratic processes, improve the productivity and competitiveness of all economic sectors, create new sources of wealth, and increase the efficiency of public services, including healthcare, education and disaster assistance. In the era of the global information economy and society, ICTs are an increasingly essential part of development policies and programs. (Implementation team on global Policy participation, 2002). Governments world-wide recognize the crucial role ICTs play in facilitating and accelerating socio-economic development; which has made a number of countries in the developing world are putting in place policies and strategic plans that will enable them to transform their economies into information and knowledge-based economies(Dzidonu, 2002).

Developing countries need to proactively evaluate the impact of ICT on existing sectors, identify the potential for ICT to create new economic and social opportunities, and address development priorities by designing and implementing comprehensive national ICT strategies. By being proactive, countries that have put in place national ICT policies that address all the relevant priority areas (ICT infrastructure and access, human capacity development, the network and regulatory environment, business and entrepreneurship, content and applications) are better able to avail themselves of the opportunities offered by the global economy. They are in a position not only to learn and creatively transform existing ways

of working, communicating and living but also to enhance the productive and human capacities of their people and to assist to reduce existing forms of economic and social inequality and exclusion. On the other hand, those erring on the side of caution risk being excluded from the benefits of the emerging networked economy. Delays in action in becoming integrated can mean a widening of economic and social development gaps, increased marginalization and social exclusion. The social and economic losses that come from not effectively deploying ICT are also likely to increase progressively (UNDP, 2001). This chapter discusses the objectives of national information and communication technology policy in developing countries,. ICT policy development process, national ICT policy development initiatives, country case studies, the role of an ICT policy in developing countries, factors affecting the formulation of national ICT policies, feature trends of the topic and conclusion..

BACKGROUND

Information and communication technology is any technology which enables communication and electronic capture, processing and transmission of information. (Parliament Office of Science and Technology, 2006 cited by Adomi, 2008). ICT has a role to play in any country's development (http://wougnet.Org/ICTpolicy/ug/ugictpolicy. html). All over the world today, governments are introducing policies aimed at integrating ICT into the national hub and strategies are evolving to enable countries exploit ICT for progress and development. ICT is fast becoming a way of life in Africa and many developing countries.

ICTs can be grouped into three categories (Adomi, 2008):

- Information technology uses computers, which have become indispensable in mod-

Figure 1. Relationship between ICT policy and other development policies (Source: UNCSTD, 1997; adapted from Marcelle 2001)

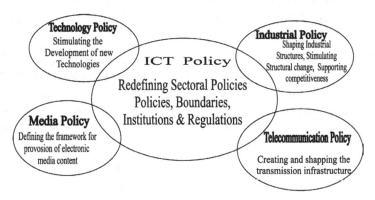

ern societies to process data and save time and effort.

• Telecommunications technologies which include telephone (with fax) and the broadcasting or radio and television often through satellites.

• Networking technologies of which the most widely known is the internet, but which has extended to mobile phone technologies; voice over IP telephone (VoIP), satellite communications and other forms of communication that are still in their infancy (p107).

The first few years of the century have witnessed a number of leaps in technology advancement and many of these leaps can be traced to advances in information and communication technology (ICT). ICTs have characterized what has become known as the information age.

A national ICT policy is an integrated set of decisions, guidelines, laws, regulation, and other mechanisms intended to direct and shape the production, acquisition, and use of ICTs. ICT policy as an official statement which spells out the objectives, goals, principles, strategies, etc intended to guide and regulate the development, operation and application of information and communication technology (Marcelle, 2001. APC, n.d. cited by Adomi & Igun, 2008). ICT

policy generally covers three main areas, like telecommunications (especially telephone communications), broadcasting (radio and television), and the internet; it may be national, regional, (and or sub regional), or international; each level may have its own decision–making bodies, sometimes making different and even contradictory policies (Adomi & Igun, 2008). According to Marcelle (2001) the ICT sector is heterogeneous, extending beyond traditional classification of industrial or services sectors and because production and diffusion of ICTs are of equal importance, national policies in the ICT sector intersect with a number of other areas of policy- making like technology, media, industrial, and telecommunication policy (Marcelle, 2001).

Figure 1 shows these areas of intersection among the various areas of policy spheres.

OBJECTIVES OF ICT POLICIES IN DEVELOPING COUNTRIES

According to UNESCAP (2009) the ICT evolution will occur with or without a systematic, comprehensive and articulated policy; that the lack of a coherent policy is likely to contribute to the development (or prolonged existence) of ineffective infrastructure and a waste of resources. The following are some aspirations that ICT policies often try to meet (UNESCAP, 2009):

- Increasing the benefit from information technology
- Helping people and organizations to adapt to new circumstances and providing tools and models to respond rationally to challenges posed by ICT
- Providing information and communication facilities, services and management at a reasonable or reduced cost
- Improving the quality of services and products
- Encouraging innovations in technology development, use of technology and general work flows
- Promoting information sharing, transparency and accountability and reducing bureaucracy within and between organizations, and towards the public at large
- Identifying priority areas for ICT development (areas that will have the greatest positive impact on programmes, services and customers)
- Providing citizens with a chance to access information technology resources
- Attaining a specified minimum level of information technology resources for educational institutions and government agencies
- Supporting the concept of lifelong learning
- Providing individuals and organizations with a minimum level of ICT knowledge, and the ability to keep it up to date
- Helping to understand information technology, its development and its cross- disciplinary impact

An ICT policy framework is recognized as an important step in order to create an enabling environment for the development of ICTs and their uses to social outcomes (.Njuguna, 2006, cited by Adomi and Igun, 2008),

There are three levels of ICT policy: infrastructural policy (dealing mostly with communication side), vertical policy (looking at the role of IT in a given sector), and horizontal policy (for which Mozambique's approach which regards ICTs as facilitators for a wide range of national activities, is an example). Some countries have focused on developing ICT as an economic sector –either to boost export (Costa Rica and Taiwan) or to build domestic capacity (Brazil, India and Korea) while others are pursuing strategies which seek to use ICT as an enabler of wider socio – economic development process. Countries which use ICT as an enabler may be further subdivided into those which have focused primarily on repositioning the country's economy to secure competitive advantage in the global economy (Malaysia, Trinidad and Tobago) and those which explicitly focus on ICT in pursuit of development goals such as those set forth in the UN Millennium Summit (Estonia and South Africa). (James, 2001, cited by Rowan, 2003).

ICT POLICY DEVELOPMENT PROCESS

The African Information Society Initiative (AISI) sets out important set of policy guidelines for national information policy in the African context. (Marcelle, 2001). Early studies of policy–making in informatics in African countries, within the ambit of AISI and its forerunners, have provided important research and analysis of the readiness of Africa Countries to undertake policy interventions in the ICT sector. One such study reviewed informatics policy in 10 African Countries; Cameroon, Congo, Cote d'Ivoire, Ethiopia, Kenya, Madagascar, Nigeria, Senegal, Tanzania, and Zimbabwe (Browne, 1996 cited by Marcelle, 2001).

The study defined national informatics policy as a plan for the development and optimal utilization of information technology (IT) and reported that limited financial resources, poor institutional capability, and inadequate access to human resources and technological know-how plague Africa's attempts to harness ICTs (Brown 1996). A recent study on national ICT policy–making revealed only very moderate success in formulating and implementing ICT policies some developing countries (Marcelle, 2001).

Policy making process normally involves the expression of conflicting interests. It is convenient to think of such interest as being represented by "actors" who engage in debate and decision –making in appropriate locations or "fora" where decisions are made (Okado, 2007). The African Information Society Initiative (AISI) recommended that the ICT policy formulation process should start with the definition of national development priorities as contained in the various documents such as the five-year plan and cabinet directions which the ICT policies should support. In practice, the process involves elaborate and prolonged processes that often cover analysis of national priorities, holding sensitization workshops, development of ICT framework, writing policy documents, taking them through legislations, establishing action plans and implementation programmes. It is particularly important to go through these steps in countries where the ICT policy awareness is low as has been the case for most countries in Sub-Saharan Africa (Okado, 2007).

Lishan & When (1988) (cited by Okado (2007) posited that the relevance of ICT policies to development can be increased by engaging all key stakeholders (actors) in national development plans and implementation of ICT policies. McQuail (2000), Okado (2007) have accordingly identified a number of actors and interests responsible for ICT policy formulation. They are:

- These actors reflect the nature of public interest in ICT PolicyPublic versus private interests
- Owners and employers versus employees
- Economic versus social cultural Organisations

- Opposition politicians versus ruling party
- National versus international Organisations
- Developed (the North) versus developing (the South) countries.

These actors reflect the nature of public interest in ICT policy. The potential of conflict between public and private interests demonstrate the struggle over who should initiate or control the expansion of ICTs. The struggle between national and international actors reflects increased globalization of ICT infrastructure as service. As a result national governments may easily find themselves at odds with multinationals. The main fora at which policies are formed are transnational, national and regional. At the international forum national governments, international ICT organizations and development agencies are the likely actors. Examples of these actors are UNESCO or the ITU. At the international level, ICT policy formulation encounters forces of globalization. This is because the global trend in ICTs and ICT policy formulation process is determined by northern forces, as a result many developing countries do not obtain fair share of the benefits of globalization and some actually suffer net losses (Khor, 1995, Okado, 2007). The North is in control not only on account of strength but also due to lack of coordination in the South. National technology policies are largely formulated by the work of global institutions and their rules and standards. It is important that developing countries participate more forcefully and effectively in these institutions. This requires policy coordination among developing countries. Figure 2 shows the various actors involved in ICT policy-making.

Figure 2. Key agent in policy-making process (Source: UNCSTED 1997; adapted from Marcelle 2005)

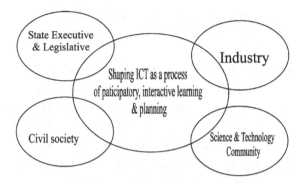

NATIONAL ICT POLICY DEVELOPMENT INITIATIVES IN DEVELOPING COUNTRIES

Nigeria

Nigeria began the implementation of its IT policy April 2001 after the Federal Executive Council approved it by establishing the National Information Technology Development Agency (NITDA), the implementation body. The policy empowers NITDA to enter into strategic alliances and joint ventures and to collaborate with the private sectors to realize the specifics of the country's vision of, making an IT capable country in Africa as well as a key player in the information society by the year 2005 through using IT as an engine for sustainable development and global competitiveness. However, this vision is yet to be fulfilled (Agyeman, 2007).

Objectives

The objectives of Nigeria's IT policy are as follows (Federal Republic of Nigeria, 2001):

- To ensure that Information Technology resources are readily available to promote efficient national development.

- To guarantee that the country benefits maximally, and contributes meaningfully by providing the global solutions to the challenges of the information age.
- To empower Nigerians to participate in software and IT development.
- To encourage local production and manufacture of IT components in a competitive manner.
- To improve accessibility to public administration for all citizens, bringing transparency to government processes.
- To establish and develop IT infrastructure and maximize its use nationwide.
- To improve judicial procedures and enhance the dispensation of justice.
- To improve food production and food security.
- To promote tourism and Nigerian arts & culture.
- To improve healthcare delivery systems nationwide.
- To enhance planning mechanisms and forecasting for the development of local infrastructure.
- To enhance the effectiveness of environmental monitoring and control systems.
- To re-engineer and improve urban and rural development schemes.
- To empower children, women and the disabled by providing special programs for the acquisition of IT skills.
- To empower the youth with IT skills and prepare them for global competitiveness.
- To integrate IT into the mainstream of education and training.
- To create IT awareness and ensure universal access in order to promote IT diffusion in all sectors of our national life.
- To create an enabling environment and facilitate private sector (national and multinational) investment in the IT sector.

- To stimulate the private sector to become the driving force for IT creativity and enhanced productivity and competitiveness.
- To encourage government and private sector joint venture collaboration.
- To enhance national security and law enforcement.
- To endeavour to bring the defence and law enforcement agencies in line with accepted best practices in the national interest.
- To promote legislation (Bills & Acts) for the protection of on-line, business transactions, privacy and security.
- To establish new multi-faceted IT institutions as centres of excellence to ensure Nigeria's competitiveness in international markets.
- To develop human capital with emphasis on creating and supporting a knowledge-based society.
- To create Special Incentive Programs (SIPs) to induce investment in the IT sector.
- To generate additional foreign exchange earnings through expanded indigenous IT products and services.
- To strengthen National identity and unity.
- To build a mass pool of IT literate manpower using the NYSC, NDE and other platforms as "train the trainer" Scheme (TTT) for capacity building.
- To set up Advisory standards for education, working practices and industry.
- To establish appropriate institutional framework to achieve the goals stated above.

Policy Strategies

Nigeria intends to achieve her ICT policy through the following strategies:

- Establishing a coordinated program for the development of a National Information Infrastructure (NII) State Information Infrastructure (SII) and Local Information Infrastructure (LII) backbone by using emerging technologies such as satellite including VSAT, fibre optic networks, high speed gateways and broad band/multimedia technologies.
- Providing adequate connectivity to the Global Information Infrastructure (GII)
- Addressing open standards for further liberalization and the fiscal measures including including incentives to substantially improve telephone teledensity and make IT more affordable to the citizenry.
- Establishing IT Parks as incubating centres for the development of software applications at national, state and local levels.

Ghana

The Government of Ghana is committed to pursuing an ICT for Accelerated Development (ICT4AD) Vision intended to improve the quality of life of the people of Ghana by significantly enriching their social, economic and cultural well-being through the rapid development and modernization of the economy and society using information and communication technologies as the main engine for accelerated and sustainable economic and social development (http://www.ict.gov.gh/html/ministerial%20ict%20policy%20statements).

The main mission of the Ghana ICT4AD Vision is: to transform Ghana into an information-rich, knowledge-based and technology-driven high-income economy and society. Some of the key sub-missions of the vision are (http://www.ict.gov.gh/html/ministerial%20ict%20policy%20statements)

- To develop Ghana's information and knowledge-based society and economy through the widespread development, deployment, and exploitation of ICTs within the society and economy.

- To transform the educational system to provide the requisite and educational and training services and environment capable of producing the right types of skills and human resources required for developing and driving Ghana's information and knowledge-based economy and society. research and development

- To develop Ghana's (R&D) capacity and capabilities with the potential to conduct and engage in advanced and cutting-edge R&D work required for supporting the development of a globally competitive information, knowledge-based and high-tech export industry and services.

The Government is committed to pursuing a number of key strategies geared towards the achievement of the stated missions of the vision. Key among them are strategies to:

- Transform Ghana into an information and knowledge-driven ICT literate nation;

- Promote the deployment and exploitation of information, knowledge and technology within the economy and society as key drivers for socio-economic development;

- Modernize Ghana's educational system using ICTs to improve and expand access to education, training and research resources and facilities, as well as the improve the quality of education, and training and make the educational system responsive to the needs and requirements to the economy and society with specific reference to the development of the information and knowledge-based economy and society and

- Improve the human resource development capacity and the Research and Development (R&D) capacity of Ghana to meet the demands and requirements for the nation's information and knowledge-based economy and society.

The Government as part of this vision acknowledges the key role that ICTs can play in educational delivery and training and the need for ICT training and education in schools, colleges and universities. The Government further acknowledges the role that ICTs can play in literacy education and needs to improve the educational system as a whole.

The ICT Policy Statement is designed to guide the process of the deployment and utilizaation of ICTs within the Ministries to support its organizational activities and operations within the framework of the national ICT4AD vision.

- Ghana is ahead of several West African countries including Nigeria telephone penetration higher, Internet bandwidth (e.g. NCS BW is bigger than many West African Telco's 10mb)

- Telecommunications assets of GT, GBC and VRA maybe strategic to development

- Private sector is becoming foreign owned e.g. South Africa 30% empowerment, 49% foreign max investment in Policy Framework, force alignment (http://www.ict.gov.gh/html/ministerial%20ict%20policy%20statements)

The Strategies

- Create the conditions for Government Ministries, Public Sector Organizations (PSO) and private sector establishments to train and up-date the skills of their personnel through in-house training, provision of regular refresher courses, study and training leaves

- Promote basic training in ICTs skills in all schools and tertiary institutions

- Ensure the development of a large pool of ICT professionals with wide range of state of-the-art ICT skills to meet the manpower needs of the country.

- Encourage the private sector, particularly key industries, to establish comprehensive training programmes for their own work-forces, especially in new technologies

- Support and facilitate the training of women in key skills required by the information and knowledge economy
- Encourage life-long learning within the working population to promote on the job training, skills update, further and continuing education and learning within the public and private sector
- Ensure that all citizens who qualify to enter tertiary education in Ghana will benefit from such education, irrespective of their socio-economic background
- Ensure the provision of support to increase access to technical, vocational education and training to all categories of persons with disabilities and other vulnerable groups
- Encourage links between the education and training systems and the industry
- Ensure that traditional apprenticeship system is reformed and strengthened to improve productivity in the informal sector
- Encourage internships, co-opts and work-study programmes in all levels within various institutions.
- Enact laws that provide equal access to physically challenged and vulnerable groups to ICT training and education.
- Ensure that training systems and mechanisms are developed to facilitate coordination and linkage between different sectors of the economy including research institutions and industries
- Promote and enforce high standards in education, vocational training and life-long learning to facilitate the development of a globally competitive quality and professional manpower to support the development of Ghana's information and knowledge-based economy and economy

- Promote world-class standards, to support nation-wide professional ICT skills accreditation
- Promote initiatives targeted at re-training and re-skilling of workers within the civil and public services as well as workers in the private sector to provide them with requisite professional skills and expertise to enable them fully and effectively participate in the development of the information and knowledge economy. (http://www.google.com.ng/search?hl=en&q=ghana+ict+policy+and+strategy+to+achieve+it&start=10&sa=N)

Pupua New Guinea

In October of 1978, the genesis of a process leading to a coherent national policy on Information and Communication Technologies (ICT) commenced in New Guinea. Since then, a structured national ICT policy has begun to emerge, under the mandate of then Secretary to the Department of Information and Communication in November 1992 (www.scrind.com/.../ICT.policy.final-Experts-Report-Papua-New-Guinea). This resulted in the National Policy on Information Communication Technology, which was tabled in parliament by the then Minister for Information and Communication Services, the Hon Martin P Thompson.

In June 2006, the National Executive Council (NEC) approved the development of an integrated government information system (IGIS) to govern the use of information services in New Guinea government departments. This decision is aligned to the focus within the Medium Term Development Strategy (2005-2010) on the use of ICTs to bridge the digital divide, and improving delivery of government service.

The Draft National Information Communication Technology Policy Frame work was approved by the NEC in 2005. In response to the draft framework, an inter-agency ICT Taskforce was established under National Executive council

Decision No. 280/2005, to take steps to formulate a national ICT policy. In the light of the above, the government's ICT policy focuses on the following areas (www.scrind.com/.../ICT.policy. final-Experts-Report-Papua-New-Guinea)

- It is the overarching objective of government to secure the social and economic benefits of an efficient ICT sector. These benefits can be obtained in such areas as education, health, national security, justice, agriculture, government administration and commerce. Increased access to information and communication technology has transformed many parts of the world and helped many developing economies.
- Pupua New Guinea must have an efficient ICT infrastructure as the backbone of ICT Policy with the use of technology appropriate to circumstances of PNG (Pupua New Guinea). This will require substantial investment to refurbish the existing network, extend its availability across the country and increase technical capabilities to support high-speed broadband.
- The government aims to substantially increase access to basic telecommunications services across PNG with service to be available at affordable prices. By reforming the ICT sector, the government intends to make telecommunications services available to ever more PNG.
- It is crucial to have a transformed and efficient Telikom PNG (Pupua New Guinea). The government's stage introduction of open competition is aimed at achieving this transformation, in combination with other efforts to improve Telikom PNG's operational capabilities.
- PGN will enjoy effective and sustainable competition to deliver market discipline and economic benefits. This competition will be built upon a clear ad achievable ICT policy with a part to take PGN from

the existing environment to competitive markets.

The government sleeks improved international capacity and connectivity to help PGN to truly become part of the international community.

- Finally, the government aims to secure the benefits that can flow from increased availability and use of the Internet (www. scribd.com/.../ICT.Policy.Final-Experts-Report-Pupua-New-Guinea).

Bangladesh

In October 2002, the government of Bangladish issued its national information and communication Technology (ICT) policy stating the importance of this sector and referring to the designation granted by the Prime Minister as a thrust sector. (ww.usaid. gov/bd/files/afe_review). With this document, the government voiced the intention to utilize the ICT sector as a tool to increase the socio-economic development of the country. As a follow up on the ICT policy, necessary organizational structure will be created exclusively for ICT industry to consolidate the various aspects of ICT now being handled by ministries of science and technology, commerce, industries, cultural affairs, law, post and telecommunications and others.

The country's ICT policy focuses on the following thematic areas:

- ICT Infrastructure
- Human Resource Development
- Industry
- Export Market Development
- Fiscal
- Legal and Regulatory
- Electronic Commerce
- E-governance
- Bangla Standard and Practices
- Others (www.fbeci-bd.org/ictp.pdf)

Trinidad and Tobago

The creation of a viable strategy for the effective deployment of information and communication technology (ICT) for national development continues to be a major challenge faced by many of the countries in the Commonwealth, especially the less developed countries and small Island States(Ramarine and Wilson, nd).Trinidad and Tobago is in a prominent position in the global information society through real and lasting improvement in social, economic and cultural development as a result of ICT deployment and usage. Amongst their admirable National policy objectives and strategies are:

- Providing all citizens in the country with affordable internet access.
- Focusing on the development of our children and adult skills to ensure a sustainable solution and a vibrant feature.
- Promoting citizen trust, access and interaction through good governance.
- Maximizing the potential within all our citizens, and acceleration of innovation to develop knowledge – based society (www. fastforward.it).

Notwithstanding the strength and dominance of the country in other areas like oil and gas, the government decided in 2002 to take the bold step of exploiting ICT for national development. This action was expected to drive the development of a vibrant knowledge based society as a major facet of the national development agenda. As a result, the National ICT (NICT) plan was crafted during 2003, launched in December of that year and accelerated into rapid implementation mode (www.fastforward.it).

FACTORS AFFECTING THE FORMULATION OF NATIONAL ICT POLICY IN DEVELOPING COUNTRIES

The importance of ICT policies is understood at the highest political level in many developing countries, and some countries have already adopted their own policies. The effectiveness of an ICT policy in one country does not guarantee that the same recipe would work in another and many developing countries face similar constraints that need to be taken into account when ICT policies are formulated (UNESCAP, 2009) The following are factors affecting ICT policies in developing countries (UNESCAP, 2009; Maclean, Souter, Deane Lilley, 2002; Wohler, Correa & Almeida, nd.)):

- **Weak ICT infrastructure**: Lack of computer and telecommunications infrastructure is a key problem in many developing countries. Therefore, National ICT policies need to be very strong in this area. A master infrastructure development plan can be supported by detailed policies for administrative sectors, geographic areas, types of service, types of educational institute, etc. Government involvement is very essential in the construction of the infrastructure in the foreseeable future in rural areas and remote locations. Only large cities are, at the present time, sufficiently attractive for most private developers, such as mobile phone and Internet service providers.
- **ICT-related goods and services are made available on suppliers' terms and low per capita purchasing power does not allow markets to mature**: Basic information technology, such as personal computers, their peripherals and software are available in major cities of developing countries. Low purchasing power, however, keeps the number of vendors down.

Government ICT policies can help the development of ICT markets by reducing red tape, reducing import taxes and creating a favorable entrepreneurial environment.

- **Continued existence of telecommunications monopolies:** Developing countries in the Asian and Pacific and other regions are mulling over the possibilities for reforming their telecommunications sectors, which are to a great extent, in the hands of government monopolies. A fair degree of liberalization has been achieved in several domestic telecommunications markets, and private Internet service providers have become commonplace. Consequently, more countries are succeeding in eradicating waiting lists for telephone services. However, the liberalization of international telecommunications is taking place painstakingly slowly, and retail prices have practically nothing to do with transmission costs. Governments are protecting their rights to collect tax-like revenue through monopolies, and attempts to change the international accounting rate settlement system (which is an additional reason for the high price of international telephone calls) have not been very successful. National ICT policies cannot afford to ignore the fact that the need for low-cost telecommunications services in developing countries is higher than ever. The policies also need adjustments because the existing market mechanism is being taken over by new modes of operation.

- **Significant variation of ICT readiness between government departments:** Departments and agencies operating in a naturally ICT-intensive field are likely to be more advanced than others. A government can assist by identifying a coordinator agency to maintain information about government ICT development ventures. Another way to benefit from the heteroge-

neity is to develop and test pilot applications in the more advanced departments before these are released for wider use within the government.

- **Public sector is a dominant employer:** The computerization of routine functions allows governments to reduce staff and simultaneously to improve the quality of their services. The effectiveness of such moves is often moderated by inflexibilities in employment contracts that limit the scope for staff retrenchments.

- **Unconducive management structures and styles:** Most failures in ICT application development are caused by poor planning and management, and not by the lack of resources or wrong technology choices. Management of ICT projects is often made more difficult by overly hierarchical organizational structures which are not conducive to innovative ideas. This can create a problem if the management is not aware, or resists becoming aware, of the benefits that could be achieved through the application of ICT. National policies should emphasize the importance of involving senior executives in ICT development and making them accountable for their organization's ICT-related performance

- **Governments are struggling to find money for basic public services:** Government budgets tend to be tight, especially in developing countries, and this can create problems for rational ICT development and obstruct the ability to react quickly to new requirements or to buy the latest technology. In order to get value for money, ICT policies should require that the specifications of systems developed or purchased are reconfirmed by third-party experts before the order is placed

- **The penetration and influence of the Internet are still minimal:** The Internet is changing the way in which data and information are collected and disseminated and how services are provided to clients. Therefore, most new systems should be developed with either immediate or future Internet connectivity in mind.

- **Governments find it difficult to recruit and retain qualified ICT staff:** A key problem for the effective application of ICT in developing countries is the inadequacy of human resources Apart from a lack of qualified ICT-system personnel, there is often high turnover of such personnel which can seriously hamper systems development or daily operations. In addition, the ICT skills of other related personnel are not very developed. These constraints can lead to delayed and uncoordinated ICT development and contribute to inadequate data security. ICT policies need to address human resource development needs in a broad educational context (www.unescap.org/stat/gc/box-ch8.asp*)*

- **Problem of lack of awareness on the importance of ICT decisions for national policies and regulations:** There is the problem of lack of awareness on the importance of ICT decisions for national policies and regulations. In a research conducted by Maclean, Souter, Deane & Lilley (2002) it was reported that was lack of awareness of the role that ICT policies can play in supporting economic and social development..

- **Lack of technical and policy capacity on ICT issues:** especially on emerging issues in such areas as the migration to IP based networks, the implementation of feature generation mobile communication systems, e–commerce applications, and protection of intellectual property rights are viewed as fundamental obstacles to effective participation by developing countries.

- **Weaknesses in ICT policy processes:** are also a barrier to effective ICT policy formulation and implementation. The barriers include lack of political leadership, absence of national ICT strategy, ineffective coordination between different government departments and agencies with ICT responsibilities, and the absence of ICT policies processes posses a great problem.

FUTURE TRENDS

The last few years have witnessed steady growth in the number of developing counties that have formulated and integrated ICT policies. This is due to growing level or awareness of the role of policy framework in facilitating deployment of ICTs for the development of different facets of the society. It is hoped that, before long, all developing countries will adopt ICT policies.

CONCLUSION

This chapter has focused on national ICT policy processes in developing countries. It examined the need for information and communication technology policy, ICT policy development process; the role of an ICT policy, case study of ICT policies in some developing countries as well as factors affecting national ICT policies. Efforts should be made by developing countries yet to formulate ICT policies in order to devise and implement them to give proper direction to ICT deployment.

REFERENCES

A study by the commonwealth Telecommunications Organization and Panos London. (n.d.). Retrieved February 17, 2009 from http://www.cto.int/publications/Loudervoicesfinalreport.pdf

Adomi, E. E. (2008). *Library and information service policies*. Benin City, Nigeria: Ethiope Publishing Corporation.

Adomi, E. E., & Igun, S. E. (2008). ICT polices in Africa . In Cartelli, A., & Palma, M. (Eds.), *Encyclopedia of information communication technology* (pp. 384–389). Hershey, PA: Information Science Reference.

Agyeman, O. T. (2007). *Survey of ICT and Education in Africa: Nigeria Country Report.* Retrieved on the 18th of August, 2009 from www.infodev. org/en/Document.422. Pdf

Browne, P. (1996). *Study of the effectiveness of Informatics policy instruments in Africa.* Addis Ababa, Ethiopia: ECA. Retrieved on the 20th of July 2009 from www.belanet.org/partners/aisi/ policy/infopol/sumbrown.htm

Building Local Capacity for ICT Policy and Regulation. A needs Assessment and Gap Analysis for Africa, the Carribean, and the Pacific. Retrieved March 21, 2009 from www.infuder.org/en.project

Considerations for ICT policy formulation in developing countries. (n.d.). Retrieved on of June 20th, 2009 papers.ssrn.com/sol3/.../SSRN_ID983944_code733084.pdf?

Dzidonu, C. (2002). *A blueprint for developing national ICT policy in Africa.* Retrieved October 1, 2009 from http://www.atpsnet.org/pubs/specialpaper/Dzidon.pdf

Federal Republic of Nigeria. (2001) Nigerian National Policy for Information Technology (IT). Retrieved April 14, 2009 from www.nita.gov.ng/ document/nigeriaitpolicy.pdf

Government of Mozambique. *Information and Communication Technology Policy Implementation Strategy: Towards the Global Information Society.* Mozambique, June. 2002. http://www. idrc.ca/en/ev-90304-201-1-DO_TOPIC.htm, Retrieved on the 7th of July 2009

http://wougnet.Org/ICTpolicy/ug/ugictpolicy. html. (n.d.). Retrieved on the 20th of June, 2009.

ICT Policy for Lesotho. Final 4. (March 2005). Retrieved April 10 2009 from www.Lesotho.gov. is/documents/lesotho-ICT-Policy-Final.p

ICT sub sector in Bangladesh. (n.d.). Retrieved on the 9th of November, 2000 from www.usaid. gov/bd/files/afe-review

Implementation team on global Policy participation. (2002). Global policymaking for information And communications technologies: Enabling meaningful participation by Developing-nation stakeholders. Retrieved October 1, 2009 from http://www.markle.org/downloadable_assets/ roadmap_report.pdf

James, T. (Ed.). (2001). *An Information Policy Handbook for southern African, A Knowledge Base for Decision-Makers.* International Development Research Centre.

Khor, M. (1995). *Globalisation and the Need for coordinated Southern policy Response, Cooperation South.* New York: UNDP.

Marcelle, G. N. (2001). *Getting Gender into African ICT Policy: A Strategic View.* Retrieved July 17th 2009 from http://www.idrc.ca/en/ev-9409-201-1-DO_TOPIC.htIm

McQuail, D. (2000). *McQuail's Mass Communication Theory.* London: Sage Publications.

Njuguna, E. (2006). *ICT policy in developing countries; Understanding the Bottlenecks.* Retrieved March 11. 2008.from http://www.ptc. org/events/ptc06/program/public/Proceedings/ Emmanuel%20Njuguna_pa-per_w143.pdf

Okado, G. (2007). *Formulation of a National ICT Policy: African Technology Policy Studies Network.* Retrieved on the7th of July 2009 from www.atpsnet.org./pubs/brief/Technopolicy%20 Brief%.pdf

Ramuarine, D., & Wilson, J. (n.d.). *Developing National ICT Strategies for Small Island States: Case Study – Trinidad and Tobago*. Retrieved on the 11th November 2009 from www.fastforward.it

Republic of Ghana. *National ICT policy and Plan Development committee: Ministerial ICT policy statements*. (n.d.). Retrieved on the 18th of August 2009 from http://www.ict.gov.gh/html/ministerial%20ict%20policy%20statements.htm

Rowan, M. (2003*). Lesson Learned from Mozambique's ICT Policy Process*. Retrieved on the 7th of July 2009 from www.idrc.ca/en/ev-46223-201-1-DO_TOPIC.html

The National ICT policy in Bolivia. (n.d.). Retrieved on the 10th of November, 2009 from http://www.ucd.org/articles/the-national-ict-policy-in-bolivia/?

UNDP. (2001). *Role of UNDP in information and communication technology For development*. Retrieved October 1, 2009 from http://www.undp.org/execbrd/pdf/DP2001CRP8.PD

UNESCAP. (2009*). Considerations for ICT policy formulation in developing countries*. Retrieved October 1, 2009 from tat/gc/box-ch8.asp http://www.unescap.org/S

Weama, T. M. (2005). *Brief history of the development of an ICT Policy in Kenya*. Retrieved from www.scridb.com/.../ict-policy-final-expert-report-pupua-new-guinea

World Bank. (2006). *2006 information and communications for development: Global trends and policies*. Retrieved October 1, 2009 from http://www-wds.worldbank.org/external/default/WDSContentServer/WDSP/IB/2006/04/20/00001200 09_20060420105118/Rendered/PDF/359240PAP ER0In101OFFICIAL0USE0ONLY1.pdf

KEY TERMS AND DEFINITIONS

Economic Development: The attaining of economic power by a country.

Governance: The act of exercising authority.

Informatics: The application of computers and statistics to the management of information.

Information and Communication Technology (ICT): Is an electronic device for collecting, storing, processing, and disseminating information.

Public Service: A service that is performed for the benefit of the public or its institutions.

Satellite: Man-made equipment that orbits around the earth or the moon.

Telecommunication: Systems used in transmitting messages over a distance electronically.

Chapter 15
Regulation of Internet Content

Esharenana E. Adomi
Delta State University, Nigeria

ABSTRACT

This chapter focuses on regulation of Internet content. It presents the arguments for and against Internet content regulation, approaches to content regulation on the Net, how Internet content is regulated in different parts of the world, issues inherent in content regulation, choice of content regulation mechanism as well as future trends.

INTRODUCTION

The growth of communications and telecommunication technology has spurred the growth in regulation of new industries. As the technologies advance and cultural and social mores change, new regulation has been needed. (Song, 2001) The Internet today has reached a level of political importance where some form of regulatory policy is needed. The problem is to determine which policies to govern which aspects of the Internet. From one perspective, the Internet does not exist: it is just a conglomeration of linked individual networks which has no formal corporeal existence. From this perspective, there is no need for any policy

save laissez-faire. What may need to be governed are specific pieces, since there is no whole. From another perspective, however, the elements of the Internet constitute a conceptual whole, the ultimate commons, where no part can function well without all other parts operating well. From this perspective, some form of overall governance is essential, which without a whole, there will be no parts(Mathiason & Kuhlman, n.d.)

There has been increasing concern in the last ten years about damaging Internet content including violence and sexual content, bomb making instructions, terrorist activity, and child pornography. Consequently, many governments around the world have sought to address the problem posed by material on the Net that are illegal under the offline laws, and those considered

DOI: 10.4018/978-1-61692-012-8.ch015

harmful to or unsuitable for minors. The nature of content of principal concern has varied greatly from political speeches to material promoting or inciting to racial hatred, to pornographic material (Wikibooks, 2008).

Essentially, to "regulate" means to monitor or control certain product, process or set of behaviours according to certain requirements, standards, or protocols. It would however seem that at least two different senses of regulation have been used in discussion involving the Internet. Sometimes regulatory discussions have focused on the content of the Internet, as in the case of whether online pornography and hate speech should be censored on the Net. And sometimes the regulatory discussion have centered on questions pertaining to which kinds of processes – rules and policies – should be implemented and enforced in commercial transactions in cyberspace. In physical space, both kinds of regulations also occur (Tavani, 2007). In this chapter, regulation of the Net is used in the sense of the former – content control/regulation.

Internet users appear puzzled by governments' intention to regulate content of the Net. Often users say the Internet is a powerful medium that will be stifled by regulation. However, the power of the Internet is the reason that governments want to regulate it.(Ang, 1997).

The concept of "content" in cyberspace encompasses anything that is created and would apply, from ordinary e-mail to websites and weblogs. The wide scope of the word "content" provides a difficulty for governments when considering content regulation on the Net. The issue of regulating Internet content is part of ongoing debate between those who believe that the state has a role to ensure that harmful content is prohibited and those who believe that the individual must have the right to choose. The gulf between community rights and individual rights divides the debate over Internet content regulation. Governments across the globe, seek to regulate access to speech on the Net so as to establish local control (Papadopoulos, Kafeza & Lessig, 2006).

As far as content regulation is concerned, it is necessary to distinguish between "illegal" and "harmful" content – these two types of content need to be treated differently. The former is criminalized by national laws while the latter is regarded as offensive or disgusting by some people but is generally not criminalized by national laws. For instance, child pornography falls under the "illegal content" category while adult pornography falls under the "harmful content" category. There are also some grey areas such as hate speech and defamation, which in some countries are considered as criminal offences and in others not (Akdeniz, n.d.)

The purpose of this chapter is to describe regulation of Internet content. It presents the arguments for and against Internet content regulation, approaches to content regulation on the Net, how Internet content is regulated in different parts of the world, issues inherent in content regulation, choice of Internet content regulation mechanism as well as future trends.

ARGUMENTS FOR AND AGAINST INTERNET CONTENT REGULATION

Several individuals have expressed views for and against control of Internet content. The following arguments have been advanced in favour of some form of regulation of Net content (Darlington, 2009):

1. The Internet is fundamentally just another communications network. According to this argument we should regulate the Internet as we regulate radio, television, and telecommunications networks.. This argument suggests that, not only is the Internet in a sense, just another network, as a result of convergence it is essentially becoming *the* network so that, if we do not regulate the it at all, effectively over time we are going to abandon the notion of content regulation.

2. There is a range of problematic content on the Internet. There is illegal content such as child abuse images; there is harmful content such as advice on how to commit suicide; and there is offensive content such as pornography. The argument goes that we cannot regulate these different forms of problematic content in the same way, but equally we cannot simply ignore it.

3. There is criminal activity on the Internet. This includes spam, scams, viruses, hacking, phishing, money laundering, identification theft, grooming of children. Almost all criminal activity in the physical world has its online analogue and again, the argument goes, we cannot simply ignore this.

4. The Internet now has users in every country totaling about 1.5 billion. This argument implicitly accepts that the origins of the Internet involved a philosophy of free expression but insists that the user base and the range of activities of the Interet are now so fundamentally different that it is a mass media and needs regulation like other media.

5. Most users want some form of content regulation or control.

However, some people argue that it would be wrong to regulate the content of the Internet because of the following reasons (Darlington, 2009):

1. The Internet was created as a totally different kind of network and should be a 'free' space. This argument essentially refers back to the origins of the Internet, when it was first used by the military as an open network which was designed to ensure that the communication always got through, and then by academics who largely knew and trusted each other and put a high value on freedom of expression.

2. The Internet is a 'pull' not a 'push' communications network. This argument implicitly accepts that it is acceptable, even essential, to regulate content which is simply 'pushed' at the consumer, such as conventional radio and television broadcasting, but suggests that it is unnecessary or inappropriate to regulate content which the consumer 'pulls' to him or her such as by surfing or searching on the Net.

3. The Internet is a global network that cannot simply be regulated. Almost all content regulation is based on national laws and conventions and of course the Net is a worldwide phenomenon, so it is argued that, even if one wanted to do so, any regulation of Net content could not be effective.

4. The Internet is a technically complex and evolving network that can never be regulated. Effectively the Web only became a mass media in the mid 1990s and, since then, developments - like Google and blogging - have been so rapid that, it is argued, any attempt to regulate the medium is doomed.

5. Any form of regulation is flawed and imperfect. This argument rests on the experience that techniques such as blocking of content by filters have often been less than perfect - for example, sometimes offensive material still gets through and other times educational material is blocked.

CONTENT REGULATION IN DIFFERENT PARTS OF THE WORLD

There have been variour efforts to regulate content of the Internet in different parts of the world. Countries regulate different content differently, using various laws and regulations (Mosteshar, 1996). These regulatory efforts have been widely documented by scholars and organizations, (Mosteshar, 1996; Electronic Frontiers Australia, 2002;Sytnyk, 2005; The OpenNet Initiative (ONI); 2007; Wikipedia, 2009) and are presented as follows.

Asia

It is natural that Asia, a region with extraordinary cultural, social, and political diversity, is home to a broad and range of approaches, policies, and practices toward Internet censorship. Many Asian and Middle Eastern countries use any number of combinations of code-based regulation to block material that their governments have deemed inappropriate for their citizens to view. China and Saudi Arabia are two good examples of nations that have achieved high degrees of success in regulating their citizens access to the internet (ONI, 2007; Wikipedia, 2009).

Malaysia, and Nepal do not use technical filtering to implement their policies on information control, but China, Myanmar, and Vietnam heavily rely on pervasive filtering as a central platform for shaping public knowledge, participation, and expression. The filtering practices of Thailand and Pakistan are more targeted, as they blocked a substantial number of sites across categories of content considered sensitive or illicit. The remaining countries in Asia tested by ONI filtered on a selective basis and on targeted topics, including India (ethnic and religious conflict), South Korea (sites containing North Korean propaganda or promoting the reunification of North and South Korea), and Singapore (pornography).

Of countries filtering *political* content, China, Myanmar, and Vietnam blocked with the greatest breadth and depth, spanning human rights issues, reform and opposition activities, independent media and news, and discrimination against ethnic and religious minorities. Thailand and Pakistan blocked political content to a much more limited degree than China, Myanmar, or Vietnam.

A narrower range of *social* content was blocked in Asian countries. Many countries, including Vietnam, cited obscene content as a major justification for engaging in technical filtering. Singapore, Thailand, China, Pakistan, and Myanmar actually blocked pornographic content to varying degrees. Pakistan filtered a number of sites posting Danish cartoon images of the Prophet Muhammad widely condemned as blasphemous, while India also blocked a limited number of sites providing more extreme viewpoints on religion. South Korea and Thailand filtered a small selection of gambling sites.

Conflict and security blocking was carried out by Myanmar, China, South Korea, India, Pakistan, and Thailand most frequently in regard to groups or movements implicated in "secessionist" or pro-independence activities, or in regard to disputed territories and border conflicts.

Myanmar, China, Vietnam, Thailand, and Singapore filtered *Internet tools*, including free Web-based e-mail providers, blog hosting services, and more frequently proxies and other circumvention tools. South Korea blocked pirated software on a nominal basis.

Each of the countries that practice extensive filtering in the region have issued ambitious regulations that aim to bring Internet users under government supervision and control, even if the feasibility of such oversight remains in doubt. Myanmar, China, and Vietnam engage in constant, unremitting supervision of and interference in other forms of media. Well-established strategies include the shuttering of reformist newspapers and Web sites, the institutionalized supervision over content, and the intimidation and harassment of dissidents, journalists, and human rights activists.

In the regulation of Internet, the corresponding phenomenon is the delegation of policing and monitoring responsibilities to Internet Service Providers (ISPs), content providers, private corporations, and users themselves. These frameworks are not structured to accommodate only voluntary self-regulation along industry lines, but rather they exact compliance with state-imposed requirements through the looming threat of shutdowns, loss of license, fines, job dismissals, and even criminal liability. Vietnam and China also apply a more direct form of censorship through the detention of cyber dissidents, while in Pakistan the Supreme Court authorized the police to register criminal cases against publishers of blasphemous content against the Prophet Muhammad, even though no one was apprehended.

China can point to a series of regulations which systematically proscribe nine to eleven types of illegal content, and this number is growing. With the aid of a legal framework where even unemployment rates and family planning statistics are state secrets, the central propaganda organ issues instructions throughout the government hierarchy to media organizations, hosts such as BBS and blog platforms, and other content providers to suppress discussion of an ever-expanding list of proscribed topics.

Whether or not a legal basis for filtering is implicit in content regulations, in many Asian countries filtering has proceeded despite the lack of clear authority to do so. This includes countries with established democratic systems and protections for the press and other forms of speech. For instance, while India is in the process of centralizing its filtering at the international gateway level and therefore improving its efficacy, many still question whether its primary legal authorization for filtering, the 2000 Information Technology Act, is valid in light of constitutional requirements for limits to the freedom of expression (ONI, 2007).

Australia and New Zealand

Australia maintains some of the most restrictive Internet policies of any Western nation, while its neighbor, New Zealand, is less rigorous in its Internet regulation. Without any explicit protection of free speech in the constitution,(Jordan, 2002, cited by ONI, 2007) the Australian government has used its "communications power" delineated in the constitution to regulate the availability of offensive content, endowing a government entity with the power to issue take-down notices for Internet content hosted within the country. A number of state and territorial governments in Australia have also passed legislation making the distribution of offensive material a criminal

offense, as the constitution does not afford that power to the national government.(Electronic Frontiers Australia, 2006, ONI, 2007)

The Australian government also promotes and finances an "opt-in" filtering program, in which Internet users voluntarily accept filtering software which blocks offensive content hosted outside of the country. There are no plans for a countrywide Internet service provider (ISP)-level filtering regime at present, though Australia's handling of hate speech, copyright, defamation, and security signal the government's desire to increase the scope of its Internet regulation.

By contrast, New Zealand is less strict in its Internet regulation. The government maintains a more limited definition of offensive content that can be investigated by a designated government entity, although—unlike in Australia—the definition includes hate speech (despite it being illegal in both countries). Furthermore, the government has not passed legislation to permit issuance of take-down notices for such content and its enforcement of Internet content regulation by prosecution almost solely focuses on child pornography. Although New Zealand Internet copyright policies have not yet been formalized, its defamation and security policies are fairly similar to Australia.

Overall, however, Australia maintains a stricter regime of Internet censorship and regulation than New Zealand and much of the Western world, though not at the level of the more repressive governments that ONI has studied.

Commonwealth of Independent States

As a former superpower—with a tradition of authoritarianism, poorly developed independent media, and lack of private rights—the Commonwealth of Independent States (CIS) would seem to be a typical setting for substantive and pervasive Internet controls. (The CIS consists of eleven countries: Armenia, Azerbaijan, Belarus, Georgia, Kazakhstan, Kyrgyzstan, Moldova,

Russia, Tajikistan, Ukraine, and Uzbekistan. Turkmenistan has been an associated member since 2005. With a strong political and economic influence over its neighboring countries, Russia remains the predominant political actor and strategic economic power in the group.) However, the reality is variegated and complex. While the CIS region is home to some of the world's most repressive measures and advanced techniques for subtly "shaping" Internet access, it also showcases examples of just how profoundly the Internet can affect social and political life.

The states within this region have a conflicted relationship with the Internet. Most of them have adopted national development strategies that emphasize information technology (IT) as a means for economic growth, with some even declaring their intent to become regional "IT powerhouses." IT development is favored because it is seen to leverage the comparative advantage of the ex-Soviet educational system with its emphasis on mathematics and engineering, and the strong tradition of innovation in the computing and technology sector. Until its demise in 1991, the Union of Soviet Socialist Republics (USSR) was one of the few countries with a "homegrown" capacity in supercomputing, cryptography/ crypto-analysis, and worldwide signals intelligence gathering. Currently many former Soviet citizens are among the leaders of the global IT industry.

CIS governments, at the same time, are wary of the civil networking and resistance activities that these technologies make possible. In recent years, Ukraine, Georgia, and Kyrgyzstan have experienced "color revolutions," where networked opposition movements (albeit movements that are more reliant on cell phones than on the Internet) have effectively challenged and overturned the results of unpopular (or allegedly fraudulent) elections. Neighboring governments fear that these challenges were made possible by opposition groups leveraging IT to organize domestic protest (often with the help of foreign-funded NGOs), and are therefore wary of leaving the sector unregulated

and without control. Many now see the Internet and other communications channels in national strategic terms, and these countries have increasingly turned to security-based arguments—such as the need to secure "national informational space"—to justify regulation of the sector.

ONI, in 2006, tested for the presence of filtering in eight of the eleven CIS countries: Azerbaijan, Belarus, Kazakhstan, Kyrgyzstan, Moldova, Tajikistan, Ukraine, and Uzbekistan. Background and baseline testing was also carried out in a further two countries: the Russian Federation and Turkmenistan, although in these two cases limitations on the testing methodology do not allow us to claim comprehensive results.

Of the eight countries in which ONI tested, results did not yield significant patterns of substantive or pervasive filtering. Only Uzbekistan pursued pervasive filtering of the kind found in China, Iran, or some parts of the Middle East. In almost all countries some degree of filtering was present, but this filtering occurred mostly on corporate networks (such as educational and research networks) where accepted usage policies (AUPs) dictated that inappropriate content was not permitted, or in "edge locations", such as Internet cafés where the reasons for filtering were more benign (conserving bandwidth) or left to the discretion of the Internet café owners themselves.

ONI observed that in all eight countries, authorities had taken steps of one kind or another to restrict or regulate their national informational space. These measures include:

- expanded use of defamation and slander laws to selectively prosecute and prevent bloggers and independent media from posting material critical of the government or specific government officials (however benignly, including, as was the case in Belarus, through the use of humor);
- strict criteria pertaining to what is "acceptable" within the national media space,

leading to the deregistration of sites that did not comply (Kazakhstan);

- efforts to compel Internet sites to register as mass media, with noncompliance then being used as grounds for filtering "illegal" content;
- national security issues (Ukraine); and,
- in some cases, government officials have "asked" ISPs — formally or informally— to temporarily suspend sites detrimental to "public order" (Tajikistan).

These sanctions (legal and quasi-legal) are intended to create overall environments that encourage varying degrees of self-censorship among ISPs, who are afraid of jeopardizing their licenses, and among individuals for whom prosecution or imprisonment is too high a price to pay for voicing criticism, which at times amounts to little more than a form of digital graffiti (ONI, 2007).

Europe

The Internet in Europe, in less than a decade, has evolved from a virtually unfettered environment to one in which filtering in most countries, particularly within the European Union (EU), is the norm rather than the exception. Compared with many of the countries in other regions which block Internet content, the rise of filtering in Europe is notable because of its departure from a strong tradition of democratic processes and a commitment to free expression. Filtering takes place in various ways, including the state-ordered takedown of illegal content on domestically hosted Web sites; the blocking of illegal content hosted abroad; and the filtering of results by search engines pertaining to illegal content. As in most countries around the world that engage in filtering, the distinction between voluntary and state-mandated filtering is somewhat not clear in Europe. In many instances filtering by ISPs, search engines, and content providers in Europe is regarded "voluntary," but is carried out with the implicit understanding that

cooperation with state authorities will prevent further legislation on the matter.

The scope of illegal content that is filtered in Europe pertains largely to child pornography, racism, and material that fosters hatred and terrorism, although more recently there have been proposals and revisions of laws in some countries that deal with filtering in other areas such as copyright and gambling. Filtering also takes place on account of defamation laws, and this practice has been criticized, particularly in the UK, for curtailing lawful online behavior and promoting an overly aggressive notice and takedown policy, where ISPs comply by removing content immediately because of fear of legal action. ISPs in Europe do not have any general obligation to monitor Internet use and are protected from liability for illegal content by regulations at the European Union (EU) level, but must filter such content once it is brought to their notice. Therefore the degree of filtering in member states depends on the efforts of governments, police, advocacy groups, and the general public in identifying and reporting illegal content.

There have been moves over the past decade to create a set of common policies and practices at the EU-level on Internet regulation. This is viewed as necessary to encourage regional competitiveness and commerce, to counter Internet crime and terrorism, and to serve as a platform to share best practices amongst nations. Notable advancements in regulation at the EU level—although not directly in the area of filtering— are the definition of ISP liability toward illegal content and obligations toward data retention (ONI, 2007).

Latin America

With the exception of Cuba, systematic technical filtering of the Internet has yet to take hold in Latin America. The regulation of Internet content largely addresses the same concerns and strategies seen in North America and Europe, focusing on combating the spread of child pornography and restricting child access to age-inappropriate material. As Internet usage in Latin America grows, so have defamation, hate speech, copyright, and privacy issues.

The judiciary in Latin America has played an important role in shaping and tempering filtering activity, a development common to North America and Europe. At the same time, there has been a wide range of legal and practical responses to regulating Internet activity. Latin American countries have relied primarily upon existing law to craft remedies to these challenges, though a growing number of Internet-specific laws have been debated and implemented in recent years. These issues have been addressed primarily through the application of cease and desist orders in conjunction with requests to have materials removed from search engine results.

The level of openness of the media environment in Latin America is reputed to be subject to considerable self-censorship, particularly in Brazil, Colombia, Mexico, and Venezuela (ONI, 2007).

Middle East and North Africa

ONI (2007) conducted in-country testing for Internet filtering in sixteen countries in the North Africa and Middle East region. It was discovered that eight of these countries broadly filter online content: Iran, Oman, Saudi Arabia, Sudan, Syria, Tunisia, United Arab Emirates, and Yemen. Another four—Bahrain, Jordan, Libya, and Morocco—carry out selective filtering of a smaller number of Web sites. ONI found no evidence of consistent technical filtering used to deny access to online content in Algiers, Egypt, Iraq, or Israel.

Most of the sites that are targeted for blocking are selected because of cultural and religious concerns about morality. Political filtering, however, is the common denominator in the region. Bahrain, Jordan, Libya, and Syria focus their filtering efforts primarily on political content. Iran, Oman, Saudi Arabia, Sudan, Tunisia, the United Arab Emirates, and Yemen, on the other hand, not only extensively filter political content but also pervasively block content that is perceived to be religiously, culturally, or socially inappropriate.

Regional and internal political conflicts are also reasons for content blocking. For instance, Syria and the United Arab Emirates block all Web sites within the Israeli domain. Morocco blocks Web sites arguing for the independence of Western Sahara.

Internet censorship in the Middle East and North Africa is multilayered, relying on a number of complementary strategies in addition to technical filtering; arrest, intimidation, and a variety of legal measures are used to regulate the posting and viewing of Internet content.

Sub-Saharan Africa

Internet penetration in sub-Saharan Africa lags behind that of much of the rest of the world due to a variety of economic, political, and infrastructural reasons. In spite of these hurdles, most countries in the region view their future success as inextricably linked to harnessing the Internet's promise for economic development. Internet regulation in Africa, as a result, primarily focuses on infrastructure and access-related issues rather than on content regulation, though countries are making moves to broaden the scope of regulation as the Internet spreads.

Given the current restrictions on the freedoms of expression and the press in sub-Saharan Africa, one would expect similar restrictions on Internet freedom. A number of countries in Africa have sought to limit the use of Voice-over Internet Protocol (VoIP) to protect incumbent telecommunications companies. However, ONI discovered evidence of systematic blocking of Internet content in only one country, Ethiopia.. Also, Uganda is alreported, by other sources, to have engaged in one temporary incidence of filtration during the past year ()NI, 2007).

United States and Canada

Though neither the United States nor Canada practices widespread technical Internet filtering at the state level, the Internet is far from "unregulated" in either state. Restrictions of Internet content take the form of extensive legal regulation, as well as technical regulation of content in specific contexts, such as libraries and schools in the United States. While there is some United States law that does control access to materials on the internet, it does not truly filter the internet. In the wake of the United States Telecommunications Act 1996 and the Bill introduced to overturn its decency provisions, it will be clear that legislators have the power to regulate the content of Internet sites. Such control was in fact effectively exercised by the German Government to ban access to certain Web sites through CompuServe because of decency concerns. The pressure to regulate specific online content has been expressed in concerns related to four problems: child-protection and morality, national security, intellectual property, and computer security. In the name of "protecting the children," the United States has moved to step up enforcement of child pornography legislation and to pass new legislation that would restrict children's access to material deemed "harmful." Legislators invoke national security in calls to make Internet connections more traceable and easier to tap. Copyright holders have had the most success in this regard by pressing their claims that Internet intermediaries should bear more responsibility—and more liability—than they have in the past. Those concerned about computer security issues, such as adware and spam, have also prompted certain regulations of the flow of Internet content. Also, in Canada, although not in the United States, there is restriction in publishing of hate speech (Mosteshar, 1996; ONI, 2007; Wikipedia, 2009).

There is heated debate on each of these restrictions. Public dialogue, legislative debate, and judicial review have led to different filtering strategies in the United States and Canada than those described elsewhere in this volume. In the United States, many government-mandated attempts to regulate content have been barred on First Amendment grounds. In the wake of these restrictions, though, fertile ground has been left for private-sector initiatives. The government has been able to exert pressure indirectly where it cannot directly censor. In Canada, the focus has been on government-facilitated industry self-regulation. With the exception of child pornography, Canadian and U.S. content restrictions tend to rely more on the removal of content than blocking; most often these controls rely upon the involvement of private parties, backed by state encouragement or the threat of legal action (see Palfrey and Rogoyski, 2006; cited by ONI, 2007). In contrast to those regimes where the state mandates ISP action through legal or technical control, most content-regulatory moves in both the United States and Canada are directed through private action (ONI, 2007)..

APPROACHES TO INTERNET CONTENT REGULATION

Because the Net is a global medium with no central control, it is not possible to monitor and remove objectionable content completely. However, it is possible to make it more difficult to access, and for barriers to be placed in the way of businesses and Individuals providing such content. (Parliamentary Office of Science and Technology, 2001). Various governments of countries that are concerned about the harm some Internet materials can cause are taking certain steps to regulate access to the Internet. Policies concerning control of Internet content may be grouped into four categories (Wikibooks, 2008; Electronic Frontiers Australia, 2002):

1. Government policy to encourage Internet industry self-regulation and end-user voluntary

use of filtering/blocking technologies. This is the approach taken in the United Kingdom, Canada, and many Western European countries. It also appears to be the current approach in New Zealand where applicability of offline classification/censorship laws to Internet content seems less than clear. In these countries, laws of general application apply to illegal Internet content such as child pornography and incitement to racial hatred. It is not illegal to make content "unsuitable for minors" available on the Internet, nor is access to the same controlled by a restricted access system. Some governments encourage the voluntary use and ongoing development of technologies that enable Internet users to control their own, and their children's, access to content on the Internet.

2. Criminal law penalties (fines or jail terms) applicable to content providers who make content "unsuitable for minors" available online. This is the approach taken in some Australian State jurisdictions and has been attempted in the USA. In these countries, in addition, laws of general application apply to content that is illegal for reasons other than its unsuitability for children, such as child pornography.

3. Government-mandated blocking of access to content deemed unsuitable for adults. This approach is taken in Australian Commonwealth law (although it has not been enforced in this manner to date) and in China, Saudi Arabia, Singapore, the United Arab Emirates and Vietnam, among others. Some countries require Internet access providers to block material while others allow only restricted access to the Internet through a government-controlled access point.

4. Government prohibition of public access to the Internet. A number of countries, like China, either prohibit general public access to the Internet, or require Internet users to be registered/licensed by a government authority before permitting them restricted access.

INTERNET CONTENT REGULATION ISSUES

There are some issues inherent in Internet content regulation .which legislators and others concerned with regulatory frameworks have to considered.. These are as follow (EURIM, 1997):

•   **The application of existing law:** The Internet does not exist in a legal vacuum. All those involved (authors, content providers, host service providers who actually store the documents and make them available, network operators, access providers and end users) are subject to their respective national laws. Those who unwittingly and despite reasonable precaution convey illegal material should have their position under the law clarified beyond reasonable doubt. The issue involves clarification of existing law and of enforcement, not of new legislation.

•   **Illegal content:** It is important to differentiate between content which is illegal and other harmful content. These different categories pose radically different issues of principle and call for very different legal and technological responses. Priority must be given to the application of resources to combat criminal content - such as clamping down on child pornography or on use of the Internet as a new technology for criminals. The task is however not easy, because the definition of what constitutes an offence varies from country to country. Moreover, where certain acts are punishable under the criminal law of one country, but not in another, practical difficulties of enforcing the law may arise.

- **Harmful content:** Various types of material may offend the values and feelings of other people (for example, content expressing political opinions, religious beliefs or views on racial matters). What is regarded harmful depends in part on cultural differences and countries differ on what is permissible or not permissible. International initiatives must therefore take such differences into consideration when exploring co-operation to protect against offensive material whilst ensuring freedom of expression.

- **Detection of breaches of the law:** While detecting breaches of the law in public applications of the Internet (for instance, the WWW) is straightforward, detection is not easy in private applications (for instance,. e-mail). Also, while enforcement of the law is relatively easy within national boundaries, it is much more difficult in an international context. There are technical problems which mean that control is most practical at the entry and exit points to the Network (the terminal used to read or download the information plus the server through which the user gains access to the Internet and the server on which the document is published). Therefore, international co-operation is required to avoid "safe havens" for content that is generally agreed to be illegal.

- **Chain of responsibility:** ISPs play significant role in giving users access to Internet content. It should not however be forgotten that the prime responsibility for content lies with authors and content providers. It is therefore essential to identify accurately the chain of responsibilities in order to place the liability for illegal content on those who create it. The widespread use of filtering devices at points of access should act as a powerful incentive to content providers to "rate" their content.

- **Trade Bodies and Regulatory Models:** In the United Kingdom, a Code of Conduct has been developed and agreed within the trade association body (Internet Service Providers Association - ISPA), with the support of the Department of Trade and Industry. But ISPA membership does not cover all major providers and although the Code is mandatory for members, the range of sanctions is limited. ISPA needs enhanced government support and encouragement if it is to achieve recognition as the authoritative voice of the industry. The primary issue, however, is not how the industry itself is organized, but whether regulation of content on the Internet should be voluntary (that is, self-regulation within existing law) or imposed (by the State or by some other recognized authority), and how such a regulatory regime should relate to the industry. In a number of Member States, information service providers have already set up systems of self-regulation: indeed, the Commission welcomes this general movement and is encouraging a European network of associations of ISPs. INCORE (Internet Content Rating for Europe), a loose association of industry, government, police and user interests, is evidence of the sort of co-operation which could further be extended to the wider international level, but it lacks formal recognition and public funding. The UK Government supports INCORE but is not itself a partner in the project. Industry self regulating bodies, which face common problems, could usefully co-ordinate their approach, in particular regarding technical solutions. The voluntary regulatory body in the UK is the Internet Watch Foundation (IWF) which is funded by a number of providers and has parallels with ICSTIS. The IWF is not independent from the providers (as is ICSTIS from the Premium Rate

Telephone Service industry) and lacks the credibility and influence which formal recognition and legal status could give. The IWF is an attempt at voluntary regulation and deserves to be afforded recognition and status similar to that of ICSTIS. The ICSTIS model is at Appendix I. It is beyond the remit of this paper to consider whether ICSTIS and the IWF should exist as separate bodies in future, or, indeed, whether their natural home might be within a restructured ITC. Suitably strengthened, the ISPA-IWF model is one which other countries could be encouraged to adopt as the foundation for a network of international regulatory regimes, offering co-operation between the authorities and providers to ensure that control measures are effective and not excessive.

- **Methodology and Technology:** There is the issue of how the universal application of codes such as Platform for Internet Content Selection (PICS) (http://www.w3.org/PICS/) content classifications are to be defined and applied. International co-operation is required if such problems are to be overcome.
- **Convergence Issues:** Whatever action is taken to regulate content on the Internet, the convergence of media and means of presentation will require a convergence of regulatory regimes. The control of audio-visual content on television, for instance, should be compatible with that of content on the Internet if the regimes are to be even-handed.

CHOOSING INTERNET CONTENT REGULATION MECHANISM

Whatever the approach to content regulation, the important consideration is that regulation must not inhibit innovation. It would seem that a hybrid between a government-regulated regime and an industry-regulated regime may be the right combination when dealing with censorship in cyberspace.(Wikibooks, 2008)

Because the Internet is universal, there is great need for an international network of hotlines governed by a framework agreement containing minimum standards on the handling of content concerns and stipulating mutual notification between hotlines. The hotline in the country where the content is located is asked to evaluate it and to take action. This mechanism will result in content providers being acted against only if the material is illegal in the host country. The mechanism also overcomes impediments in the complex diplomatic procedures necessary for cross-border cooperation of law enforcement authorities.

Essentially, no regulatory mechanism can work independently of an education and awareness campaign. The Internet industry should have a continuous online and offline program to develop general awareness of self-regulatory mechanisms such as filtering systems and hotlines. Schools should provide the necessary skills for children to understand the benefits and limitations of online information and to exercise self-control over problematic Internet content. The Internet is itself a process, an enormous system for change and response, feedback and transformation. Like the Internet, the legal system and regulatory mechanisms around it must incorporate similar practices of learning and changing (Wikibook, 2008).

FUTURE TRENDS

It is now well recognized that the Internet as we know it today, defies total traditional content regulatory practices. The main reasons are associated with the blurring of the concepts of territory and sectors. But as we consider the future of the Internet, we see even greater challenges ahead, with many questions related to privacy, security and regulation of the Internet. It is also the moment to initiate a global reflection on an improved, more effective and inclusive Internet.(Reding, 2009). The debate on the legitimacy or otherwise of internet content regulation will still continue in the future. Governments and other bodies will continue the efforts to apply regulatory measures. Illegal, offensive and harmful content will not disappear from the Net. However, the number of countries controlling Net content will increase. There is need for studies on how regulation of Internet content can be made more effective and widespread.

CONCLUSION

The Internet as a mass medium of communication not only contains content considered beneficial but objectionable content. Different measures have been adopted in different countries to regulates content considered illegal, offensive and harmful. However, these efforts have not resulted in total control of such content. This could be due to absence of international laws and measures to regulate Internet content, failure of regulation technique (such as blocking of content by filters) to black access to harmful content among others. There is need for developing and adopting international framework for regulation of Internet content for it to be effective.

REFERENCES

Akdeniz, Y. (n.d.). *Speech 3: Controlling internet content: Implications for cyber-speech*. Retrieved November 4, 2009 from http://portal.unesco.org/ci/en/files/18278/11086401441statement_akdeniz.doc/statement_akdeniz.doc

Ang, P. H. (1997). How countries are regulating Internet content. Retrieved April 30, 2009 from http://www.isoc.org/INET97/proceedings/B1/B1_3.HTM

Darlington, R. (2009). *How the Internet could be regulated*. Retrieved May 8, 2009 from http://www.rogerdarlington.co.uk/Internetregulation.html

Electronic Frontiers Australia. (2002). *Internet Censorship:Law & policy around the world*. Retrieved May 12, 2009 from http://www.efa.org.au/Issues/Censor/cens3.html

Electronic Frontiers Australia. (2006). *Internet censorship laws in Australia*. Retrieved May 11, 2009 from http://www.efa.org.au/Issues/Censor/cens1.html

Jordan, R. (2002). *Free speech and the constitution*. Retrieved May 11, 2009 from http://www.aph.gov.au/LIBRARY/Pubs/RN/2001-02/02rn42.htm

Mathiason, J. R., & Kuhlman, C. R. (n.d.). *An International Communication Policy: The Internet, international regulation & new policy structures*. Retrieved November 4, 2009 from http://www.un.org/esa/socdev/enable/access2000/ITSpaper.html

Mosteshar, S. (1996). *Some legal aspects of internet regulation*. Retrieved November 15, 2009 form http://www.mosteshar.com/sig.html

OpenNet Initiative. (2007) *Global Internet Filtering*. Retrieved May 8, 2009 from http://map.opennet.net/filtering-pol.html

Palfrey, J., & Rogoyski, R. (2006). The move to the Middle: The enduring threat of harmful speech to the end-to-end principle. *Washington University Journal of Law and Policy, 21*, 31–65.

Papadopoulous, M., Kafeza, I., & Lessig, L. (2006). *Legal considerations for regulating content on the Internet*. Retrieved from http://www.marinos.com.gr/bbpdf/pdfs/msg52.pdf

Parliamentary Office of Science & Technology. (2001). *Regulating Internet content.* Retrieved May 8, 2009 from http://www.parliament.uk/post/pn159.pdf

Reding, V. (2009). *Internet of the future: Europe must be a key player.* Retrieved May 12, 2009 from http://ec.europa.eu/commission_barroso/reding/docs/speeches/2009/brussels-20090202.pdf

Song, B. K. (2001). *Content regulation on the Internet.* Retrieved May 1, 2009 http://iml.jou.ufl.edu/projects/fall01/song/page5.html

Sytnyk, O. (2005). *Control of internet content: Purposes, methods, legislative bases.* Retrieved November 4, 2009 from http://www.personal.ceu.hu/students/04/Oleksandra_Sytnyk/testfile.pdf

Tavani, H. T. (2007). Regulating cyberspace: Concepts and controversies. *Library Hi Tech, 25*(1), 37–46. doi:10.1108/07378830710735849

URIM. (1997). *EURIM briefing no 19: The regulation of content on the Internet.* http://www.eurim.org/briefings/brief19.htm

Wikibooks. (2008). *Legal and Regulatory Issues in the Information Economy/Censorship or Content Regulation.* Retrieved May 8, 2009 from http://en.wikibooks.org/wiki/Legal_and_Regulatory_Issues_in_the_Information_Economy/Censorship_or_Content_Regulation

Wikipedia. (2009). *Cyberlaw.* Retrieved November 15, 2009 from http://en.wikipedia.org/wiki/Cyberlaw

KEY TERMS AND DEFINITION

Breaches of the Law: Refer to violations or infractions, transgression of the law.

Internet Content Censorship: Is the suppression of speech or removal of material which is considered illegal, objectionable, harmful, sensitive, or inconvenient to the government on the Internet.

Local Control: Is the exertion of regulatory measures by a government of a country over Internet content.

Offline Laws: Refers to laws not connected to a computer or computer network or the Internet

Problematic Content: Refers to material on the internet considered harmful, illegal, and offensive to users. Examples include pornography, content that promote racism, etc.

Regulatory Mechanism: Has to do with the approach or measure adopted to control Internet content.

Restricted Access: Is denied access to material on the Internet. It also refers to Internet content to which access has been denied.

Chapter 16
The New Zealand Response to Internet Child Pornography

David Wilson
Researcher, New Zealand

ABSTRACT

New Zealand's approach to regulating illegal material on the Internet varies from other comparable countries. A single law governs the legal classification of Internet content, commercial films, printed material and a wide variety of other media and covers legal and illegal content. A Crown agency rather than the judiciary determines the legality of material. A specialist, non-police, enforcement agency deals those who possess or distribute illegal material, particularly child pornography. This agency actively seeks out child pornographers and has a high success rate in prosecuting them. This chapter describes the history, development and operation of the New Zealand censorship system, as it applies to Internet content. It is likely to be of interest to policy-makers, law enforcement officers and media regulators in other countries.

COUNTRY PROFILE

New Zealand is a parliamentary democracy located in the South Pacific Ocean, some 2000 kilometres east of Australia. It is composed of two main islands and has a similar land size to the United Kingdom or Italy. New Zealand's population of four million people is primarily descended from mid-19th century British and Irish settlers, though there are significant numbers of indigenous Maori

(15 percent), and later immigrants from the Pacific Islands and Asia. A former British colony, New Zealand is a member of the Commonwealth. New Zealand ranks highly on the Human Development Index (Human Development Report Office, 2008) and consistently is measured as having the lowest levels of perceived corruption in the world (Transparency International, 2008). New Zealand has a high rate of Internet use with 3.36 million users (79 percent of the total population) and a high rate of Internet connectivity – being ranked 31st in the world despite being ranked 122nd for population

DOI: 10.4018/978-1-61692-012-8.ch016

size. Young New Zealand adults have one of the highest Internet usage rates in the world, on a par with Canada, Sweden, Netherlands, Norway and Barbados (UNICEF, 2008).

This high level of Internet use brings with it New Zealand's share of online criminals. One area in which the nation has been innovative in responding to changing technology is in addressing the distribution of child pornography on the Internet. This chapter examines how New Zealand's censorship laws have responded to the challenges of the Internet and identifies particular features of the New Zealand system likely to be of interest to policy makers, academics and law enforcement officials in other countries.

NEW ZEALAND'S APPROACH TO CHILD PORNOGRAPHY

The creation, possession, distribution, importation and provision of access to child pornography are illegal in New Zealand. Child pornography is broadly defined as material that promotes or supports the sexual exploitation of children or young people or that exploits their nudity. A child or young person is considered to be someone 16 years of age or less. Decisions about the legal status of material alleged to be child pornography are made by a specialist, non-judicial classification body, the Office of Film and Literature Classification. The Censorship Compliance Unit of the Department of Internal Affairs (an organisation separate from the New Zealand Police) is primarily responsible for investigating and prosecuting offenders within New Zealand.

The New Zealand approach to child pornography will be detailed in this chapter. But before it does so, it is useful to consider approaches taken in other developed nations.

OVERSEAS APPROACHES TO CHILD PORNOGRAPHY

New Zealand's approach to the detection and legal classification of objectionable material differs from other comparable countries. In Australia state laws prohibit making, possessing or distributing material that describes or depicts a person who is, or appears to be, under 16 years of age in a manner that would offend a reasonable adult (Krone, 2005). Fictitious depictions of the sexual abuse of children or young people are covered, in addition to depictions of real people and the laws apply to a wide variety of media, including electronic material. The state and federal police investigate child pornography offences and the courts determine the legal status of material alleged to be child pornography. Material may be submitted for classification to the Office of Film and Literature Classification and the Office's decision may be used in evidence (Office of Film and Literature Classification, 2006). The legislation governing child pornography and other illegal material is the federal Classification (Publications, Films and Computer Games) Act 1995 and censorship or general criminal law in the eight states and territories.

In England and Wales the law relating to child pornography is spread across five statutes (Gillispie, 2005). It is illegal to make, distribute or possess an indecent image of a child less than 18 years of age. The police investigate and prosecute cases involving 'obscene' material and the courts determine whether material meets the obscenity test (Andrews, 2003). The British Board of Film Classification may classify obscene films and DVDs but it does so in consultation with the Crown Prosecution Service and the police (Perkins, 2009).

In Germany it is illegal to offer, transmit, procure or possess child pornography under the Strafgesetzbuch (German Penal Code). The penal code defines child pornography as pornographic publications that deal with the sexual abuse of children less than 18 years of age. The criminal courts determine whether material is child pornography. The police of each state, many of which have developed specialist online investigative units, carry out enforcement of the law. The criminal investigation department of the federal police works closely with the state police and co-ordinates nationwide investigations (Heuck, 2005). Different specialists undertake investigation, forensic analysis and prosecution of offenders.

The law on child pornography and associated enforcement activity in the United States of America has been the subject of a large body of writing and research. Currently, U.S. law defines child pornography as a visual depiction of a minor engaged in sexually explicit activity and a visual depiction of an actual minor engaging in sexually explicit conduct (Loftus, 2008). The U.S. Constitution protects free speech to an extent unrivalled in most developed countries. The Supreme Court has stated that it "has long held that obscene speech—sexually explicit material that violates fundamental notions of decency—is not protected by the First Amendment."[1] A 2008 decision of the Supreme Court confirmed that virtual child pornography is illegal only if the person in seeking to obtain it believes it to depict real children (United States v Williams, 2008). The decision upheld current law that has been the subject of considerable amendment after earlier provisions had been struck down as unconstitutional. Material such as a work of fiction is protected by the First Amendment. Simulated child pornography is generally not illegal, unless offered as if real children were involved, or if the material is deemed to be obscene (Cohen, 2003). The test of obscenity is complex and involves, in part, applying contemporary community standards to determine if "material appeals to the prurient interest in sex" (Ferraro & Casey, 2005). American child pornography laws apply across a range of media, including the Internet. The law is enforced by local and state police as well as the F.B.I., U.S. Customs and the Postal Inspection Service at the federal level.

In the Republic of Ireland the production, dissemination, handling or possession of child pornography is illegal under the Child Trafficking and Pornography Act 1998. Child pornography is defined as a representation of a person who is, or is depicted as being, under 17 years of age involved in sexual activity. Irish law also outlaws depictions, for sexual purposes, of the genitals of a child (O'Donnell & Milner, 2007). The Gardai, the police force, enforce the law and the courts decide, in the course of a prosecution, whether the material constitutes child pornography.

The paragraphs above outline five jurisdictions with different legal systems and different laws governing child pornography. However, each country has in common that images that record the sexual abuse of children are illegal, the courts generally determine the legal status of alleged child pornography and the police in each country enforce that law. Although Australia, England, Germany and the United States of America are much larger countries than New Zealand they have in common that they are developed democratic nations with high levels of Internet use. They are also countries with which New Zealand frequently co-operates in international efforts to combat child pornography. Ireland has a similar population size to New Zealand and is a developed democratic nation with a high level of Internet use.

A SINGLE NON-JUDICIAL DECISION-MAKING BODY

The overseas jurisdictions outlined above leave it to the courts to determine the legal status of alleged child pornography. In 1993 New Zealand took a different approach in creating the Office of Film and Literature Classification (Classification Office) as the sole body permitted to determine the legal status of a wide variety of media. It was given jurisdiction over the films, videos and literature previously regulated by the three different censorship bodies. Provision was also made for the classification of a wide range of other material, collectively called "publications". This included any word, image or representation recorded in any way that makes it capable of being reproduced. The Classification Office was to have been part of a government department but the parliamentary committee that scrutinised the legislation prior to its enactment recommended that it be established as a Crown entity, a state sector organisation that would operate independently of the government of the day to prevent political censorship. The Crown Entities Act 2004 classed the Classification Office as an 'independent Crown entity', a body with legal protection from government influence that is able to operate without regard to government policy in discharging its duties.

One of the notable features of New Zealand's censorship law is the way it combines a wide variety of media into one law. The Films, Videos, and Publications Classification Act 1993 provides that it applies to any 'publication'. The Act defines 'publication' broadly to include:

- films, books, sound recordings, newspaper, photograph, print or writing; and
- any 'thing' that has images, representations, signs, statements or words printed, impressed, recorded or stored on it.

Since 1995 the courts had recognised computer files as 'publications', within the meaning of the Act (Manch and Wilson, 2003). To avoid any future doubt, the law was amended in 2005 to make specific reference to computer files as publications. Electronic files comprised 82 per cent of all the publications found to be illegal by the Office of Film and Literature Classification in the 2008 (Office of Film and Literature Classification, 2008).

Concern in the early years of censorship centred on addressing the "danger to moral health" posed by films (Christoffel, 1989). New Zealand's censorship system has evolved considerably since that time and now aims to balance the right to free

expression enshrined in the New Zealand Bill of Rights Act 1990 with protecting the "public good" from harm. Central to the classification regime is whether or not a publication is 'injurious to the public good', an issue that is determined by statutory criteria. The focus on 'injury' is a deliberate departure form earlier statutes that used terms such as 'offence', 'indecency' or 'obscenity' as the test for classification. This was the overt intention of the legislation when it was introduced (Bather, 2002). A publication that is not injurious is not subject to any legal restriction. If a publication is found to be injurious then the Classification Office must consider how to remedy the injury. The most common 'injury' is the likely harm caused by exposing children to graphic violent or sexual material that may adversely effect their development. The injury is addressed by restricting the publication to people of an age where they are able to process the content and put it in context without being adversely effected by it.

Where a publication is found to be injurious to such an extent that an age restriction is not able to remedy the harm it will cause then the publication is classified as 'objectionable' and is banned. Certain publications are automatically 'objectionable' under the Films, Videos, and Publications Classification Act 1993 and must be banned. These automatically banned publications include those that promote or support the sexual exploitation of children, sexual violence, necrophilia, bestiality and torture. The legislation also governs the classification of material that degrades or demeans people, promotes discrimination, or depicts self-harm, though such material is not automatically banned.

The wide purview of the Films, Videos, and Publications Classification Act 1993 allows the Classification Office to deal with any publication from an animated children's film to a graphic horror film to child pornography images under the same legal criteria. This allows remarkable consistency in decision-making. While the Classification Office is obliged only to provide written reasons for its decisions to ban publications, its practice of producing written reasons for all classification decisions enables considerable public scrutiny of its activities. Anyone who is dissatisfied with a classification decision is able to have it reviewed by the Film and Literature Board of Review, composed of nine expert members of the public. The Classification Office has 30 staff located in the nation's capital, Wellington. In 2008, the Classification Office deemed 300 publications to be objectionable and half of these banned publications were child pornography. Most objectionable material is submitted for classification by law enforcement agencies (Office of Film and Literature Classification, 2008). Most of the publications received by the Classification Office are DVDs intended for commercial release but anyone can submit a publication for classification so the Classification Office deals with a wide variety of material.

Consistent decision-making is also aided by the fact that all classification decisions in New Zealand are made by the same professional body. The courts have no jurisdiction to classify a publication. If the question of the legality of a publication arises in the course of a prosecution, the courts must refer the matter to the Classification Office. The Classification Office applies the legislation to classification decisions on a daily basis. It made almost 3,000 decisions in 2008, across a wide variety of media and classifications. The courts deal with approximately 100 objectionable material cases each year, meaning that many judges would preside over no more than a single case of that type each year. If they were responsible for classifying material, judges would have little experience on which to base their decisions.

The Classification Office carries out its own research and gathers external research to better inform its classification decisions. It also provides a service to law enforcement agencies that no court could offer – training in application of the law. Staff from the Classification Office provide training to investigators to assist them in identifying objectionable material. They also speak to a wide variety of audiences as part of a community education programme that the Classification Office is required by law to provide.[2] In addition to receiving training from the Classification Office, law enforcement agencies may submit publications to it for classification at any stage in an investigation. This means that an investigator can have legal certainty that material is illegal before obtaining a search warrant or before laying charges against an offender.

New Zealand's censorship system focuses primarily on classifying material intended for public consumption. It does not concern itself with the private viewing choices of adults, except in relation to objectionable material. As a result, there is no mandatory filtering of Internet access in New Zealand. However, those who choose to download or distribute objectionable material face the risk of detection, prosecution and imprisonment. This approach contrasts with that of Australia, which has a similar media classification scheme but prohibits sexually explicit content on the Internet including X18 (legal, sexually explicit content) and RC (refused classification because it contain sexual violence, child abuse, criminal instruction or extreme violence). Such content, if hosted in Australia, is subject to 'take-down' notices issued by the Australian Communications and Media Authority and to prosecution for failure to comply with the notice (Brown and Price, 2006). Australia is currently trialling ISP-level mandatory filtering of the same prohibited Internet content. At the same time New Zealand has commenced voluntary ISP-level blocking of known child pornography sites.

A DEDICATED ENFORCEMENT BODY

Censorship law in New Zealand is primarily enforced by the Inspectors of Publications from the

Censorship Compliance Unit of the Department of Internal Affairs. This is notable for the fact that the Inspectors are not police officers and is the other major way in which New Zealand's approach to censorship varies from many other nations. The Department had been responsible for inspecting cinemas for many decades and had also carried out inspections of video stores to ensure they complied with classification decisions. When the Films, Videos, and Publications Classification Act 1993 was passed, the Department continued to hold the primary responsibility for investigation of crimes involving objectionable material. The Department established a Censorship Compliance Unit to carry out this function. Inspectors from the Unit have the power to prosecute offenders and to carry out search and seizure, under a search warrant. While such powers are commonplace amongst law enforcement officers, the Censorship Compliance Unit's operational model is unusual in that the investigators also carry out the forensic analysis of seized computers and appear as prosecution witnesses in Court. One of the strengths of the model is that the officers who carry out the 'field-work' (detection, investigation and seizure) also carry out the 'lab-work' (forensic analysis). This makes them very knowledgeable witnesses for the prosecution. Their technical knowledge also helps when interviewing suspects because they are aware of the details of alleged offences and the exact nature of the electronic evidence they have obtained. Another strength of this 'unified' model is the fact that the chain of custody for evidence is much easier to maintain when the same officers are involved at every stage of the investigation (Carr 2006). The high quality of their investigations, knowledge as prosecution witnesses and reliable chain of custody of evidence has resulted in a conviction rate of more than 99 percent. Only two people charged by the Censorship Compliance Unit have ever escaped conviction. This is an enviable record by any measure.

Investigators capable of carrying out all stages of an online investigation are uncommon. Law enforcement agents involved in investigating child pornography offences in other countries will often be divided into those who detect offenders, those who carry out search and seizure and those who analyse seized computers. The Inspectors work closely with the New Zealand Police and are accompanied by police officers on every search warrant they execute on offenders physical addresses. Co-operation with the Police is important because of the likelihood of child pornography offenders also carrying out physical sexual offending against children and because only the Police have the power to arrest. The Censorship Compliance Unit also works closely with the New Zealand Customs Service that, the border protection agency that investigates the importation of objectionable material. The three agencies work together under a formal agreement. The Police, although they have the powers of Inspectors to investigate censorship offences, do not duplicate that work but, instead, focus on investigating other online crimes such as adults grooming children for sexual offending (Van Der Stoep, 2009).

The Censorship Compliance Unit has developed working relationships with overseas law enforcement agencies, particularly in Canada, the United States of America, Norway, Germany, the United Kingdom and Australia. Since 2007, 30 per cent of the Unit's prosecutions have arisen from intelligence on New Zealand offenders provided by overseas law enforcement agencies. Almost all other cases are detected through the Unit's online covert surveillance of offenders, though some (13 per cent) have arisen from public complaints, including complaints from computer repair stores (correspondence with the Department of Internal Affairs, 2009). This situation contrasts with some other countries. A United States' study found that only 3 per cent of child pornography cases came to light because of online investigations, while the vast majority (87 per cent) became known as a result of conventional child sexual abuse cases (Wolak, et. al, 2005). This difference may reflect the different approaches taken to investigations by the Censorship Compliance Unit in New Zealand, which has a mandate focussed on child pornography, and police departments in the United States, with a much wider mandate, including child pornography but focussed on physical sexual offences. Because the New Zealand Police do not aim to replicate the work of the Censorship Compliance Unit, and, instead focus on child sexual abuse investigations, the majority of child pornography cases prosecuted by the Police stem from such investigations.

LAWS REFLECT CHANGING TECHNOLOGY

The Censorship Compliance Unit has focussed on investigating two offences primarily – the possession of child pornography and the making and distribution of child pornography. While its mandate is to investigate and prosecute all types of censorship offending including material promoting sexual violence, necrophilia and bestiality, child pornography is its focus because the making of such material often involves the sexual abuse of children. Online child pornography offenders are able to socialise through a virtual community with thousands of members. Some researchers suggest that this can normalise deviant behaviour and encourage the avoidance of individual responsibility by providing behavioural reinforcement and anonymity (Taylor et al. 2001, Lanning 1992, Quayle et al. 2000, Shelley 1998, O'Connell 2001).

The way that objectionable material is obtained and exchanged has changed significantly in the past two decades. When the Department of Internal Affairs began its investigations most child pornography and other illegal material was contained on videotapes, magazines and photographs. A relatively small amount of this material was in circulation due to New Zealand's geographic isolation and the work of the Customs Service in detecting it at the border (Wilson, 2002). The first

online investigations were into computer bulletin boards being used to exchange objectionable images in 1995. The Censorship Compliance Unit was established in 1996 to focus on Internet offenders and, at that time, more than half of its prosecutions involved online child pornography (Department of Internal Affairs, 1997). In the years since 1996, the majority of the unit's investigative effort has come to be spent on such cases.

While some early online offending involved downloading material from child pornography pay websites, most offending occurred in dedicated child pornography chat rooms (Department of Internal Affairs 1998). In these chat rooms, offenders would openly solicit child pornography and offer it to other offenders in return for new material. Investigators from the Censorship Compliance Unit would join the chat rooms, pose as offenders seeking new material, and engage New Zealand offenders. When those offenders sent objectionable images to the investigators their unique Internet Protocol (IP) address would be captured. A search warrant executed on the offender's ISP would identify the person to whom the IP address was allocated and this information would provide the vital connection between the online identity and the real-life offender. A second search warrant executed on the physical address supplied by the ISP would enable the Censorship Compliance Unit to interview the offender and seize computer equipment for subsequent forensic analysis. Confronted with a search by officers who held detailed information about the offending and were likely to obtain damning evidence from the seized computer, a large proportion of suspects (86 per cent) simply confessed to the their offending (Carr 2004).

The Censorship Compliance Unit found that, between 2004 and 2005, there was a distinct movement away from chat rooms towards peer-to-peer applications (Wilson and Andrews, 2004; Sullivan, 2005). A smaller number of offenders continued to post images or links to images in newsgroups, participated in chat rooms or operated websites from which people could download material. But the majority operated passive distribution systems such as file-servers or peer-to-peer networks where other people could open folders on computers and download material without the need for the possessor of the material to actively transmit it. This development was consistent with overseas trends noted in contemporary international research (Ferraro & Casey, 2005; Koontz, 2003). The use of peer-to-peer networks for offending made it easier for investigators to seek out objectionable material since it could be searched for in the network through the use of key words with out the need to engage with offenders. However, the huge amount of material discovered made it difficult to identify New Zealand offenders, without manually checking their IP addresses. To address this problem the Censorship Compliance Unit developed its own software to automatically search peer-to-peer networks and locate New Zealand-based offenders. The use of this software allows investigators to better focus their efforts on local offenders, rather than those located overseas over which they have no jurisdiction. When they do locate overseas offenders, the details are passed on to the relevant law enforcement agency through Interpol.

In parallel with these changes in investigative technique came changes to the law to ensure it took account of developments in technology and patterns of offending. Historically, penalties for censorship offending had been low. The maximum penalty for possession of objectionable material was a $2,000 fine. The penalty for distributing or supplying objectionable material, knowing that it was objectionable (the most serious offence under the Act), was a maximum of one year's imprisonment. There had been calls for change, to toughen penalties and align them with overseas jurisdictions (ECPAT 2003, Department of Internal Affairs 2002a), in a climate of international concern about online child abuse and growing public and political dissatisfaction with the sentences given to child pornographers in New Zealand. In March 2003 the Minister of Justice stated that the penalties were

"clearly inadequate and fail to reflect the fact that the production of child pornography involves the actual abuse of children". The new penalties were to be a maximum of 10 years' imprisonment for supply and distribution of child pornography and a maximum of two years' imprisonment for possession of child pornography (Goff, 2003). Anti-child abuse organisations Stop Demand Foundation and ECPAT advocated a higher penalty for possession offences. These groups argued that the possession offence should be treated as seriously as the supply offence since the demand for child pornography led to child abuse. The result was a revision of the maximum penalty for possession offences to five years' imprisonment.

Important definitional changes were made in parallel to the increased penalties. The focus of these amendments to the Films, Videos and Publications Classification Act 1993 was on the sections of the Act that defined the offence of distribution of objectionable material. The law originally had required elements of monetary or material gain in order to prove a distribution charge. However, New Zealand experience showed that very few 'traders' in objectionable material aimed to do anything other than increase the size or range of their collection of objectionable material by exchanges with other 'traders'. The Act was amended in 2005 so that 'distribution' included delivering, giving, offering or providing access to a publication. The 'passive' distribution of objectionable material by providing access to it but not actively transmitting it was caught by the broader definition of 'distribution'.

When the amendments to the legislation were before a parliamentary committee some groups, including ISPs and the Internet Society of New Zealand, expressed concern that 'providing access to' objectionable material could be taken to apply to businesses that provide the networks through which material is distributed such as ISPs and postal services. The legislation was amended to specifically exclude such services. This has an important implication for libraries, universities and other organisations that provide Internet access to the public. In order to be held liable for the use of its computers to download illegal material, the organisation would have to know of the nature of the material and that it was being downloaded rather than simply provide the facilities that could be used to commit an offence (Wilson, 2008).

A recent High Court decision in an appeal against convictions for distributing child pornography showed that the law changes had had the intended effect. The judge hearing the case dismissed the appeal that was based on the argument that downloading objectionable material and saving it into a shared folder accessible to others on a peer-to-peer network did not constitute 'distribution' of the material. The judge noted that the offender could have saved the images in a personal folder that others could not access if distribution was not intended.

Instead, [he] elected to download them onto files which, as the [trial] Judge pointed out, are networks which exist for the very purpose of trading objectionable material... The statute expressly provides that a person distributes by providing access to an objectionable publication. That is the act that Parliament has expressly proscribed within measures taken to prohibit the use of computer technology as a means of disseminating child pornography.[3]

Clarity around the parameters of the offence provisions was important given that the maximum penalty for distribution of objectionable material was increased to 10 years' imprisonment in 2005.

Possession of objectionable material is an offence under the Films, Videos, and Publications Classification Act 1993. When the law was initially enacted, possession of objectionable material was made a 'strict liability' offence, meaning that it did not have to be established that an offenders knew that material was objectionable in order to be convicted of possessing it. In 2002, the High Court found that a defendant did not need to know

a publication was objectionable to be prosecuted for possession, but did need to be aware of the presence of the publication and have the ability to exercise control over it (for example, by saving or deleting it).[4] This confirmed Censorship Compliance Unit prosecution decisions and earlier court judgments over what constituted possession (Manch and Wilson, 2003). The real weakness of the possession offence was the very low penalty - a maximum $2,000 fine. Under the same legislation, the maximum penalty for selling a DVD that was not correctly labelled was a fine of $3,000. There had been calls to toughen penalties, and align them with overseas jurisdictions. In 2005 a new, related possession offence was created which carried a maximum sentence of five year's imprisonment. However, the new offence requires proof that the offender knew that the material in their possession was objectionable. This is a reasonable safeguard in light of the severity of the penalty. The material that is the subject of prosecutions for possession is so clearly objectionable that proving that the person in possession of it knew it was illegal is not a significant barrier to conviction. This conclusion is supported by the fact that 90 per cent of prosecutions in the last two years have involve convictions for the new offence of possession of child pornography, knowing it to be objectionable (correspondence with the Department of Internal Affairs, 2009).

An important contribution to case law about the possession of child pornography occurred in 2004 when a school teacher was convicted after opening objectionable images, viewing them and then closing them. Although he did not save the images, they were cached on the school computer's hard drive. The judge in the case found that the defendant sought out and came into possession of the material knowingly and exercised control over the material, albeit for a short period of time. The judge held that any reasonable interpretation of the law that inhibited the use of material that sexually exploits children was to be preferred.[5] The decision made it clear that even the briefest period of possession of objectionable material was an offence.

APPLICATION OF THE LAW TO NEW ZEALANDERS OVERSEAS

The Internet has changed the way that governments should think about their borders. The Internet belongs to no one country. It is possible for a company in the United States to operate a server from Jamaica that contains images or information accessible to people in almost any country in the world. Furthermore, material that may be perfectly legal in the country in which it was created, or from which the server storing it operates, may be prohibited elsewhere. This issue was addressed in New Zealand in 2005 when a man was convicted on charges of making available objectionable publications as a result of a Censorship Compliance Unit investigation. The offender had hosted the pictures on an overseas server but controlled the content of the server from New Zealand. He appealed his convictions. One of the grounds of appeal was that the objectionable computer images in question were not made available in New Zealand, since they were hosted on a server in the United States and were intended for an overseas audience. He argued that making the images available in New Zealand should require the person to direct the activity towards New Zealand. The High Court did not accept the appellant's submission and noted that the files were placed on a website server so that they would be available world-wide, including to people in New Zealand. The fact that an Inspector was able to access the files in New Zealand was accepted by the Court as evidence of the availability of the files in that country. The Court found that:

It is irrelevant that the appellant may have believed his primary market to lie elsewhere. It is likewise irrelevant that the server utilised by the appellant was situated in the USA. While in New Zealand, *the appellant undertook certain steps to display or make available the images on his website.*[6]

The court also pointed to section 7 of the Crimes Act 1961 which provides that where any element of an offence occurs in New Zealand, the offence shall be deemed to be committed in New Zealand. The decision was significant because it largely eliminated a legal defence available to New Zealanders who based their child pornography operations oveseas to avoid prosecution.

SENTENCING SINCE THE 2005 LAW CHANGES

In the two years following the enactment of the 2005 law changes the proportion of offenders receiving a custodial sentence remained low, compared to other countries, at 40 per cent. This was because the courts were still sentencing for some offences committed before the law change and were only able to sentence in accordance with the law at the time of offending.[7] Since 2007, 60 per cent of convicted offenders have received a custodial sentence. The average prison sentence imposed is increasing. In the two years immediately following the law change, the average sentence was 13 months. Since 2007, the average prison sentence has increased to 18 months (correspondence with the Department of Internal Affairs, 2009).

NEW ZEALAND'S LAW IN THE INTERNATIONAL CONTEXT

Amendments to the censorship offence provisions in 2005 significantly strengthened New Zealand's efforts to combat child pornography and other objectionable material. The maximum penalties for offences involving this material are similar to those in the United Kingdom, Australia, Canada, Ireland and the United States.

It is difficult to establish objective criteria by which to analyse the effectiveness of a law, particularly in an international context. The International Centre for Missing and Exploited Children (2008) has established some criteria against which to assess the responses to child pornography of the 187 Interpol member nations. The 10 criteria are:

1. Defining "child" for the purposes of child pornography as anyone under the age of 18, regardless of the age of sexual consent;

2. Defining "child pornography," and ensuring that the definition includes computer and Internet specific terminology;

3. Creating offences specific to child pornography in the national penal code, including criminalizing the possession of child pornography, regardless of one's intent to distribute, and including provisions specific to downloading or viewing images on the Internet;

4. Ensuring criminal penalties for parents or legal guardians who acquiesce to their child's participation in child pornography;

5. Penalizing those who make known to others where to find child pornography;

6. Including grooming provisions;

7. Punishing attempt crimes;

8. Establishing mandatory reporting requirements for workers likely to come into contact with child pornography;

9. Addressing the criminal liability of children involved in pornography; and

10. Enhancing penalties for repeat offenders, organized crime participants, and other aggravated factors considered upon sentencing.

New Zealand's law meets nine of the 10 criteria, because it does not require mandatory reporting of offences. Only five countries in the world have this requirement and thereby meet all of the standards set by the report's authors (International Centre for Missing & Exploited Children, 2008). A mandatory requirement to report crime is not a normal feature of New Zealand criminal law. However, New Zealand's privacy laws permit a person to breach another's privacy in order to report a crime.

New Zealand's laws against child pornography have some significant advantages over many other jurisdictions. They enable a wide variety of material that promotes or supports child sexual abuse to be classified as objectionable, whether or not a real child was victimised in the production of the publication. This includes works of fiction, cartoons, computer-generated ('morphed' or 'pseudo') images and images depicting sexualised nudity. The New Zealand definition of child pornography is broader than that used by the United Nations which defines it as "any representation, by whatever means, of a child engaged in real or simulated explicit sexual activities or any representation of the sexual parts of a child for primarily sexual purposes".[8]

PROFILING OF OFFENDERS

In 2000 the Censorship Compliance Unit identified a lack of knowledge about the common characteristics of offenders who distributed objectionable material on the Internet. It initiated a research project aimed at better understanding the nature and offending behaviour of censorship offenders. The research was based on a questionnaire completed by the Censorship Compliance Unit, based on its observations and interviews of offenders. The questionnaire collected social, demographic and behavioural information about offenders (Carr, 2004).

In the study, the offenders were overwhelmingly male (only one was a woman); mostly white, likely to be middle class and adept at using the Internet. The offenders' ages, at the time they were investigated, ranged from 14 to 67 with the average age being 30. But almost a quarter of the offenders were aged less than 20 years. The most common occupations among offenders were student or information technology professional. Of greatest interest were the 13 per cent of offenders who had been convicted for physical sexual offences in addition to their child pornography offending. Two similar studies found that 12 per cent of American offenders (Wolak et, al., 2005) and 5.1 per cent of Irish offenders (O'Donnell and Milner, 2007) had previous convictions for sexual offences. The results of these studies do not, by themselves, prove that collectors of child pornography necessarily go on to commit child sex offences but it appears that they are significantly more likely to do so than the rest of the population. All three studies found that most child pornography offenders had never been in trouble with the law previously. This result supports the approach taken by the Censorship Compliance Unit in searching out child pornography offenders on the Internet, since most will not come to the attention of law enforcement authorities for any other reason or until they have carried out physical sexual offending.

The New Zealand research also examined the extent to which online offenders had access to the subjects of their image collections. It found that 40 per cent had frequent access to children in their daily lives. Most offenders used no security mechanisms to protect their collections of objectionable images from discovery and only five per cent encrypted them. Offenders valued ease of access to their collection more highly than protecting it from unwanted scrutiny. This profiling research provided valuable insights into offender behaviour, contributing to increased efficiency in investigation, prosecution and treatment as well as providing a sound empirical base on which to develop censorship policy. Similar research has since been undertaken in other countries. The Censorship Compliance Unit publishes updates on key offender statistics each year.

FUTURE TRENDS

Child pornographers will continue to change their methods of offending as information and communication technology changes. New Zealand has witnessed these changes as hard copy photographs and video tapes largely have been replaced by digital media. There has been a progression from bulletin boards to chat rooms and then to peer-to-peer networks. Offenders have adopted new technology to avoid detection and better access objectionable material. But their changes in method are not necessarily linear or exclusive of other methods. It would be a mistake to assume that because many offenders use peer-to-peer networks to collect child pornography, that they no longer use chatrooms, for example. It would also be a mistake to think that offenders will not use other methods, employing technology not yet available or reverting to previous methods. The point is well illustrated by a group of New Zealand offenders who downloaded child pornography images from the Internet but circulated them by handing a hard drive from one person to the next.

Technology for storing data continues to decrease in size and increase in capacity. This development bring its own challenges to those charged with searching for evidence of offending. Offenders will always balance their desire to avoid detection with a wish to easily access and disseminate child pornography. Thus, encryption of material or hiding storage devices is weighed against accessibility. The dissemination of child pornography to others involves offenders making themselves visible, in varying degrees, to law enforcement agencies. Those agencies must continue to adapt to the changes in the uses of technology, in the pursuit of offenders. The key to keeping pace with methods of online offending is in gathering intelligence about offenders and sharing it globally. But effective enforcement also requires the support of effective and fair legal and administrative frameworks.

CONCLUSION

It is always challenging for legislation to keep pace with emerging technology. New Zealand's approach to date has not been to legislate for a wide variety of new Internet-based offences. Rather, policy-makers have generally viewed new technology as a different means to carry out already-existing offences. It has also approached new information technology as beneficial, but with the potential to be used for criminal activity. Child pornography was made and exchanged in New Zealand prior to the widespread availability of Internet access. It is not possible to accurately state what effect, if any, Internet availability has had on rates of offending. However, New Zealand's censorship law is well-placed to deal with current and future offending. Its offence provisions are expressed in broad and permissive terms to cover existing technology and technology not yet available. The classification provisions keep the law tightly focussed on banning harmful material while allowing unfettered access by adults to legal material. A single law to deal with legal and illegal publications of all types leads to a consistent decision-making regime.

New Zealand's censorship laws are able to be effective because of the agencies that operate under them. The Censorship Compliance Unit of the Department of Internal Affairs is an effective law enforcement agency with a firm focus on detecting and prosecuting child pornographers. The Office of Film and Literature Classification is an expert classification body that makes decisions based on research, precedent and experience in classifying tens of thousands of publications. These agencies and the government will need to remain vigilant to ensure that their processes and legislation are not eclipsed by technology. Based on experience to date, this is a challenge they are well equipped to face.

The New Zealand response to Internet child pornography works well in the New Zealand context. As a New Zealander who has played a part in shaping that response I hope I have done it justice without being overly effusive in my praise. I do not advocate that other jurisdictions simply attempt to replicate the New Zealand system within their own settings but I hope that the approach described in this chapter may highlight that there are different ways of approaching a global problem. A national classification organisation to adjudicate all illegal material works effectively in a country with a small population. It would not work well in a populous country, if we assume a similar level of offending and enforcement activity, without significant financial and human resources. Similarly, a single investigative body may be problematic in nations with strong state government. However, the most recent study of child pornography laws by the International Centre for Missing and Exploited Children (2008) found that only 29 of the 187 Interpol member countries have legislation sufficient to combat child pornography offences while 93 have no legislation at all that addresses child pornography. The approach outlined in this chapter may be of most interest to those policy makers and law enforcement officials in countries wishing to start tackling child pornography.

REFERENCES

Andrews, C. (2003). *The problem of child pornography on the Internet and the ability of the British Police to regulate it*. Unpublished MA thesis, University of Hull, Hull, UK.

Brown, R., & Price, S. (2006). *The Future of Media Regulation in New Zealand: Is There One?* Retrieved March 19, 2009, from http://www.bsa.govt.nz/publications/BSA-FutureOfMediaRegulation.pdf

Burrows, J., & Cheer, U. (2005). *Media Law in New Zealand* (5th ed.). Melbourne, Australia: Oxford University Press.

Carr, A. (2004). *Internet Traders of Child Pornography and other Censorship Offenders in New Zealand*. Wellington, New Zealand: Department of Internal Affairs.

Carr, A. (2006). *Internet censorship offending: a preliminary analysis of the social and behavioural patterns of offenders*. Unpublished PhD thesis, Bond University, Queensland.

Cohen, H. (2003). Child Pornography Produced without an Actual Child: Constitutionality of Congress Legislation. In W. Holliday (Ed.), *Governmental Principles and Statutes on Child Pornography*. New York: Nova Science Publishers.

Department of Internal Affairs. (1997). *Report of the Department of Internal Affairs for the year ended 30 June 1997*. Wellington, New Zealand: Author.

Department of Internal Affairs. (1998). *New Zealand Censorship Compliance Unit*. Unpublished paper presented at the Australian Institute of Criminology conference, Internet Crime, Melbourne University, 16-17 February.

Department of Internal Affairs. (2002a). *Briefing to the Incoming Minister of Internal Affairs*. Wellington, New Zealand: Author.

Ferraro, M., & Casey, E. (2005). *Investigating Child Exploitation and Pornography*. San Diego, CA: Elsevier.

Heuck, J. (2005). *Comparative Report on Child Pornography on the Internet in Germany and New Zealand*. Auckland/Freiburg. ECPAT Deutschland e. V and ECPAT New Zealand.

Human Development Report Office. (2008). *Human development indices*. Retrieved February 11, 2009, from http://hdr.undp.org/en/media/HDI_2008_EN_Tables.pdf

International Centre for Missing & Exploited Children. (2008). *Child Pornography: Model Legislation & Global Review*. Retrieved on February 15, 2009, from http://www.icmec.org/en_X1/pdf/ModelLegislationFINAL.pdf

Koontz, L. (2003). File-Sharing Programs: Child Pornography is Readily Accessible over Peer-to-Peer Networks. In Holliday, W. (Ed.), *Governmental Principles and Statutes on Child Pornography*. New York: Nova Science Publishers.

Lanning, K. (1992). *Child Molesters: A Behavioural Analysis*. Washington, DC: National Centre for Missing and Exploited Children.

Loftus, R. A. (2008). Disconnecting Child Pornography on the Internet: Barriers and Policy Considerations. *Forum on Public Policy Online*, Spring. Retrieved on 15 June 2009 from http://www.forumonpublicpolicy.com/archivespring08/loftus.pdf

Manch, K., & Wilson, D. (2003). *Objectionable material on the Internet: Developments in enforcement*. Retrieved February 18, 2009, from www.netsafe.org.nz/Doc_Library/netsafepapers_manchwilson_objectionable.pdf.

Morris, J., Haines, H., & Shallcrass, J. (1989). *Report of the Ministerial Inquiry into Pornography*. Wellington, New Zealand: New Zealand Government.

ECPAT New Zealand. (2003). *Williamson's Child Pornography Sentence Too Light*. media release, 31 July.

O'Connell, R. (2001). *The Structural and Social Organisation of Paedophile Activity in Cyberspace: Implications for Investigative Strategies*. Accessed October 21, 2002, from http://www.uclan.ac.uk/facs/science/gcrf/crime1.htm

O'Donnell, I., & Milner, C. (2007). *Child Pornography Crime, Computers and Society*. Portland, OR: Willan Publishing.

Office of Film and Literature Classification. (2007). *Publication of Objectionable Computer Files*. Retrieved February 18, 2009, from http://www.censorship.govt.nz/thelaw-occasional-objectionable-computerfiles.html

Office of Film and Literature Classification. (2008). *Report of the Office of Film and Literature Classification for the year ended 30 June 2008*. Wellington, New Zealand: Office of Film and Literature Classification.

Office of Film and Literature Classification [Australia]. (2006). *Enforcement Package*. Retrieved 16 June 2009 from http://www.oflc.gov.au/www/cob/rwpattach.nsf/VAP/(084A3429F-D57AC0744737F8EA134BACB)~896.pdf/$file/896.pdf

Perkins, M. (2009). Prime Cuts. *Index on Censorship*, *38*(1), 129–139. doi:10.1080/03064220802712258

Quayle, E., Holland, G., Lineham, C., & Taylor, M. (2000). The Internet and offending behaviour: a case study. *Journal of Sexual Aggression*, *6*(1/2).

Shelley, L. (1998). Crime and corruption in the digital age. *Journal of International Affairs*, *51*(2).

Sullivan, C. (2005). *Internet Traders of Child Pornography: Profiling Research*. Retrieved February 24, 2006, from http://www.dia.govt.nz/pubforms.nsf/URL/Profilingupdate2.pdf/$file/Profilingupdate2.pdf

Taylor, M., Quayle, E., & Holland, G. (2001). Child pornography, the Internet and offending. *ISUMA*, *2*(2). Retrieved May 27, 2002, from http://www.isuma.net/v02n02/taylor/taylor_e.shtml

Transparency International. (2008). *2008 Corruption Perceptions Index*. Retrieved February 5, 2009, from http://www.transparency.org/news_room/in_focus/2008/cpi2008/cpi_2008_table

UNICEF. (2008). *The State of the World's Children 2009*. New York: United Nations Children's Fund.

United States v Williams. No. 06–694. (2008). Retrieved 26 June 2009 from http://www.supremecourtus.gov/opinions/07pdf/06-694.pdf

Van Der Stoep, L. (2009). Police target web predators. *Sunday Star Times*, 22 February 2009. Retrieved March 3, 2009, from http://www.stuff.co.nz/stuff/thepress/4856032a26834.html

Watson, C., & Shuker, R. (1998). *In the Public Good? Censorship in New Zealand*. Palmerston North, New Zealand: Dunmore Press.

Wilson, D. (2002). Censorship in New Zealand: the policy challenges of new technology. *Social Policy Journal of New Zealand*, *19*, 1–13.

Wilson, D. (2007). Responding to the Challenges: Recent Developments in Censorship Policy in New Zealand. *Social Policy Journal of New Zealand*, *30*, 65–78.

Wilson, D. (2008). Censorship, new technology and libraries. *The Electronic Library*, *26*(5). doi:10.1108/02640470810910710

Wilson, D., & Andrews, C. (2004). *Internet Traders of Child Pornography and Other Censorship Offenders in New Zealand*. Retrieved February 24, 2009, from http://www.dia.govt.nz/pubforms.nsf/URL/profilingupdate.pdf/$file/profilingupdate.pdf

Wolak, J., Finkelhor, D., & Mitchell, K. (2005). The Varieties of Child Pornography Production . In Quayle, E., & Taylor, M. (Eds.), *Viewing Child Pornography on the Internet. Lyme Regis*. Dorset, UK: Russell House Publishing.

KEY TERMS AND DEFINITIONS

Objectionable Material: Material that is illegal under New Zealand law, generally for promoting sexual or violent crimes.

Child Pornography: In New Zealand the term refers to any publication that promotes child sexual abuse or exploits child nudity.

Peer-to-Peer Networks: A network of computers that connect directly to each other (as 'peers') rather than through a central server. Commonly used to share music and image files.

Chat Room: A website where live, real-time keyboard conversations with other people occur, usually on a specific topic.

Publication: Under New Zealand law, any thing that has words, statements, signs, representations or images impressed or stored on it and capable of being reproduced identically. Includes films, DVDs, digital image files, still images and text files.

Possession (of Objectionable Material): Under New Zealand law, an offence committed by knowingly possessing and controlling objectionable publications.

Distribution (of Objectionable Publications): Under New Zealand law, an offence committed by delivering, giving, offering or providing access to an objectionable publication.

IP Address: A numerical identification that is assigned to computers participating in a network that utilises the Internet for communication. It is expressed in a string of numbers such as 121.72.27.60. Each IP number is unique and relates to one Internet connection.

ENDNOTES

1. Miller v California 314 U.S. 15 (1972).
2. See section 88 of the Films, Videos and Publications Classification Act 1993.
3. Espinosa v Department of Internal Affairs [2008] CRI 2008-404-233.
4. Goodin v Department of Internal Affairs [2002] unreported, AP11/01.
5. Department of Internal Affairs v Young [2004] DCR 231, 234.
6. Batty v Department of Internal Affairs [2005] CRI 2005-404-313
7. s. 6(1), Sentencing Act 2002.
8. Optional Protocol to the Convention on the Rights of the Child on the Sale of Children, Child Prostitution, and Child Pornography, G.A. Res. 54/263, Annex II, U.N. Doc. A/54/49, Vol. III, art. 2, para. c, *entered into force* Jan. 18, 2002.

Chapter 17
Towards Electronic Records Management Strategies

Basil Enemute Iwhiwhu
Delta State University, Nigeria

ABSTRACT

Records are a vital business resource and are key to the effective functioning and accountability of the organization. Efficient management of records is essential in order to support organization's core business activities, to comply with legal and regulatory obligations, and to provide a high quality service to individuals. Electronic records management programme ensures that the organizational business activities are well documented, organized and managed, accessible, protected from unauthorized access and disposed off (either destroyed or archived). Credible and dependable information systems are desired to achieve this. Also, adequate skills sets are required by personnel working with and managing electronic records. The relevance of all these are articulated in this chapter.

INTRODUCTION

Records are documentary evidence of transactions made or received in pursuance of legal obligations regardless of the physical form or characteristics of the media. They are a class of information identified by the particular functions they perform in support of business, accountability and cultural heritage. They substantiate who did what, where, and when. It is an essential business function with responsibility cascading to every level of

decision-making. Many record keeping functions have been automated or streamlined so they are not as labour intensive (and therefore costly) as they were in the traditional record keeping environment. The advent of automated records, often called electronic records or e-records, has dominated most offices today.

Electronic records (e-records) are information generated electronically and stored by means of computer technology. Electronic records are fragile in nature due to change in hardware and software used for their storage, processing and use, making them prone to high risks. If damaged

DOI: 10.4018/978-1-61692-012-8.ch017

or deteriorated, restoration is a difficult task, if not impossible. Electronic records have short life expectancy dependent on the average service life of the hardware and software required to read and process them. As a result, creating, using and managing e-records in a digitized and globalize world are a challenging venture that must be well thought-out, planned and implemented. This can only be achieved with a well articulated records management programme in place. Though, there seems to be an international standard for the management of e-records as prepared by the International Records Management Trust (IRMT) and the World Bank for Commonwealth Countries (National Archives of Australia, 2002), records management activities generally and e-records management in particular are still issues to contend with. Most governments today have their records in electronic form. Moore (2000) stated that as much as two-thirds of the world's information was 'born digital' in the late 1990s and 2000, meaning that its original occurrence was in a digital format generated from computers. While this is true for most developed countries, developing countries are still behind with attendant problems of infrastructure, policy, technical-know-how, electricity generation and sustenance, hardware and software availability among others. As a result, most developing nations adopt the hybrid nature, where records are created and managed both in the paper and in electronic formats. Keeping records in this hybrid nature has implication for the management of e-records, as accessing and using readily available records may not be easily realized (Mazikana, 1999). This chapter will examine the electronic record management programmes from creation, use and maintenance, preservation and storage, distribution and disposition, e-recordkeeping systems requirements, the basics of good electronic records management, strategic considerations and appropriate skills sets among records staff.

BACKGROUND

As the volume of records produced began to grow in earnest during the 20th century, specialised facilities for the storage of records (in records repositories) grew in number and in the quality of environmental and security controls to ensure the availability as well as integrity of these primary media-centric (paper or films) records. Records with historical or permanent value are therefore, disposed, which included destruction and transfer of discrete sets of records to an archival facility and administered under careful document and audit procedures. With most of today's business information created in digital format, the rules of the records management game have changed dramatically. Research suggests there has been 30% increase in the amount of stored information (includes four types of physical media: paper, film, magnetic and optical) each year, from 1999-2002, with 92% of new information stored on magnetic media primarily hard disks (Lyman & Varian, 2003). As a result, organisations need efficient and suitable ways to capture and protect their electronic records, as well as efficient and veritable means to dispose of or destroy information and medium when there are no longer requirements to keep them.

The risks associated with poor or non formal records management programme have been well documented in the mainstream media in recent years in terms of legal and regulatory compliance risks and costs, as well as the overall effectiveness of business operations (The Association of Information and Image Management, 1991; Cohasset Associates, 2005). The discipline of records management has not been particularly exciting despite all the sensational media attention in recent years. Guided by laws, regulations and the organisation's culture and structure, corporate records management programme has traditionally been responsible for establishing proper recordkeeping guidelines, policies and procedures. For decades, the protection of significant documents, files, contracts and physical objects has been performed by clerks relying on trial and error methods and tools such as filing cabinets, file folders, imaging, indices and secure storage facilities to meet the organisation's basic recordkeeping requirements. Many organisations often consider records management as a back-office cost centre with little business benefits (Cohasset, 2005). Records managers have had to rely on the creators and users of business records and their administrative support staff to comply with retention rules for hard copy documents and hoped that drafts and other non-records were systematically destroyed (Cohasset, 2005). The records management function served a custodial role for vast stores of inactive records.

Citizens and workers do not treat records with some sanity in the past because there were no adequate policies, technical-know-how, infrastructure and the public-will. Records managers were preoccupied with managing the records in central filling systems or records' store while archivists were used to defining their function within the walls of archives. Access and retrieval of records were difficult. The Association of Information and Image Management (1991) stated that in the United States:

- office staff spent almost 2.5 hours per day retrieving and putting away paper files;
- 50 minutes a day was spent searching for missing or misfiled papers;
- 17 per cent of files needed were not readily accessible; and
- one to six information searches were impeded.

In the same vein, the 2003 Electronic Records Management survey conducted by Cohasset Associates (2005) on the management of electronic records revealed that:

- 41% of records managers responded that electronic records were not included in the organization's records management program.
- More than two-thirds (71%) reported that IS/IT had primary responsibility for the day-to-day management of their organization's electronic records.
- An overwhelming majority (93%) believed the process by which electronic records will be managed will be important in future litigation, but by a ratio of nearly 2:1, a majority (62%) were less than confident that the organization could demonstrate its electronic records are accurate, reliable and trustworthy – many years after they were created.

The above is associated with the poor management of records; and the situation can better be imagined in developing countries like Nigeria. Studies have revealed several problems associated with the management of records in Africa (Mnjama (2005); Kemoni et al. (2003); Khamis (1999). Thurston and Smith (1986) in their study identified the following:

- inability by researchers to use archival information;
- inadequate finding aids;
- lack of recognition by national governments of the role played by archival institutions;
- outdated archival legislation which impacts negatively on access to the collections;
- inadequate number of professional archivists;
- lack of adequate archival training schools;
- poor systems of archival arrangement and description;
- understaffing of archival services;
- poor storage facilities for records; and
- inadequate retrieval tools.

Meeting the unique challenges associated with electronic records management require new ways of thinking and a renewed spirit of collaboration to work across domains and disciplines. According to the International Council on Archives, the users of electronic records, over the long-term, are necessarily speculative and can be diverse and differing in their needs (ICA, 2005). Users may include:

- government and municipal administrations needing records for reasons of corporate memory or accountability;
- prosecuting authorities and lawyers who need records as evidence for their clients and cases;
- scholars carrying out research based on historical sources;
- teachers using historical sources in their teaching;
- students;
- those who are working on cultural projects – including employees of cultural institutions;
- journalists;
- genealogists;
- people who need records as evidence of their rights, or to document events which have a direct bearing on them personally.

ELECTRONIC RECORDS MANAGEMENT PROGRAMME

The inclusion of records management programme (RMP) in daily organisational business operations is of extreme importance. Most organizations face the problem of what to do with the multifarious volumes of records/files in their disposal, storage space, as well as decisions on which records to retain or which to discard. Instituting RMP in that organisation can assist offices and departments to take decision and overcome the problems. One of the primary roles of RMP is to provide services that support the organisation's ability to meet their information needs (Cruickshank, 2006). It

is in the best interest of organizations to address e-records management issues as soon as possible. This is in order to assist the organization to achieve economy and efficiency in creation, use and maintenance and disposal of records. The ISO standard provides the following three principles for records management programmes:

- Records are created, received and used in the conduct of business activity. To support the continuing conduct of business, comply with the regulatory environment, and provide necessary accountability, organisations should create and maintain authentic, reliable, and usable records, and protect the integrity of those records for as long as required;
- Rules for creating and capturing records and metadata should be incorporated into the procedures governing all business processes for which there is a requirement for evidence of that activity; and
- Business continuity planning and contingency measures should ensure that records that are vital to the continued functioning of the organisation are identified as part of risk analysis and are protected and recoverable when needed.

Developing a records management programme (RMP) therefore, will ensure that:

- High value information is trustworthy and protected;
- Low value information is routinely purged and appropriately destroyed;
- Information system technology investments are aligned to business priorities and gaols and meet corresponding information and record life cycle management requirements; and
- Employees have the support and tools they need to comply with retention and disposi-

tion rules for managing the e-records that they create and receive (ICA, 2005).

Records management programme (RMP) is needed in organisations in order to ascertain economy, efficiency, legal protection, meeting statutory requirements, etc. Achieving all these require a top management team commitment, setting up RMP committee, defining the objectives of the RMP, communicating staff and management, describing file location, specifying data to be collected, preparation of inventory form, establishment of work schedule and selection of personnel to conduct the survey. The records management programme consist of different phases as follows;

Electronic Records Creation

Records used in organisational activities can either be born (created in the office) or received from other department or organisation offices. By this, the record is created or the act is referred to as records creation. Electronic records could be born digital or made digital. Those records created directly with a computer system are born digital. while those converted from paper format to electronic format are made digital. At this stage the physical form which may be paper, electronic media, magnetic tape or disc, photographic media, etc. and information contents are established in terms of value. Hence, the need to be more careful in creating records in the organisation. Taking the right decision at this point of creation with regard to the medium and classification enhance the management of the records through its life and improves its effectiveness. Adequate records must be created where there is need to be accountable for decisions, actions, outcomes or processes. Records should be created and maintained in a manner that ensures they are clearly identifiable, accessible and retrievable. The creation of records is a fundamental aspect of the management of any business operation. As such, organisations should determine how and why e-records are being cre-

ated. This is important because the capture of appropriate content, creation of metadata, declaration of record type, etc. are best addressed at the record creation stage. E-records management procedures are most effective when carried out at the point of creation or very shortly thereafter. In other words, records should be created when there is need to have evidence that is credible and authoritative to protect the rights of the organization, its staff and anyone else affected by its activity. Records should be created when there is a requirement:

- To provide evidence of a transaction;
- To prove that policies, procedures, rules are followed in arriving at a decision or outcome;
- To defend against possible claims or future legal actions, for example workers compensation, breach of contract; or
- For others to know what action occurred, what was decided, when it occurred, who was involved and the sequence of actions.

Generally, records are created as part of the organization's business activities. However, records may be created indiscriminately when there is no need for evidence that something has been done, a record duplicates evidence or information in one or more other records, and a redundant duplicate of a record is created (for example via a multipart form such as a receipt). Also, the hardware and software environment will affect the capture of e-records in the organisation. Records could be captured through:

- The user interface layer;
- Modification of the applications software;
- The operating system;
- The application programme interface (API), or
- The front end to a corporate filling system.

The organizational environment will also influence the point at which records are captured.

This include perceptions about what constitutes a record, assignment of responsibility, organisation requirements to create records, and staff understanding of the technology involved. It must be able to identify specific information objects (e.g. documents, e-mail messages, database entries) as records and somehow distinguish between the types of records to which different business and retention requirements must be applied (ICA, 2005)

Records Use and Maintenance

After creation/capture or receipt, records would be used and their maintenance is paramount for the preservation of the records for posterity. Different kinds of electronic office systems are used in records creation and management. They can focus on information retrieval (e.g. document management systems) or on supporting the business process of an organisation (e.g. workflow systems). Electronic systems can also include stand-alone, non networked systems. In a modern office context, however, they are usually embedded in distributed networked environments on different levels. In order to provide evidence, tools are needed to preserve records and make them available for use. A recordkeeping system should be an instrument that governs records management functions through the entire life cycle/records continuum (ICA, 2005).

A recordkeeping system (RKS) is an information system that has been developed for the purpose of storing and retrieving records, and is organised to control the specific functions of creating, storing, and accessing records to safeguard their authenticity and reliability (ICA, 1997). They are the systems that capture, manage and provide access to records overtime. RKS guarantees the maintenance and preservation of authentic, reliable, and accessible records overtime. Records are kept to provide evidence of functions, activities and transactions.

RKS maintain linkages to the activities they document and preserve the content, structure, and context of the records, which make them differ from generic information system. They often accommodate records that exist in more than one format and should be able to identify all records, active and inactive, and the version of the computer software that supports access. They should be able to identify records stored off-line and off-site and all media. RKS should document all successful outcomes that can be sustained within the organisation overtime and accurately reported. The systems should meet their accountability requirements without detracting from the benefits provided by modern technology and organisational change (ICA, 2005).

The longer it takes before records are maintained, the more difficult it becomes to fully maintain their content, structure and context. To maintain record content, systems should be in place to ensure that:

- The identity of a record's creator is verified (through the use of a password and possibly encryption);
- Permission to both read and write file is appropriately restricted;
- Periodic system audits are conducted;
- Data transmission includes data error checking and correction;
- Data are regularly backed up, and
- Data on off-line media such as magnetic tape are regularly refreshed to avoid catastrophic loss of data due to medium degradation.

Data should also be encoded in such a way that the bits will continue to be readable overtime. The records of business processes may span different media and multiple systems. Records creation may be restricted to certain media, which should be clearly articulated and communicated to staff. RKS should be designed to enable access to the complete record without hindrance. Where multiple recordkeeping systems are in place, links should be provided for records that span these multiple systems.

RKS need to capture and maintain information about the structure of records as an integral part of the metadata associated with the records in separate formal documentations. Since structure is more difficult to maintain than content, it is often neglected. The simpler the record structure, the easier it is to preserve it overtime. It is also nice for records structure to be based on open standards such as Standard Generalised Markup Language (SGML) and eXtensible Markup Language (XLM).

The value of a record is severely diminished or lost if its content is separated from key information about the agency and person(s) who made it, place, the time and reasons for its creation, and its relationship to other records. Though the content of a record may still be of interest, it will have no value as evidence if not placed in context. This is best achieved when records are filed. Once record quantities pass a critical mass, they cannot be found efficiently if they are kept in a big heap or saved electronically into a single folder. Contextual information, therefore, is information about the records and the administrative environment in which they were created and maintained. It can range from high-level information such as the name and location of the organisation that created the record to more detailed information such as the date it was made. The ideal in the electronic environment is to link to records the metadata and contextual information necessary to read and understand them (National Archives of Australia, 2002).

RKS need to maintain and provide access to information about the business and administrative context in which records were created and used. For computer systems developed by information technology professionals, system design documentation, data dictionaries and related business documentation are fundamental to providing context for records that are held in those systems.

Active data dictionary - lists of all files in a database management system, the number of records in each file, and the names and types of each field - and computer-aided software engineering (CASE) tools - software that provides a common development environment for programming teams - automate much of the process of keeping metadata authentic. Maintaining the context of records created and managed outside of systems developed by information technology professionals is more difficult. The ubiquity of personal computers allows records to be created, modified, copied, transmitted and deleted, often with little regard for business and legal records management requirements. Even if records are managed appropriately on an individual workstation, their existence may not be known to other users, and the contextual information may be inadequate for future retrieval. Considerations need to be given to assigning and preserving meaningful document names, author, work groups and organisational identifiers designating whether records are draft or final versions and linking them to other documents or information objects. Off-the-shelf software exists to address these problems. Contextual information needs to be collected, structured, and maintained from the time records are created. This involves identifying and labelling (or tagging) records and linking them to contextual information (i.e. keeping records about records). In some cases, this can be achieved by embedding key contextual information into the metadata or e-records themselves. The more e-records can be made self-describing, the less need there is for mounting separate information. While contextual information is absolutely necessary for long-term retention of e-records, it can also improve the quality of records in active use, support information sharing, and enhance their quality as evidence.

ISO Standard 15489 contains an extensive list of policy issues and suggested requirements that archivists can use as basis for cooperation with records managers in promoting good records management as for the basis for creation and preservation of sound archival records. Some life cycle/continuum issues to be addressed include:

- The development of new systems so that they can identify those which will create records of archival value and ensure that those systems will support their preservation and continued accessibility;
- The operation of systems in which archivists need to monitor systems management to ensure that all parts of the archival record (e.g. the records themselves, related metadata, and documentation of how the system operated) are properly maintained and so that no changes are made to the system that would affect the archival quality of the records;
- Decisions concerning modification, upgrading, migration, and other changes to the system (e.g. changing hardware or software platform) could affect the authenticity and integrity of the records, the ability of the system to preserve the records, and the ability of the archives or the creating organisation to provide for long-term accessibility to the records, and
- Decisions to discontinue with systems containing archival records or to remove archival records from those systems.

Electronic Records Disposition

Effective record keeping systems require good information management and retention schedules in order to be able to find the requested information and to ensure the information is kept for the appropriate period (Cruickshank, 2006). Records can be stored for a period of few days to several years. However, the type of record and its intended use determine the length of time a record should be kept. Users may reference the record frequently and need quick access to it, requiring the records to be maintained in the office area (Cruickshank, 2006). Cruickshank (2006) further stated that

records activity must b e weighed against factors of urgency, multi-user access and geographic dispersion. Generally, highly active records require a medium that facilitates ease of irretrievability.

Electronic Recordkeeping Systems

Since effective management of e-records depends so heavily on the information systems involved, identifying recordkeeping requirements when new systems are designed or when existing systems are upgraded is of paramount importance. If recordkeeping requirements are identified during process analysis, effective procedures and automated routines can be built into the revised processes to handle records more effectively. Several aspects of recordkeeping should be considered during the system design and procurement process. If the system is expected to support electronic recordkeeping, then some customisation of commonly available software may be needed. It may be necessary to establish special permissions which give different individuals authority to create, alter, and view records based on their authority and responsibility within a business or administrative process, for instance. Recordkeeping requirements should be considered when information system are being replaced or upgraded. The recordkeeping aspects of the system that is being phased out should be reviewed by analysts and use the analysis to identify opportunities for improvement. Also, important is whether any of the electronic records stored in the old system need to be retained and migrated into the new system and if they are readily identifiable and well described (National Archives of Australia, 2002).

The complexity of today's information systems environments makes it difficult for employees to control or direct the movement, storage, access/retrieval, preservation or deletion of many of the business records they produce or receive from internal or external sources. As a result, information is widely dispersed on desktops, laptops, applications, shared servers, backup tapes, etc.

Dependence on technology, which has created such enormous challenges in the management of e-records, is also a key part of proactive and sustainable solutions. Capabilities that enable assured levels of recordkeeping performance must be demonstrably trustworthy in the 'chain of preservation', the systems of controls that extend over the entire life cycle of records and ensure the identity and integrity of the records as they are maintained in storage repositories and whenever they are retrieved or reproduced. In this manner, information systems/technology (IS/IT) provides critical support to the creators and users of business records to meet the organisation's mission and comply with its information governance guidelines (ICA, 2005).

Key characteristics of information systems which create, transmit, archive and dispose of e-records include (National Archives of Australia, 2002):

- Reliability over time and in the normal course of business;
- Control measures to monitor, verify, authorise and secure access
- Compliance with changing business, regulatory and legal requirements;
- Ability to manage media-centric and content-centric records produced by the complete range of business activities for the organisation;
- Organisations should imbed into its policies, procedures and tasks as rules, systematic creation, maintenance and management through the design and operation, distribution, protection and disposition requirements for the organisation's e-records and applied across all technical and business systems and processes.

Developing and maintaining a recordkeeping, system requires (National Archives of Australia, 2002):

- Preliminary investigation: basic information concerning the legal, administrative, and economic environment generated, gives a general view of the strengths and weaknesses of the records.
- Analysis of business activity: the hierarchical structured view and description of the functions, activities and transactions of an organisation is provided at this stage. Analysis needs to go as deep as is necessary to show the stages in the business process when records are regularly created or received in the normal course of business. This step provides a useful framework for organising records.
- Identification of a recordkeeping requirements: this stage reflect which records to be captured and maintained, reason for capturing the records, the length of time the records will be maintained, and other characteristics of records required, which should be implemented.
- Assessment of existing systems;
- Identification of a recordkeeping strategy;
- Designing of recordkeeping system(s);
- Implementation of recordkeeping system(s), and
- Ongoing management and review

The Basics of Good Electronic Records Management

Electronic records are the products of computers and applications software; by definition, they do not include the software used in creating them and in the recordkeeping process. E-records are distinct from digitised images as they are created as electronic documents and not converted from another form of a digitised picture (Oregon State University, 2003). E-records are always machine dependent formats; thus they are accessible and readable only with the assistance of digital processors. Electronic records or files exist in different forms and formats. Some distinct types of e-records, which their disposition can be easily determined, are as follows (OSU, 2003):

- **Text Format:** Text documents are now often created for use only in machine readable form, though they had been used traditionally to prepare hard copy records through word processing and desktop publishing software. Textual records consist of drafts and copies of correspondence, memoranda, reports, and publications distributed in hard copy form. Also, spreadsheets and database management programmes can be used to create text documents.
- **Database Formats:** Databases contain large amounts of information organised in data fields which may contain text, numbers, graphics, or mixed character elements. These data elements are organised and stored so that they can be manipulated or extracted to serve diverse applications. However, the database as a whole is managed electronically and independently of the special application.
- **E-mail Format:** Electronic mail (e-mail) consists of any memo, letter, note, report, or communication between individuals and groups that is stored and/or transmitted in a format that requires an electronic device to capture and access (OSU, 2003). E-mail is perhaps the most common electronic record format found in most offices in organisations.
- **Voice Mail Format:** This consists of messages recorded on the organisations' computerised telephone message system. Voice mail allows the recipient to hear and respond to telephone messages at a later time.
- **Electronic Publications Format:** Public access servers provide access to electronic files to anyone, worldwide, who has compatible systems and appropriate connections. These servers usually consist of

software that link remote and networked personal computer workstations to a central file server and through that to similar file servers located elsewhere.
- **Graphic Format:** There are software packages that allow users to create graphics ranging from simple figures and tables to extremely complex images. Digitizing scanners and video conversion hardware also allow for the direct conversion of visual images into digital format for electronic manipulation and storage.

According to Cruickshank (2006), good electronic records management (ERM) requires:

- **Understanding:** The need to have a clear understanding of both the nature of e-records and the information that the organisation captures as e-records is of importance to the management of that organisation. Hence, the need to appoint an experienced record manager to provide the necessary expertise, to develop and maintain strategy and procedures, and to monitor compliance with them.
- **Policy:** There should be a corporate policy formally endorsed by top management for maintaining electronic evidence as corporate records. This provides a framework, which ERM procedures and practices are developed and maintained, together with an authentic basis for the Record Manager to exercise powers and perform duties, responsibilities and functions.
- **Strategy:** Developing an ERM strategy ensures that government-wide standards are adhered to and that corporate e-records remain both accessible and usable for as long as they are needed (including during prolonged computer system failure).
- **Procedure:** This provides means of routinely capturing intimation, which should be designed into the electronic system that

generates e-records. They should be easy to understand and used.

- **ERM System:** This should be designed to manage authentic records and to ensure that their authenticity can be confirmed easily if necessary. A 'process control records' should be considered in the design of the system.

- **Maintenance Capability:** maintaining records of value requires that policies and procedures will spell out appropriate appraisal, scheduling and disposal actions to e-records.

- **Culture and Training:** Training is of importance in other to promote a culture of good recordkeeping practices in the organisation. This will make the requirements stated above to be easily realised.

- **Monitoring Compliance:** This is a vital component of the requirement of a good electronic record management. It ensures that organisation adheres to its own and to externally imposed recordkeeping requirements. Monitoring compliance by individuals and business units is essential to provide confidence that the organisation as a whole complies. Equally, there is little point in establishing internal policy and standards if there are no way of knowing if - and to what extent - they are observed. In either case, the Records Manager should function as an auditor and provide the Board with periodic assurance of compliance.

STRATEGIC CONSIDERATIONS

To ensure the success of electronic records management, four basic issues must be considered. These are (ICA, 2005):

- The key to a successful programme is having a clear strategic vision, a realistic understanding of the programme's abilities, and the flexibility to adjust to changing priorities and customer needs;

- To succeed, archivists must be opportunistic and interventionist;

- Archivists must add value and provide services to their customers;

- Neither archival nor records management concerns will stand well on their own as separate business priorities. Instead, they should be understood and promoted as essential if organisations are to attain their wider gaols.

Developing a strategic vision for electronic records must take into account the archival vision of what it hopes to accomplish and the reality of organisational context.

DEVELOPING APPROPRIATE SKILLS SETS

Adequately managing electronic records requires some skills sets - archival electronic records, technical, and 'soft'. The first three sets of skills are necessary to give the programme credibility; the final is needed to be effective in influencing governmental partners and customers and in promoting the archival agenda (ICA, 2005).

- **Archival Skills:** The knowledge of basic archival and records management principles and techniques is vital to all work with electronic records. Archivists and record managers should have these skills and can also apply them to the questions and problems the organisation faces. Some of the challenges are; what constitutes sufficient documentation of business activities, how can that documentation be created and maintained most effectively and efficiently, when can records be destroyed, and what needs to be preserved?

- **Electronic Records Skills:** The fundamental archival principles and practices and how those principles are changed, reformatted, and/or expanded to deal with electronic records must be understood. The personnel should posses the ability to:
- Understand and articulate what it means to undertake electronic recordkeeping;
- Understand and articulate what it means to preserve electronic records overtime, including preserving each of the components of electronic records (e.g. data, software, documentation), and successfully migrating records to new platform;
- To determine system requirements for electronic recordkeeping and preservation of electronic records; and
- To walk programme staff through the process of determining what is and should be an archival record in an electronic context.

These skills are best acquired or developed in-house.

- **Technical Skills:** technical skills are required in the aspect systems design, data management and software development. They need to know the software that would be best for meeting recordkeeping requirements. There is need therefore, to train systems developers in records management principles so that they are able to develop effective recordkeeping solutions for their clients.

POLICY SUGGESTIONS

For organizations to effectively and efficiently provide information to users, they need to establish a systematic record keeping systems that will address the issue of:

- Creating new records
- Organising records in a logical order
- Actively managing the records (where the period of active management can be several years)
- Inactively managing records that are not frequently used
- Appraising old records and either destroying those no longer required, or permanently storing those of legal or historical value.

Essentially, systems should be incorporated with record keeping processes such as:

- Records creation and capture
- Registration
- Classification
- Storage and handling
- Access and use
- Tracking
- disposal

According to Cruickshank (2006), these processes occur at varying stages of a records lifecycle and require consistently applied policies and procedures to 'operationalise' or bring to life an organisations records management programme.

FUTURE TRENDS

At this point let me reiterate the fact that records are intrinsically linked to the day-to-day business activities and fundamental to business viability, and everybody has a role to play in recordkeeping. Records management programme when fully integrated with electronic recordkeeping systems in place, organizations will engage in;

- Creating electronic records and capturing them into electronic recordkeeping systems

- Designing, building and using electronic systems that keep records
- Maintaining and managing electronic records over time
- Making electronic records accessible

CONCLUSION

Records keeping are the life wire of any meaningful organisation. Without reliable records nobody believes you, work has to be redone and things can go wrong. A consistent approach across the whole of a records life cycle is fundamental to providing transparent and reliable records. If an effective records management programme is consistently applied, transactions will be transparent, officials will be accountable and issues regarding information loss are less likely to arise, and are less likely to be viewed with suspicion by stakeholders. This will no doubt, bring about reputation to Africa in the eyes of the world.

REFERENCES

Association for Information and Image Management (AIIM). (1991). Public records survey in USA. *Records Management Bulletin, 45*, 45.

Australian Standard for Records Management. (2002, July). AS ISO 15489. National Archives of Australia, *Commonwealth Recordkeeping Publication*. Retrieved May 26, 2007 from www.naa.gov.au

Cohasset Associates. (February, 2005). *Assured records management: establishing and maintaining a new level of performance*. White paper. Retrieved June 13, 2008 from www.cohasset.com

Cruickshank, J. (2006). *Good practice guidelines for records and record keeping: a report prepared within the SAFEGROUNDS Project*, Draft 3.1. 18 October.

Heslop, H., Davis, S., & Wilson, A. (2002). *An Approach to the preservation of digital records* (pp. 1–10). Australia: Commonwealth of Australia and National Archives of Australia.

International Council on Archives (ICA). (1997 February). guide for managing electronic records from an archival perspective. *ICA Committee on Electronic Records*. Retrieved October 10, 2006 from http://www.ica.org/biblio/cer/guide_eng.html

International Council on Archives (ICA). (April 2005). Electronic records: a workbook for Archivists. Committee on Current Records in an Electronic Environment. *ICA Study 16*.

International Records Management Trust (IRMT). (2004). *The e-records readiness tool*. Retrieved October 15, 2006, from www.irmt.org

Kemoni, H. N., Wamukoya, J., & Kpilang'at, J. (2003). Obstacles to utilization of information held archival institutions: a review of literature. *Records Management Journal, 13*(1), 40–41. doi:10.1108/09565690310465722

Khamis, K. (1999). Making the transition: from backlog accumulation to a new order. Paper presented at XV Bi-annual Eastern and Southern African Regional Branch of the International Council on Archives (ESARBICA) General Conference, Zanzibar, 26-30 July.

Mazikana, P. C. (1999). Editorial. *ESARBICA Journal, 18*, 4–6.

Mnjama, N. (2005). *Archival landscape in Eastern and Southern Africa*.

Moore, F. (2000, October). Digital data's future: you ain't seen nothin' yet. *Computer Technology Review*.

National Archive of Australia. (2001). *ISO 15489-1 Information and Documentation-Records Management, Part 1: General,* 1, 3. Amsterdam: International Standards Organization. Retrieved July 10, 2008 from www.naa.gov.au

National Archives of Australian. (2002). A new approach to recordkeeping. *Commonwealth of Australia.* Retrieved May 25, 2007 from www.naa.gov.au

Oregon State University. (2003). *Archives & records management handbook.* Oregon: OSU Library. Retrieved April 2, 2007 from http://osulibrary.oregonstate.edu/archives/handbook/chapter2/electronic.html

World Bank. (February, 2005). Fostering trust and transparency through information systems. *PREM notes, 97.*

KEY TERMS AND DEFINITIONS

Archival Values: Those values—administrative, fiscal, legal, evidential and/or informational—which justify the indefinite or permanent retention of government records.

Corporate Records: Are records of an organization showing the daily activities of the organization, which is peculiar to them.

Electronic Records: Are records generated electronically and stored by means of computer technology.

Electronic Records Keeping Practices: Is the act of creating and maintaining complete, accurate and reliable records as evidence of business transactions.

Life Cycle: Is a records life span, from its creation to its final disposition.

Permanent Records: Are records appraised as having sufficient historical or other values to warrant their continued preservation or retention.

Recordkeeping System: Is the framework to capture, maintain, and provide access to evidence over time.

Retention Period: Is the length of time that a record must be kept before it can be destroyed.

Chapter 18

Uganda's Rural ICT Policy Framework:
Strengths and Disparities in Reaching the Last Mile

Carol Azungi Dralega
Western Norway Research Institute, Norway

ABSTRACT

This chapter investigates the Ugandan ICT policy approach to promoting access to and the empowerment of the poor majority, remote and "under-accessed" communities in Uganda. The chapter highlights the strengths of the policy framework while at the same time draws attention to it's weaknesses. For instance, while the chapter acknowledges the fact that the ICT policy framework recognises and has pursued strategic approaches to expanding access to remote areas, a closer scrutiny indicates disparities that may delimit its pragmatism. These disparities, it is argued, mainly emanate from the fact that the policy framework is not entirely holistic nor forwardlooking in its outlook, not only because the processes (of policy making) left out the rural users, it also fails to address the gender dynamics and most urgently, the media convergencies notably between broadcasting and telecommunication. In addition to divorcing itself from political and democratic aspects imperative for development, the policy framework seems shorthanded on sustainability fundamentals that are conjectured to restrict its propitiousness at the grassroots.

INTRODUCTION

Since the beginning of the 1990s, Uganda has joined global trends towards liberalisation. In consonance with liberalizing the economy, the country has undertaken strategic steps to develop an Information and Communication Technology

DOI: 10.4018/978-1-61692-012-8.ch018

(ICT) policy framework that can 'leap frog Uganda into the information economy' (ICT policy 2003). The Ugandan ICT policy has 'development' at the centre of its framework as it recognises the role of information, particularly ICTs, in enhancing economic development. The national ICT Policy framework also recognises that, although Uganda as a whole is underdeveloped with a high demand for ICT propagation, some parts (70%)

are more underdeveloped than others. As a result, a specific policy framework has been developed: the *Rural Communications Development Policy* (RCDP) supported by the *Rural Communications Development* Fund (RCDF) to target underserved rural areas. The aim of this chapter is to outline and interrogate the strengths and weaknesses of the RCDP/Fund and to make suggestions that can engender a more pragmatic policy framework by and for the people it's meant for - rural, impecunious and the marginalized special groups within these communities such as women and youth.

BACKGROUND

Africa, ICT Policy and Development

"With the physical boundaries that separated nations melting off due to the emergence of a boundryless Information Society, any people-group, nation or region that does not line up with the expectations of the New Economy – which is primarily driven by Information and Communication Technologies – will experience lonely moments on the island of insignificance" Dr. *Phillip Emeagwali*[1]

Africa's active participation in the Information Society started with the establishment of the African Information Society Initiative (AISI) later adopted by the United Nations Economic Commission for Africa (ECA) in Addis Ababa 1996 and the subsequent endorsement by high level Ministerial and Organizational of the then Organization of African Union (OAU), Heads of States and Governments' meeting including the 1997 G8 Summit. The role of AISI was to serve as a mechanism for achieving the Millennium Development Goals[2] in the AISI framework document[3] which also recommends the mainstreaming of Information and Communication Technologies. This informed the strong support that the AISI

gives to the development of National ICT policies and strategies through its National Information and Communication Infrastructure (NICI) plan, which helps nations link to national, regional and global development goals – including MDGs (Chiumbu, 2008; Sesan, 2004).

Although today, most African nations have developed a NICI informed ICT policy with national specific considerations, the AISI/NICI continental policy framework has come under heavy criticism for being driven more by foreign actors and ideas to the detriment of an local and grassroots participation and pragmatism (Chiumbu 2008; Dralega 2008). With an exception of a few countries (like Uganda and Kenya), who have adopted grassroots policies, several countries are still struggling to come to terms with the market driven strategies that tend to exclude marginalized and peripheral communities (Gillwald & Stork, 2008; Etta & Elder, 2005; Ogbo 2003) a matter I take up in this chapter.

National ICT Policy Ambitions and Advancements

In line with its policy of liberalisation, beginning in the early 1990s, the Ugandan government embarked on the liberalisation of its telecommunications sector and the pursuit of an ICT policy that would catalyze the role of information in strengthening the national development planning process (ICT Policy, 2003). Within this policy framework, ICTs were seen to support specific national development initiatives, including *Vision 2025*[4], a project that describes its national aspirations, and the 1997 *Poverty Eradication Action Plan (PEAP, 2000)*[5], revised in 2000 and which became Uganda's Comprehensive Development Framework. That is to say it is a national framework to guide detailed medium term sector plans, district plans and the budget process. Other similar frameworks embedded in the ICT policy ambition include: *The Uganda Information Infrastructure Agenda*, the *Plan for the Modernization of Agriculture[6]*, the National Agricultural Advisor Services (NAADS)[7]; the *Rural Electrification and Transformation Project[8]*, *Universal Primary Education* and *Health Improvement and Delivery (Health policy, 1999[9])*. These national planning initiatives, in varying stages of completion, show poverty eradication as an overarching and fundamental goal for the government which recognises and promotes the role of ICT in achieving this.

In recognising the relation between ICT and development, the National ICT Policy focuses on three mutually inclusive areas: information as a resource for development, ICT as a mechanism for accessing information for development, and ICT as an industry including e-business, software development and manufacturing (National ICT Policy. 2003). Here, ICTs are recognised as economic drivers that can propel Uganda into the global economy. The contention within the policy framework is that, although the majority of the population is still dependent on traditional information delivery systems, especially radio, new ICTs can enhance the efficiency of these systems in delivering development information (ICT Policy, 2003).

It all started in 1997 when parliament passed a significant bill - *The Communications Act* (Uganda 1997, Section 3). The main objective of the Act was to increase penetration of telecommunications services through private sector investment (a market forces approach that hinges on competition) instead of government intervention. It was from this act that the ICT Policy (and subsequently the Rural Communications Development Policy) emerged. The Act has since been criticised for aiming more at increasing investment in the telecom sector than providing access to communication facilities (Ofir, 2003).

The Ugandan ICT policy framework operates as a partnership between government, private investors and development partners[10]. The mandate to oversee media and information management falls under the Directorate of Information in the President's office, and that of overseeing telecommunications from 2006, shifted from the Ministry of Works, Housing and Communications to the newly created Ministry of ICT.

As mentioned above, the steps towards liberalisation and deregulation of its public sector, including the communications and telecommunications sector, signalled an important shift away from government monopolies and towards the private sector, who have been entrusted to transform the low levels of ICT in the country through competition within three areas: voice and data network operators; Internet service provision

Figure 1. One of many middlemen selling call cards on a mobile payphone in Kampala (Source: Litho (2006))

and the development of 'value-added' service providers, "the middlemen" like those in Figure 1.

The process began with the licensing of the first mobile cellular service – Celtel in 1993. A second cellular service provider, the South African Mobile Telephone Networks (MTN) started operation in 1998, while the third, the government owned Uganda Post and Telecommunication, which had a monopoly over all telecoms, was broken down into two and privatised (ICT Policy, 2004). In December 2007, the Saudi Arabian Warid Telecom was launched as the fourth telecom company. Since then, Orange Uganda Limited majority owned by France telecoms was launched in 2008 and others are expected to join the scene.

As a result, Uganda's earnings have been enhanced as manifested in the estimated 240UGX billion investment in telecommunications in 2007 alone, and in the sector's contribution to gross national product, reportedly up to 9%, generating 300,000 jobs directly or indirectly, compared to 3,000 in 1998 (The New Vision Supplement, 26 February, 2008).

This framework has seen an increase in the number of Internet subscription, fixed telephone lines subscribers, with the largest increase coming in mobile telephony, as explained further and shown in the tables below:

Figure 2. Telephone subscriptions March 2008 to March 2009

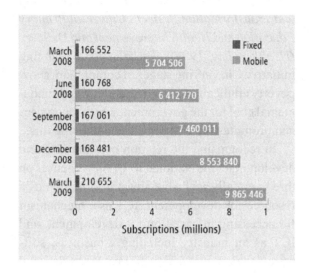

By early 2008, Uganda had 5.7 million mobile phone subscribers, a number that had almost reached 10, million by March 2009 as shown in Figure 2 (Uganda, Communications Commission 2009):

The Internet was introduced in Uganda in 1993, although limited - commercial e-mail services did not become available until 1994. By 2000 there were an estimated 60,000 Internet users in Uganda, although actual subscribers were less than 10,000 (Mwesige, 2004). By mid 2007, Internet subscribers had grown by 30% with 11,000

Table 1. Telecommunication status by 2006

Telecommunication status between 1999-2006	1996	2006
National Telecommunication Operators	1	2
Mobile Cellular Operators	1	3
Internet Service providers	2	17
Fixed Telephone Lines	45,145	108,140
Mobile Cellular Subscribers	3000	2,008,818
Payphones	-	11,082
Private FM Radio Stations	14	145
Private Television Station	4	34
National Postal Operators	1	1

Source: Ministry of Works, Housing and Communications 2003, Uganda Communications Commission (2006)

to 15,500 subscribers between 2006-2007 (Masambu 2007). The biggest challenge to Internet penetration in Uganda, as in the rest of Sub-Saharan Africa, has been the very low telecommunication infrastructure. By 2004 less than 2% of the population had a telephone line and the vast majority of these remain in the capital Kampala and a few other major towns and urban centres (Mwesige, 2004). An even smaller percentage of people own a personal computer. The cost of computers remains high, while private Internet access has been exorbitant (Masambu, 2007).

For instance, available statistics in 2007 monthly fees for dial-up Internet access were about US$30, while Internet Protocol (IP) address access connection annual fees were US$ 1500 or a monthly fees of US$250 (Masambu, 2007). However, there is optimism that this scenario will drastically change after the launch of the Seacom broadband project[11] in July 2009 linking East Africa to Europe and Asia.

However, despite these national developments in the telecom sector, 70% of communication services are still concentrated in elitist urban areas[12], leaving rural areas with the least access to the vital communication services (National ICT Policy, 2004; also see endnote ix on extent of Seacom reach by end of 2009). The information and communication systems and services have remained to a large extent urban-centred, top-down and elitist in nature, leaving the largest percentage of the population (the labour force) uninformed, illiterate and with no effective media for accessing relevant information.[13]. The desire for their participation in the Ugandan development process has been taken up in the country's ICT Policy, (2004) through the Rural Communications Policy with the aimed to extend that much needed access to remote areas. Now, let have a closer look at the policy in question.

The Rural Communications Development Policy

The aim of the Rural Communications Development policy (2001) was to address the imbalances within the national ICT policy that led to the concentration of more than 70% of communication services in urban areas, leaving rural areas, with bigger populations, who are after all contributors of a significant portion of the country's wealth, with the least access to these vital communication services. The policy required all operators to extend services to the 930 sub-counties around the country. However, citing political conflict in the north and the high degree of remoteness and poverty in rural areas, the operators (MTN, Celtel and Uganda Telecom) could only serve 154 of the 930. To address this gap the Uganda Communications Commission (a Ugandan government regulatory body) encouraged bids for licenses that would go to underserved sub-counties. It was from this that the Rural Communications Development Fund was established (National ICT policy 2001; Ofir 2003; UCC 2001).

The Rural Communication Development Fund

One of the outcomes of the RCDP is that since 2003, when the RCDF was officially inaugurated, subsidies and grants have been competitively awarded to service providers to facilitate the provision of communications services in rural areas in accordance with the 2006 policy ambitions and disbursement ratios shown in the tables below:

The Fund, run on a 1% levy on telecommunication companies, grants from donors and government funds, offers subsidies and grants to private investors who set up communication services in rural areas, thus serving as an incentive to serve the total of 154 underserved sub-counties identified. By February 2008 (New Vision Supplement, February, 2008), the Fund is reported to have established Internet Points of Presence in 54

Table 2. Rural Communications Policy on distribution of communication services throughout the country

Number	Activity
20	Internet Points of Presence, one per district[14]
52	Internet Cafes, one per district
54	ICT Training Centres
56	District Information Portals
316	Payphones
154	Telephone networked sub counties
20	Multipurpose Community Telecentres

Source: Uganda Communication Commission (2006)

districts of Uganda including the 24 newly created districts. Kampala, Jinja, Mukono and Masaka had already been served by private investors. In terms of its universal access to telephony figures from the November, 2004 mid-term review note that the World Bank agreed to provide additional funds to enable the Government to meet its new objective of having one public access point for 2,500 inhabitants, as opposed to the original target of 5,000 indicating an improvement in outreach.

In addition, the Fund subsidised the establishment of 54 ICT training centres and 50 Internet cafes all over the county. The supplement also states that the fund also helped create information portals for all the districts, enabling them to share information with local communities, development partners and the outside world. These information portals ascribe domain names for each district, e.g. www.[district name].go.ug provide information on health, agriculture, education and commerce and are also linked to the national portal (www.ugandaweb.net). Other activities of the RCDP are the installation of public payphones in 316 selected sub-counties and the establishment of 25 rural post offices under the Expansion of the Postal Network Project.

Table 3. General Disbursement Ratios

Item	RCDF Investment (%)
Public telephony infrastructure	40
User rural packages	3
Internet Point of Presence and wireless access	12
Internet Exchange Point (IXP) initiative	1
Vanguard Institutions/organisation ICT projects (one per district)	10
ICT start-ups and training (including in schools)	20
Rural post franchise support costs	8
ICT training capacity investment	3
ICT awareness and content creation projects	3
Total	100

Source: UCC (2001)

So Just How Pragmatic is the Policy Framework?

While the policy efforts mentioned above suggest a positive trend in the inclusion of the rural majority of Uganda into ICT discourses, several issues of concern are apparent. In reference to the policy priorities in the above table, the policy thrust seems to be towards access "to meet customer demand" rather than providing an integrated approach that fosters sufficient and meaningful access, participation and people-centred communication.

The preoccupation with promoting access as shown by the biggest percentages, for example the 40% aimed at increasing public telephony, 12% at increasing Internet points of presence and the 20% at setting up Vanguard Institutions and ICT start ups, is at the expense of the preferred integrated approaches that promote processes for meaningful ICT assimilation for development, as shown by the 3% allocated for such as important aspect as awareness and content creation projects. Not to mention that the new Seacom project mentioned above, which is not yet accounted for in the policy frames aims at infrastructural expansion (see website under endnote vii). Proponents of ICT for development point out the limited role of ICTs if they do not socially constructed to integrate and involve users in the technological processes (Hughs 2003; Jacobson and Servaes 1999; Melkote and Steeves 2001). Rural communities, it is argued are embedded in local cultures and indigenous knowledge, and have needs and capacities that must be channelled in technological developments. Contemporary development (communication) theory insists on the recognition, participation and integration of these endogenous dimensions of local communities in any development initiative – including policy (Dralega, 2008; Hughs, 2003; Servaes & Jacobson, 1999; Melkote & Steeves, 2001; Nicol, 2001).

Within the context of every developing country, it is imperative to recognize the marginalised within the already marginalised rural communities, such as women and youths (Dralega, 2007; Dralega, 2008; Dralega, 2009). Both the National and Rural Development Communication Policy frameworks make passing statements without specifying or giving guidelines on how to go about empowering them. In addition, basing on the comprehensive evaluative report on the processes of ICT policy making in Uganda, that was commissioned by IDRC, women, youth, disabled and other minorities within rural areas did not participate in the policy-making processes, which consisted of mainly the private sector, the government, donor communities (See: Ofir, 2003).

As mentioned before, the RCDF derives its funding from a 1% percentage levy on service providers and grants from multilateral partners and UCC, which is arguably too little for the task at hand: i.e. that of rapidly expanding access and effective use for the majority of the country's population, which is also the portion at the least developed levels. In addition, for those communities that are impecunious, the market approach currently dominating the policy agenda is fraught with sustainability issues, something that has to be addressed if the policy is to be effective. This is compounded by the poor government commitment to ICT if one were to consider that the ICT Ministry accounts for a meagre 1% share of the national budget (see endnote ix).

Although, influenced by the more "traditional policy frameworks" such as The Press and Journalist Statute (1995) [15] and The Electronic Media Status (1996)[16], the ICT policies seem to work in parallel with these "traditional" media policies, which may perhaps explain why the ICT policies focus on access issues rather than usability aspects. This regulatory omission provides loophole when it comes to practicalities and disjunctures in the access and utilisation of "old" and "new" ICTs which in reality work together in rural setting – see for example the current trends in the convergence between telecoms and broadcasting in Dralega (2009) in which community media practitioners in the small village of Nakaseke are experimenting new technologies such as mobile telephony, internet and the radio to reach a diverse group of people. In this case the policy framework needs to "catch up" be synchronised to suit the changing times and the budding media convergences and experiments on the ground – a good reason too to involve "non-conventional" grassroots stakeholders.

It is also important to note that ICT policy emphasises the role of ICTs for development, and although a worthy ambition, it fails to acknowledge the role of ICTs for democratic engagement. The development of district portals as governance resources is not sufficient in promoting active civic engagement (they provide mainly district administrative information in addition to economic information). Democracy is an integral part of development and should be given equal footing in rural-aimed initiatives, including policies (Dralega, 2009[17]).

FUTURE TRENDS

It is not a cliché to conjecture that the new information and communication technologies are not just here to stay, they are evolving and fast. Any policy move especially aimed at achieving distributive justice, must look to the past, present but most importantly to the future in order to strategically position itself so as to capture its full potential. In this brief section on future trends I will focus on the potential of the fibre optic projects yet in their budding stages. The Seacom project has been instituted (while the TEAMS and Eassy projects are not yet launched) and as mentioned earlier Uganda has initiated a four phased nationwide expansion starting with the main capitols. Broadband expansion should not be seen merely as more bandwidth and in technical terms, it should be perceived as a paradigm shift in the use of ICTs to propel the intelligence and ingenuity of the networks because according to Mureithi (2009) "The limitation is your capacity

and not the network". A new policy should adopt measures that promote affordable, widespread access to a full range of services.

Since national operator-commercial-interests driven models are fraught with efficacy and sustainability issues, it is imperative for local communities to take the initiative to extend the fibre to communities from the bottom-up (Mureithi, 2009; Ingjerd; Grøtte Strand, forthcoming). These counter networks (Mosse 2005) forged by communities should develop must take up the reigns by exploring their own community needs, determine appropriate ownership models that encourages partnerships and develop innovative services for their respective local communities.

A private-public sector synergy to combine utilities such as power, water and transport utilities with fibre and the implementation of Internet Points of Presence (PoPs) along the fibre would be a forward looking synchronized rural development approach. While local communities have a central role to play in this, so does civil society organizations – who must create awareness of the opportunities and feasibility of community fibre networks for local development, establish forums for community networks to share ideas and experiences of community based networks as well as create advocacy forums to lobby for enabling environment for community fibre networks (Ingjerd et al, forthcoming).

There is a significant role for research and academia in terms of developing the appropriate knowledge base and engaging the issues. In this matrix, development partners also play a significant role in lending support to community fibre initiatives as demonstrated by SIDA pilots in Kenya and Tanzania (Mureithi, 2009) but most importantly the development of grassroots policy for and by the people it is meant to serve is of a crucial importance – more on this in the conclusions below.

CONCLUSION

The information provided above shows that the Ugandan ICT policy has development concerns at the centre of its framework as it recognises the role of information, particularly ICTs, in enhancing development. It is also recognised that, although Uganda as a country is underdeveloped with a high demand for ICT propagation, the policy framework acknowledges that some parts are more rural that others. This has been exacerbated because the initial operations by the licensed operators have created geographical pockets that are unreachable. As a result, a specific policy framework has been developed: the Rural Communications Development Policy supported by the Rural Communications Development Fund to target underserved rural areas.

However, it is argued that although noble, the ICT policy initiative's concern with widening access to include rural areas are driven by market forces and are largely technologically deterministic as policy completely ignores the fundamental issues of how this access should be used to effectively promote the development proclaimed. In essence, the policy omits to specify local content development or people-centred and participatory processes in ICT use that have been attributed to effective ICT for development results (Hughs, 2003).

The issue of participation, it has been argued also arises when considering that the policy-making process omitted rural users but instead featured the private sector, the government, development partners and academics (Ofir 2003). This is in view of conventional prerequisites in policy-making for the inclusion of beneficiary populations in any policy-making activities (Nicol, 2001; Chiumbu, 2008) In order to ensure rural participation, local stakeholders such as local councillors', local leaders, and target group representatives such as women, youth, religious leaders, teachers, farmers from the grassroots should be included in policy making and review processes so it can better re-

flect the needs of their constituencies[18]. A good example would be the Community Multi Media Centre model that requires the establishment of community representatives (i.e. farmers, women, youth, religious leaders, people with disabilities, teachers etc) in locally owned and sustained ICT initiatives[19]

The chapter argues that the policy is gender and youth insensitive. Something of a grave situation in any country's development where women make up large rural populations and work force while youths also represent current and future stakeholders – both groups formidable within the development agenda. Through the sort of representational participation mentioned above, these groups can be instrumental in the development of more pragmatic policy development and effective and sustainable practices.

To deal with high cost of implementation, there is a need to encourage investments in hardware and user-generated software as a policy prerequisite. The challenges of lowering internet costs seem to have been covered by the newly launched Seacom broadband project, however, this new development calls for the revisitation of the policy framework to stipulate guidelines that reflect the needs and limitations of the under-served rural areas in addition to keeping up with new developments. A policy move regarding the Seacom project could provide a condusive environment for local and regional institutions to actively participate in the expansion of the broadband infrastructure.

I also argue that the policy's thrust towards development is a worthy ambition, although the fact that it does not mention the democratic role of ICTs is somewhat disconcerting. For a developing country with large poor communities, development issues must be subject to a market place of ideas, a democratic prerequisite that must not be divorced from development concerns. However, it must be acknowledged that the policy framework started off with noble ambitions and strategies, but the more propitious policy framework will largely depend on the ability of policy makers to consider and to include the different grassroots realities in a *holistic* way with intended users at the forefront, actively promoting content development that engenders the said development more than just focusing on access. Such a holistic approach, driven from below, would include the nuances embedded in a multidimensional approach – for instance improving ICT access and capacity development within the rural education system – which currently reflects a geographical digital divide with urban schools which have better access to computers and training services, tackling corruption within the frameworks of the grassroots ICT projects; developing software that suites the needs and capacities of local people (Adomi et al. 2005; Dralega, 2007) are just a few approaches that ought to be considered within holistic grassroots targeted ICT policies.

REFERENCES

Adomi, E. E., Adogbeji, B. O., & Oduwole, A. A. (2005). The use of internet service providers by cyber-cafes in Nigeria: an update . *The Electronic Library*, *23*(5), 567–576. doi:10.1108/02640470510631281

Chiumbu, H. S. (2008). *Understanding the Role and Influence of External Actors and Ideas in African Information, Communication and Technology Policies*. Oslo, Norway: UNIPUB.

Dralega, C. A. (2007*).* Rural Women's ICT Use in Uganda: Collective Action for Development in ICTs, Women Take a Byte. *Durban Agenda Journal*, (71).

Dralega, C. A. (2008). *ICT Based Development of marginal Communities: Participatory Approaches to Communication, Empowerment and Engagement in Uganda*. Oslo, Norway: UNIPUB.

Dralega, C. A. (2009) *Multimedia Experiments and Disjunctures in Community Media in Uganda*. Equid Novi, 1(30).

Elder, L., & Etta, F. (2005). *At the Crossroads: ICT Policy making in East Africa*. Nairobi, Kenya: East African Educational Publishers.

Gillwald, A., & Stork, C. (2008). *Towards the African e-Index: ICT access and usage in 16 African Countries*. Johannesburg, South Africa: The Link Centre. Retrieved September 03, 2008 from http://www.researchictafrica.net/images/upload/Cairo.pdf

Hughs, S. (2003). Community Multimedia Centres: Creating Digital Opportunities for all . In Girard, B. (Ed.), *The One to Watch – Radio, New ICTs and Interactivity*. Bonn, Germany: The Friedrich Ebert Stiftung.

Ingjerd, S., & Grøtte, I. P. (in press). [Rural Infrastructural development as a Stages model: a roadmap to new rural infrastructure.]. *Strand*.

Jacobson, T., & Servaes, J. (Eds.). (1999). *Theoretical Approaches to Participatory Communication*. Geneva, New Jersey: International Association for Media Communication Research.

Litho, P. (2005). *ICTs, Empowerment and Women in Rural Uganda: A Scott Perspective*. A paper presented at the "To think is to experiment" SSMA, Centre for Narrative Research, UEL, 22nd April 2005.

Masambu, P. (2007). *A Review of the Postal and Telecommunicatios Sector: June 2006-June 2007*. Kampala, Uganda: Communications Commission.

Melkote, S. R., & Steeves, H. L. (2001). *Communication for Development in the Third World: Theory and Practice for Empowerment* (4th ed.). London: Sage.

Minges, M. (2001). *The Internet in an African LDC: Uganda Case study*. International Telecommunication Union Study.

Mwesige, P. (2004). Cyber Elites: A Survey of Internet Cafes in Uganda. *Telecommunications: Development in Africa*, *21*(1), 83–101.

New Vision Supplement. (2006, 26th February). Retrieved from Http://newvision.co.ug/D/8/12/613650

Nicol, C. (2003). *ICT Policy: A Beginners Handbook*. Johannesburg, South Africa: Association for Progressive Communication.

Ofir, Z. M. *(2003)*. Information and Communication Technologies for Development (ACACIA) The Case of Uganda: Final Evaluation Report prepared for IDRC Evaluation Unit.

Ogbo, O. (2003). *Strengthening National ICT Policies in Africa: Governance, Equity and Institutional Issues. Acacia II*. IDRC.

PEAP. (2000). *Poverty Eradication Action Plan.*

Policy, I. C. T. (2003) *The National Information and Communication Policy*. Retrieved from http://ucc.co.ug/nationalIctPolicyFramework.doc

Ssewanyana, K. J. (2007). ICT Access and Poverty in Uganda. *International Journal of Computing and ICT Research, 1*(2), 10 - 19. Retrieved from http:www.ijcir.org/volume1-number2/article2.pdf

UCC. (2006). *Uganda Communication Commission*. Retrieved from www.ucc.co.ug

KEY TERMS AND DEFINITIONS

Empowerment: There are several meanings to this term, here, it is meant to address members of groups that social discrimination processes have excluded from decision-making processes through - for example - discrimination based on disability, race, ethnicity, religion, or gender, geographic location.

Internet Protocol (IP): Is the primary protocol in the Internet Layer of the Internet Protocol Suite and has the task of delivering distinguished protocol datagrams (packets) from the source host to the destination host solely based on their addresses.

Marginalization: Is the social process of becoming relegated or confined to a lower social standing or outer limit or edge, as of social standing. Material deprivation is the most common result of marginalization when looking at how unfairly material resources (such as food and shelter, technologies) are dispersed in society. Along with material deprivation, marginalized individuals are also excluded from services, programs, and policies (Young, 2000).

Participation: Is a term associated with the importance of involving wider groups of people in decisions and decision making processes, services and design.

Point of Presence (PoP): Is an artificial demarcation point or interface point between communications entities. POPs, as understood in this study, are also located at Internet exchange points and colocation centres.

Social Construction of Technology (SCOT): Is a theory within the field of Science and Technology Studies. Advocates of SCOT — that is, social constructivists — argue that technology does not determine human action, but that rather, human action shapes technology. They also argue that the ways in which a technology is used cannot be understood without understanding how that technology is embedded in its social context. SCOT is a response to technological determinism and is sometimes known as technological constructivism.

Technological Determinism: Is a reductionist theory that presumes that a society's technology drives the development of its social structure and cultural values.

ENDNOTES

[1]   Philip Emeagwali is an Igbo Nigerian-born engineer and computer scientist/geologist who was one of two winners of the 1989

Gordon Bell Prize, a prize from the IEEE, for his use of the Connection Machine supercomputer – a machine featuring over 65,000 parallel processors – to help analyze petroleum fields. The quote was from Sesan G. (2004) Africa, ICT policy and the Millenium Development Goals.

[2] http://www.un.org/millenniumgoals/

[3] E-Strategies: National. Sectoral and Regional ICT Policies, Plans and Strategies. UNECA, 2003.

[4] For more, see: Government of Uganda (1998), Vision 2025: A Strategic Framework for National Development: Vol. 1: Main Document, Ministry of Finance, Planning and Economic Development, Kampala and Government of Uganda (1998), Vision 2025: A Strategic Framework for National Development: Vol. 2: Background Papers, Ministry of Finance, Planning and Economic Development, Kampala.

[5] Also see: Uganda's Progress in Attaining the PEAP Targets – in the context of the Millennium Development Goals. Background paper for the Consultative Group Meeting, Kampala, 14-16 May 2003. Prepared by the Ministry of Finance, Planning and Economic Development.

[6] Details can be obtained here: http://www.pma.go.ug/

[7] NAADS recognizes the role of ICTs in the promotion of sustainable agriculture. This has been followed up through the targeted delivery of information packages to Community Multimedia Centres of Nabweru, Buwama and Nakaseke among other ICT centres. However, this arrangement (NAADS) has recently been riddled with corruption and inefficiency putting to question the effectiveness and continuity of this important collaboration.

[8] Funded by the World Bank (2001-2009).

[9] Ministry of Health (1999) Uganda Health Policy, 1999.

[10] The main development partners include: International Development Research Centre (IDRC); UNESCO, International Telecommunication Union (ITU), Uganda Government -

[11] http://www.seacom.mu/index2.asp

[12] Mwesige´s 2004 study of Internet access in Uganda also indicates this trend, pointing out that most of the users, whom he calls "cyber Elites", are educated, young and middle-class (Mwesige 2004).

[13] Even the newly launched Seacom broadband project has been marred by corruption as only one of four phases has been unsatisfactorily completed connecting just 4 towns in the central region see details here: Broadband the Inside story (The Independent): http://www.independent.co.ug/index.php/business/business-news/54-business-news/1424-broadband-for-uganda-the-inside-story. Accessed: 2.12.2009.

[14] IPP – Internet Points of presence are meant to establish internet connectivity in all the 80 districts of Uganda.

[15] Press and Journalist Statute is an extension of Article 29(1) of the Constitution on Freedom of Expression in print media. It also crated the Media Council, the National Institute of Journalists of Uganda and Disciplinary Committee within the media Council which is responsible for regulating eligibility for media ownership...

[16] The Electronic Media Statute created a licensing system under the broadcasting Council for radio, television stations, cinemas and video businesses. Under the Uganda Broadcasting Corporation (UBC) Act of 2005, there was provision of free spectrum coverage for community radio up to 10kilometres.

[17] Dralega, C. A. (2009) ICTs, Youth and the Politics of Participation in rural Uganda in Mudhai, O.; Tettey W. & Banda, F. (Eds.)

African Media and the Digital Public Sphere. Palgrave Macmillan, NY.

[18] For examples see Dralega, C. A (2009) Bridging the Digital Divide: Exploring the principles of the Community Multimedia Centre model in Uganda in Ronning H. and Orgeret, S. (2009) The power of Communication: Changes and Challenges in African Media. Unipub, Oslo.

[19] Dralega, C.A. (2009) Bridging the digital divide: exploring the principles of the Community Multimedia Centre Model in Uganda in Ronning, H. & Orgeret K, S. (Eds.) The Power of Communication: Changes and challenges. Unipub, Oslo.

Compilation of References

Aalberts., B.P., & Van der Hof, S. (1999). *Digital Signature Blindness Analysis of Legislative Approaches to Electronic Authentication*, (pp. 32). Retrieved on Nov.29, 2008 from; http://www.buscalegis.ufsc.br/arquivos/Digsigbl.pdf

Abichandani, R. K. (2004). *Controvertial Copyright Issues*, National Judicial Academy, Bhopal, Retrieved on August, 21, 2009, from http://gujarathighcourt.nic.in/Articles/roundtable.htm.

About, N. A. S. S. C. O. M. (2009). Retrieved March 18, 2009, from http://www.nasscom.in/Nasscom/templates/NormalPage.aspx?id=5365

Abu-Musa, A. A. (2007). Evaluating the security controls of CAIS in developing countries: An examination of current research. *Information Management & Computer Security, 15*(1), 46–63. doi:10.1108/09685220710738778

Adedibu, L. A. (2005/2006). Collection Development Policy: The Case of University of Illorin Library. *Nigerian Libraries, 39*, 79–91.

Adewopo, A. (2009, June 9). Pay subscription broadcasting: rights acquisition and infringement. *The Guardian*, (p.96).

Adeyeye, M., & Iwela, C. C. (2005). Towards an Effective National Information and Communication Technologies in Nigeria. *Information Development, 21*(3), 202–208. doi:10.1177/0266666905057337

Adomi, E. E., & Igun, S. E. (2008). Combating cybercrime in Nigeria. *The Electronic Library, 26*(5), 16–725. doi:10.1108/02640470810910738

Adomi, E. E. (2009). *Library and information resources*. Benin City, Nigeria: Ethiope Publishing Corporation.

Adomi, E. E. (2008). *Library and information service policies*. Benin City, Nigeria: Ethiope Publishing Corporation.

Adomi, E. E., Adogbeji, B. O., & Oduwole, A. A. (2005). The use of internet service providers by cybercafes in Nigeria: an update. *The Electronic Library, 23*(5), 567–576. doi:10.1108/02640470510631281

Adomi, E. E., & Igun, S. E. (2008). ICT polices in Africa. In Cartelli, A., & Palma, M. (Eds.), *Encyclopedia of information communication technology* (pp. 384–389). Hershey, PA: Information Science Reference.

Adomi, E. E. (2007), Overnight Internet browsing among cybercafé users in Abraka, Nigeria. *Journal of Community Informatics, 3*(2). Retrieved October 30, 2009 from http://ci-journal.net/index.php/ciej/article/viewPDFInterstitial/322/351

Afele, J. S. (2003). *Digital bridges: developing countries in the knowledge economy*. Hershey, PA: Idea Group.

Africa Policy Monitor, A. P. C. (2009). Retrieved on March 27 2009 from http://africa.rights.apc.org/?apc=he_1&w=s&t=21873

African Business. (2009, Thursday, January 1). Retrieved on March 27, 2009 from http://www.allbusiness.com/media-telecommunications/11745598-1.html retrieved on 27/03/2009

Aghatise, E. J. (2006). *Cybercrime Definition. Computer Crime Research Center*. June 28, 2006. Available online at www.crime-research.org

Agyeman, O. T. (2007). *Survey of ICT and Education in Africa: Nigeria Country Report*. Retrieved on the 18th of August, 2009 from www.infodev.org/en/Document.422. Pdf

Akdeniz, Y. (n.d.). *Speech 3: Controlling internet content: Implications for cyber-speech*. Retrieved November 4, 2009 from http://portal.unesco.org/ci/en/files/18278/110 86401441statement_akdeniz.doc/statement_akdeniz.doc

Alampay, E. A. (2006a). Beyond access to ICTs: measuring capabilities in the information society. *International Journal of Education and Development Using Information and Communications Technology, 2*(3), 4–22.

Alampay, E. A. (2006b). The capability approach and access to information and communications technologies. In Minogue, M., & Cariño, L. (Eds.), *Regulatory governance in developing countries* (pp. 183–205). Cheltenham, UK: Edward Elgar.

Alexander, B. (2004). Going nomadic: Mobile learning in higher education. *EDUCAUSE Review, 39*(5), 29–35.

Aljifri, H. A., Pons, A., & Collins, D. (2003). Global e-commerce: A framework for understanding and overcoming the trust barrier. *Information Management & Computer Security, 11*(3), 130–138. doi:10.1108/09685220310480417

Alkire, S. (2005). Why the capability approach? *Journal of Human Development, 6*(1), 115–133. doi:10.1080/146498805200034275

American Bar Association. (1996). *Digital signature guidelines: Legal Infrastructure For certification Authorities and Secure Electronic Commerce*, (2nd Ed.). Chicago: Author. Retrieved on Oct. 30, 2008, from American Bar Association website; http://www.abanet.org/scitech/ec/isc/dsg.pdf

Anand, P. B. (2007). Capability, sustainability, and collective action: an examination of a river water dispute . *Journal of Human Development, 8*(1), 109–132. doi:10.1080/14649880601101465

Anderson, D. (1974). *Universal Bibliographic Control: A long term policy, a plan for action*. München, Germany: Verlag Dokumentation.

Andrew, T. N., & Petkov, D. (2003). The need for a systems thinking approach to the planning of rural telecommunications infrastructure. *Telecommunications Policy, 27*(1-2), 75–93. doi:10.1016/S0308-5961(02)00095-2

Andrews, C. (2003). *The problem of child pornography on the Internet and the ability of the British Police to regulate it*. Unpublished MA thesis, University of Hull, Hull, UK.

Ang, P. H. (1997). How countries are regulating Internet content. Retrieved April 30, 2009 from http://www.isoc.org/INET97/proceedings/B1/B1_3.HTM

Ani, O. E. (2005). Evolution of Virtual Libraries in Nigeria: Myth or Reality? *Journal of Information Science, 31*(1), 67–70. doi:10.1177/0165551505049262

Ani, O. E. (2007). Information and Communication Technology (ICT) Revolution in African Librarianship: Problems and Prospects. *Gateway Library Journal, 10*(2), 111–118.

Ani, O. E., & Ahiauzu, B. (2008). Towards Effective Development of Electronic Information Resources in Nigerian University Libraries. *Library Management, 29*(6/7), 504–514. doi:10.1108/01435120810894527

Ani, O. E., & Biao, E. P. (2005). Globalization: Its Impact on Scientific Research in Nigeria. *Journal of Librarianship and Information Science, 37*(3), 153–160. doi:10.1177/0961000605057482

Ani, O. E., Esin, J. E., & Edem, N. (2005). Adoption of Information and Communication Technology (ICT) in Academic Libraries: A Strategy for Library Networking in Nigeria. *The Electronic Library, 23*(6), 701–708. doi:10.1108/02640470510635782

Anyaegbunam, C., Mefalopulos, P., & Moetsabi, T. (1999). *Participatory rural communication appraisal (PRCA) methodology, the first mile of connectivity*. Rome: FAO. Retrieved May 29, 2006, from http://www.fao.org/sd/CDdirect/CDan0024.htm

APC WNSP. (2005). *New gender and policy monitor help women make ICT policy a priority*. Retrieved 2nd November 2009. from: http://www.genderit.org/en/index.shtml?w=9&x=91258

Arnold, J., Guermazi, B., & Mattoo, A. (2007). Telecommunications: the persistence of monopoly . In Mattoo, A., & Payton, L. (Eds.), *Services, trade and development: the experience of Zambia* (pp. 101–153). Washington, DC: The International Bank for Reconstruction and Development/World Bank.

Arnstein, S. R. (1969). A ladder of citizen participation. *Journal of the American Planning Association. American Planning Association, 35*(4), 216–224. doi:10.1080/01944366908977225

Association for Information and Image Management (AIIM). (1991). Public records survey in USA. *Records Management Bulletin, 45,* 45.

Atkinson, R. D. (2001). The New Growth Economics: How to Boost Living Standards through Technology, Skills, Innovation, and Competition. *Blueprint Magazine.* Retrieved on 22 Feb 2009 from http://www.dlc.org/ndol_ci.cfm?contentid=2992&kaid=107&subid=123

Aubeelack, P. (2004). *Mauritius. Proceedings of workshop of the regional consultation on national e-government readiness in Gaborone, Botswana.* From 14-16 April 2004. Retrieved July 16, 2005, from http://www.comnet-it.org/news/CESPAM-Botswana.pdf

Australian Standard for Records Management. (2002, July). AS ISO 15489. National Archives of Australia, *Commonwealth Recordkeeping Publication.* Retrieved May 26, 2007 from www.naa.gov.au

Avram, H. D. (1984). Authority control and its place. *Journal of Academic Librarianship, 6*(9), 331–335.

Awe, J. (n.d.). *Fighting cybercrime in Nigeria.* Retrieved October 30, 2009 from http://www.jidaw.com/itsolutions/security7.html

Bae, K. S. (2001). Korea's e-commerce: Present and future. *Asia-Pacific Review, 8*(1), 77.

Baeza-Yates, R., & Ribeiro-Neto, B. (1999). *Human-computer interaction.* Retrieved September 15, 2006, from http://www.ischool.berkeley.edu/~hearst/irbook/10/node3.html

Ball, K. S. (2001). Situating workplace surveillance: ethics and computer based performance monitoring. *Ethics and Information Technology, 3*(3). doi:10.1023/A:1012291900363

Banou, C., Kostagiolas, P., & Olenoglou, A. (2008). The Reading Behavioural Patterns of the Ionian University Graduate Students: Reading Policy of the Academic Libraries. *Library Management, 29*(6/7), 489–503. doi:10.1108/01435120810894518

Bantekas, I., & Nash, S. (2007). *International Criminal Law.* Abingdon, UK: Routledge-Cavendish.

Barja, G., & Gigler, B.-S. (2005). The concept of information poverty and how to measure it in the Latin American context . In Galperin, H., & Mariscal, J. (Eds.), *Digital poverty: perspectives from Latin America and the Caribbean.* Ottawa: IDRC.

Barrett, S., & Fudge, C. (Eds.). (1981). *Policy and Action.* London: Methuen.

Basu, S., & Jones, R. P. (2002). *Legal Issues Affecting E-Commerce A Review of the Indian Information and Technology Act 2000,* 17th Annual BILETA Conference, Amsterdam. Retrieved on Oct. 22, 2008; from http://works.bepress.com/subhajitbasu/33

Batchelor, S., & Norrish, P. (2004). *Framework for the assessment of ICT pilot projects: beyond monitoring and evaluation to applied research.* Washington, DC: InfoDev. Retrieved July 21, 2006, from http://infodev.org/en/Publication.4.html.

Batchelor, S., & Sugden, S. (2003). *An analysis of InfoDev case studies: lessons learned.* Washington, DC: InfoDev. Retrieved May 26, 2006, from http://www.sustainableicts.org/infodev/infodevreport.pdf

Batuk lal. (2006). *The Law of Evidence,* Allahabad, India: Central Law Agency.

Beaudiquez, M., & Bourdon, F. (1991). *Management and use of name authority files (personal names, corporate bodies and uniform titles): evaluation and prospects.* München, Germany: Saur.

Bebbington, A. (2004). Social capital and development studies 1: critique, debate, progress? *Progress in Development Studies, 4*(4), 343–349. doi:10.1191/1464993404ps094pr

Bellah, R. N., Madsen, R., Sullivan, W. M., Swidler, A., & Tipton, S. M. (1985). *Habits of the heart.* Berkeley, CA: University of California Press.

Biggeri, M., Libanora, R., Mariani, S., & Menchini, L. (2006). Children conceptualizing their capabilities: results of a survey conducted during the first Children's World Congress on Child Labour. *Journal of Human Development, 7*(1), 59–83. doi:10.1080/14649880500501179

BITPIPE. *COM.* (2008). Retrieved on August 15, 2007 from http:/www.bitpie.com.

Bojanc, R., & Jerman-Blaic, B. (2008). Towards a standard approach for quantifying an ICT security investment. *Computer Standards & Interfaces*, *30*(4), 216–222. doi:10.1016/j.csi.2007.10.013

Bonder, G. (2002). *From access to appropriation: women and ICT policies in Latin America and the Caribbean. A Report from the United Nations Division for the Advancement of Women (DAW) Seoul Republic of Korea II to IV November 2002.* Retrieved December 23, 2009 from http://www.un.org/womenwatch/daw/egm/ict2002/reports/Paper-GBonder.PDF

Bonsor, K. (2000). How DNA Computers Will Work. Retrieved on July 6, 2010 from HowStuffWorks.com. <http://computer.howstuffworks.com/dna-computer.htm>

Botswana, Ministry of Communications, Science and Technology (2008). *Maitlamo: National Information and Communications Technology Policy.* Gaborone, Botswana: Government Printer.

Bouguettaya, A., Ouzzani, M., Medjahead, B. & Cameron, J. (2001). Helping citizens of Indiana: Ontological approach to managing state and local government databases. *IEEE Computer,* February.

Bourdon, F. (1993). *International cooperation in the field of authority data: an analytical study with recommendations.* München, Germany: Saur.

Bourne, R. (Ed.). (1992). *Seminar on Bibliographic Records: proceedings of the Seminar held in Stockholm, 15-16 August 1990.* München, Germany: Saur.

Bozimo, D. O. (2005/2006). ICT and the Ahmadu Bello University Libraries. *Nigerian Libraries*, *39*, 1–20.

Bridges.org. (2001). *Comparison of E-readiness Assessment Models: Final Draft.* Retrieved July 16, 2003, from http://www.bridges.org/eredainess/tools.html

Brown, R., & Price, S. (2006). *The Future of Media Regulation in New Zealand: Is There One?* Retrieved March 19, 2009, from http://www.bsa.govt.nz/publications/BSA-FutureOfMediaRegulation.pdf

Browne, P. (1996). *Study of the effectiveness of Informatics policy instruments in Africa.* Addis Ababa, Ethiopia: ECA. Retrieved on the 20th of July 2009 from www.belanet.org/partners/aisi/policy/infopol/sumbrown.htm

Buchinski, E. J., Newman, W. L., & Dunn, M. J. (1976). The automated authority subsystem at the National Library of Canada. *Journal of Library Automation*, *4*(9), 279–298.

Building Local Capacity for ICT Policy and Regulation. A needs Assessment and Gap Analysis for Africa, the Carribean, and the Pacific. Retrieved March 21, 2009 from www.infuder.org/en.project

Burch, S. (1997). *Latin American women take on the Internet.* Retrieved 2nd November, 2009. Retrieved from http://www. apcwomen.org/netsupport/articles/art.01.html

Burrows, J., & Cheer, U. (2005). *Media Law in New Zealand* (5th ed.). Melbourne, Australia: Oxford University Press.

Buyya, R. (2001). *Parallel processing: architecture and system overview.* Retrieved on March 27 2009 from http://www.cs.mu.oz.au/678/ParCom.ppt#352,1,Parallel Processing: Architecture and System Overview.

Bynum, T. W. (2009). *Ethics in the information age, (Tech.Rep.).* New Haven, CT: Southern Connecticut State University.

Bynum, T. W., & Rogerson, S. (Eds.). (1996). *Global information ethics.* USA: Opragon press.

Byrne, E., & Sahay, S. (2007). Participatory design for social development: a South African case study on community-based health information systems. *Information Technology for Development*, *13*(1), 71–94. doi:10.1002/itdj.20052

Cantoni & C. McLoughlin (Ed.). *Proceedings of World Conference on Educational Multimedia, Hypermedia and Telecommunications2004,* (pp. 4729-4736).

Capuro, R. (2006). Towards an ontological foundation of information ethics. *Ethics and Information Technology*, *8*(2), 175–186. doi:10.1007/s10676-006-9108-0

Carbo, T. (2007). Information rights: trust and human dignity in e-government. [from http://www.i-r-i-e.net]. *International Review of Information Ethics*, *7*(9), 1–7. Retrieved December 8, 2008.

Carbo, T. (2006). *Understanding information ethics and policy: integrating ethical reflection and critical thinking into policy development*. Retrieved February 28, 2009, from http:/www.ethicspolicy.pitt.edu

Carlisle, A. & Lloyd. (2002). *Understanding PKI: Concepts, Standards, and Deployment Considerations*, (pp. 11-17). London: Addison-Wesley.

Carr, A. (2004). *Internet Traders of Child Pornography and other Censorship Offenders in New Zealand*. Wellington, New Zealand: Department of Internal Affairs.

Carr, A. (2006). *Internet censorship offending: a preliminary analysis of the social and behavioural patterns of offenders*. Unpublished PhD thesis, Bond University, Queensland.

Caruso, J. B. (2003). Information technology security policy keys to success. *Center for Application Research Bulletin, 2003*(23). Retrieved January 2009 from http://www.educause.edu//library//ERB0323.pdf

Cerina, P. (1998). The New Italian Law on Digital Signatures. 6 *CTLR,* 193.

Cert-in is the official website of the Indian Computer Emergency Response Team. (n.d.). Retrieved from http://www.cert-in.org.in/

Chan, A. H., & Gligor, V. (2002). *Information Security, (LNCS)*. Berlin: Springer. doi:10.1007/3-540-45811-5

Chang, S. E., & Chin-Shien, L. (2007). Exploring organizational culture for information security management. *Industrial Management & Data Systems, 107*(3), 438–458. doi:10.1108/02635570710734316

Chawki, M. (2006). *Anonymity in Cyberspace: Finding the Balance between Privacy and Security*. Revista da Faculdade de Direito Milton Campos.

Chawki, M. (2006). *Le Droit Penal à l'Epreuve de la Cybercriminalité*. Lyon, France: University of Lyon III.

Chawki, M. (2008). *Combattre la Cybercriminalité*. Perpignan, France: Editions de Saint Amans.

Chawki, M. & Wahab, M. (2006). Identity Theft in Cyberspace: Issues and Solutions. LexElectronica, *11*(1).

Chawki, M. (2005). A Critical Look at the Regulation of Cybercrime. *The ICFAI Journal of Cyberlaw, 4*(4).

Chawki, M. (2009) Nigeria Tackles Advance Fee Fraud. *Journal of Information, Law and Technology.* Retrieved May 21st, 2009. From: http://www2.warwick.ac.uk/fac / soc/law/elj/jilt/ 2009 _1 /chawki/chawki.pdf

Chetty, M. (2005). Information and Communications Technologies (ICTs) and Africa's development. Retrieved on March 5, 2009 from http://www.nepad.org/2005/files/documents/124.pdf.

Chiumbu, H. S. (2008). *Understanding the Role and Influence of External Actors and Ideas in African Information, Communication and Technology Policies*. Oslo, Norway: UNIPUB.

Chukwuemerie, A. (2006). Nigeria's Money Laundering (Prohibition) Act 2004: A Tighter Noose. *Journal of Money Laundering Control, 9*(2). doi:10.1108/13685200610660989

Chwee Kin Keong and Others v. Digilandmall.com Pvt. Ltd, (2004) 2 SLR 594.

Clement, E. A. H. (1980). The automated authority file at the National Library of Canada. *International Cataloguing, 4*(9), 45–48.

Cloud Corp. v. Hasbro Inc, 314 F.3d 289 (7th Cri. 2002).

Cohasset Associates. (February, 2005). *Assured records management: establishing and maintaining a new level of performance*. White paper. Retrieved June 13, 2008 from www.cohasset.com

Cohen, H. (2003). Child Pornography Produced without an Actual Child: Constitutionality of Congress Legislation. In W. Holliday (Ed.), *Governmental Principles and Statutes on Child Pornography*. New York: Nova Science Publishers.

Comin, F. (2001 June). *Operationalising Sen's capability approach*. Paper prepared for the conference Justice and Poverty, examining Sen's capability approach, Cambridge, UK. Retrieved August 20, 2006 from http://www.st-edmunds.cam.ac.uk/vhi/sen/papers/comim.pdf.

Common Market for Eastern and Southern Africa, Association of Regulators of Information And Communications for Eastern and Southern Africa. (2004). *Policy Guidelines on Universal Service/Access.* Retrieved on September 22, 2009 from http://programmes.comesa.int/attachments/118_Policy_Guidelines_Universal_Service.pdf.

Common Market for Eastern and Southern Africa. (2009). *COMESA Regional e-Government Framework.* Retrieved on September 23, 2009 from http://programmes.comesa.int/attachments/144_COMESA%20e-GOV%20Framework%20PresentationKampala.pdf.

Community Research and Development Information service. (CORDIS, 2009). *ICT - Future and emerging technologies: a continuing well established successful ICT scheme.* Retrieved on September 22, 2009 from http://cordis.europa.eu/fp7/ict/programme/fet_en.html

Conklin, W. A. (2007). Barriers to adoption of e-government. In *Proceedings of the 40th Annual Hawaii International Conference on System Sciences.* Washington, DC: IEEE Computer Society.

Considerations for ICT policy formulation in developing countries. (n.d.). Retrieved on of June 20th, 2009 papers.ssrn.com/sol3/.../SSRN_ID983944_code733084.pdf?

Corbridge, S. (2002). Development as freedom: the spaces of Amartya Sen. *Progress in Development Studies, 2*(3), 183–217. doi:10.1191/1464993402ps037ra

Corea, S. (2007). Promoting development through information technology innovation: The IT artifact, artfulness, and articulation. *Information Technology for Development, 13*(1), 49–69. doi:10.1002/itdj.20036

Cruickshank, J. (2006). *Good practice guidelines for records and record keeping: a report prepared within the SAFEGROUNDS Project*, Draft 3.1. 18 October.

Csete, J., Wong, Y., & Vogel, D. (2004). Mobile devices in and out of the classroom. In L.

Cullen, R. (2002). Addressing the digital divide. *Online Information Review, 25*(5), 311–320. doi:10.1108/14684520110410517

Dabeesing, T. (2006). *National ICT Strategic Plan Emerging Technologies and Standards Working Group.* Retrieved on September 22, 2009 from http://www.gov.mu/portal/goc/ncb/file/Presentation_NWGET.pdf.

Daniel, J., & West, P. (2006). *E-learning and free open source software: The key to global mass higher education*? International Seminar on Distance, Collaborative and eLearning: Providing Learning Opportunities in the New Millennium via Innovative Approaches, Universiti Teknologi Malaysia Kuala Lumpur, Malaysia, 4-5 January 2006.

Darlington, R. (2009). *How the Internet could be regulated.* Retrieved May 8, 2009 from http://www.rogerdarlington.co.uk/Internetregulation.html

Dart, J., & Davies, R. (2003). A dialogical, story-based evaluation tool: the most significant change technique. *The American Journal of Evaluation, 24*(2), 137–155.

Data Security Council of India. (2008). Retrieved March 18, 2009, from http://www.nasscom.in/Nasscom/templates/NormalPage.aspx?id=51973

Davies, P. B. (2002). *Information systems: an introduction to Informatics in organizations.* New York: Palgrave.

Davis, F. D. (1989). Perceived usefulness, perceived ease of use, and user acceptance of information technology. *Management Information Systems Quarterly, 13*(3), 319–340. doi:10.2307/249008

Delone, W. H., & Mclean, E. R. (1992). Information system success: The quest for the dependent variable. *Information Systems Research, 3*(1), 61–95. doi:10.1287/isre.3.1.60

Delsey, T. (1980). IFLA Working Group on an International Authority System: A progress report. *International Cataloguing, 9*(January/March), 10–12.

Delsey, T. (1989). Authority control in an international context. *Cataloging & Classification Quarterly, 3*(9), 13–28. doi:10.1300/J104v09n03_02

Delsey, T. (2004). Authority records in a networked environment. *International Cataloguing and Bibliographic Control, 4*(33), 71–74.

Department for International Development (DFID). (2006). *Financing ICT for Development: the EU approach.* Retrieved on March 27, 2009 from http://www.dfid.gov. uk/pubs/files/eu-financ-wsis-english.pdf.

Department for International Development (DFID). (2008). *Millennium Development Goals.* Retrieved on March 27, 2009 from http://www.dfid.gov.uk/mdg/

Department of Internal Affairs. (1997). *Report of the Department of Internal Affairs for the year ended 30 June 1997.* Wellington, New Zealand: Author.

Department of Internal Affairs. (2002a). *Briefing to the Incoming Minister of Internal Affairs.* Wellington, New Zealand: Author.

Department of Internal Affairs. (1998). *New Zealand Censorship Compliance Unit.* Unpublished paper presented at the Australian Institute of Criminology conference, Internet Crime, Melbourne University, 16-17 February.

Department of IT eTechnology Group (India). (2003). *Assessment of central ministries and departments: e-governance readiness assessment 2003: Draft Report 48.*

Department of Public Service and Administration. (1996). *Green paper transforming public service delivery.* Pretoria, South Africa: GCIS.

Dhillon, G., & Torkzadeh, G. (2006). Value-focused assessment of information system security in organizations. *Information Systems Journal, 16,* 293–314. doi:10.1111/j.1365-2575.2006.00219.x

Dholakia, R. R., Miao, Z., Dholakia, N., & Fortin, D. R. (2000). *Interactivity and revisits to websites: a theoretical framework.* RITIM Working paper. Retrieved December 15, 2008, from http://ritim.cba.uri.edu/wp/.

Diebold Systems Pvt. Ltd. *v.* The Commissioner of Commercial Taxes, (2006). 144STC59(Kar).

Diffie, W., & Hellman, M. E. (1976). New Directions in Cryptography. *IEEE Transactions on Information Theory, IT-22*(6), 644–654. doi:10.1109/TIT.1976.1055638

Dirks, K. T., & Ferrin, D. L. (2002). Trust in leadership: meta-analytic findings and implications for research and practice. *The Journal of Applied Psychology, 87*(4), 611–628. doi:10.1037/0021-9010.87.4.611

DIT & NCAER (Department of Information Technology & National Council of Applied Economic Research). (2006). *India: e-readiness assessment report 2006 - for States/Union Territories.*

Divjak, S. (2008). Mobile phones in the classroom. *International Journal on Hands-on Science, 1*(2), 92–94.

DOC. (2009). Launch of the National e-Skills Dialogue - From Agrarian, Industrial to Information Society. Retrieved July 2009, from http://www.doc.gov.za/index.php?option=com_content&task=view&id=309&Itemid=457.

Dong, L., Neufeld, D., & Higgins, C. (2009). Top management support of enterprise systems implementations. *Journal of Information Technology, 24*(1), 55–80. doi:10.1057/jit.2008.21

Dralega, C. A. (2008). *ICT Based Development of marginal Communities: Participatory Approaches to Communication, Empowerment and Engagement in Uganda.* Oslo, Norway: UNIPUB.

Dralega, C. A. (2007). Rural Women's ICT Use in Uganda: Collective Action for Development in ICTs, Women Take a Byte. *Durban Agenda Journal,* (71).

Dralega, C. A. (2009) *Multimedia Experiments and Disjunctures in Community Media in Uganda.* Equid Novi, 1(30).

Drevin, L., Kruger, H. A., & Steyn, T. (2006). Value-focused assessment of ICT security awareness in an academic environment. In S. Fischer-Hubner, K. Ranneberg, L. Yngstrom, & S. Lindskog (Eds.), *IFIP International Federation for Information Processing. Volume 201, Security and Privacy in Dynamic Environments,* (pp. 448-453). Boston: Springer.

Duggal, P. (2005). Cyber-crime in India: The Legal Approach . In *R.G. Broadhurst & P.N. Grabosky, (n.d.), Cyber-crime: The Challenge in Asia* (pp. 183–196). Hong Kong: Hong Kong University Press.

Duncombe, R. (2006). Using the livelihoods framework to analyse ICT applications for poverty reduction through microenterprise. *Information Technologies and International Development, 3*(3), 81–100. doi:10.1162/itid.2007.3.3.81

Dunsire, G. (2009). *The Semantic Web and expert metadata: Pull apart then bring together: Presented at Archives, Libraries, Museums 12 (AKM12), Poreč, Croatia, 2008.* Retrieved March 25, 2009, from http://cdlr.strath.ac.uk/pubs/dunsireg/akm2008semanticweb.pdf

Durance, C. J. (1978). "What's in the name?" Summary of the workshop . In *What's in a name? Control of catalogue records through automated authority files* (pp. 225–227). Vancouver, Toronto: University of Toronto Library Automation System.

Dzidonu, C. (2002). *A blueprint for developing national ICT policy in Africa.* Retrieved October 1, 2009 from http://www.atpsnet.org/pubs/specialpaper/Dzidon.pdf

Earl, S., Carden, F., & Smutylo, T. (2001). *Outcome mapping: building learning and reflection into development programs.* Ottawa, Canada: IDRC.

Easton, D. (1965). *A systems analysis of political life.* New York: John Wiley.

Economic and Financial Crimes Commission (Establishment) Act. (2004). Retrieved 10th May, 2009. From http://efccnigeria.org/index.php?option=com_docman&task=doc_download&gid=5

ECPAT New Zealand. (2003). *Williamson's Child Pornography Sentence Too Light.* media release, 31 July.

Edewor, N. (2008). Freedom of Information Bill: issues, imperatives and implications for Nigerian libraries. *Ozoro Journal of General Studies, 1*(1), 34–40.

Editorial, (2008). The ranking of our Universities. *ThisDay Newspapers* (Tuesday 9, December).

Elder, L., & Etta, F. (2005). *At the Crossroads: ICT Policy making in East Africa.* Nairobi, Kenya: East African Educational Publishers.

Electronic Frontiers Australia. (2002). *Internet Censorship: Law & policy around the world.* Retrieved May 12, 2009 from http://www.efa.org.au/Issues/Censor/cens3.html

Electronic Frontiers Australia. (2006). *Internet censorship laws in Australia.* Retrieved May 11, 2009 from http://www.efa.org.au/Issues/Censor/cens1.html

Encyclopaedia Britannica. (2009). Very large-scale integration. Retrieved on March 27, 2009, from *Encyclopædia Britannica Online* from http://www.britannica.com/EBchecked/topic/626791/very-large-scale-integration

Erikson, E. H. (1963). *Childhood and society* (2nd ed.). New York: W.W. Norton.

Ernest, H., & Shetty, R. (2005). Impact of Nanotechnology on Biomedical Sciences: Review of Current Concepts on Convergence of Nanotechnology With Biology. *Journal of Nanotechology online.* Retrieved on July 6, 2010 from http://www.azonano.com/details.asp?ArticleID=1242

Erwin, G. J., & Taylor, W. (2006). Assimilation by Communities of Internet Technologies: Initiatives at Cape Peninsula University of Technology, Cape Town . In Marshall, S., Taylor, W., & Yu, X. (Eds.), *Encyclopedia of Developing Regional Communities with Information and Communication Technology* (pp. 40–46). Hershey, PA: Idea Group Publishing.

Etim, F. (2006). Resource Sharing in the Digital Age: Prospects and Problems in African Universities. *Library Philosophy and Practice, 9*(1), 12–19.

European Commission. (2002). *eEurope 2005 action plan: e-Inclusion.* Retrieved December 20, 2008, from http://europa.eu.int/information_society/eeurope/2005/all_about/einclusion/index_en.htm

European Commission. (2005). *Transforming public services.* Report of the Ministerial eGovernment Conference, Manchester, UK. Retrieved December 12, 2008, from http://www.egov2005conference.gov.uk/documents/pdfs/eGovConference05_Summary.pdf

Evans, J., & Ninole, M. (2004). Minding the information gap in Papua New Guinea: a view from Wewak. In R.F. Garcia, R.F. (Ed.), *Divide and connect: perils and potentials of information and communication technology in Asia and the Pacific.* (pp. 49-60). Mumbai: Asian South Pacific, Bureau of Adult Education (ASPBAE).

Falch, M., & Anyimadub, A. (2003). Tele-centres as a way of achieving universal access: the case of Ghana. *Telecommunications Policy, 27*(1-2), 21–39. doi:10.1016/S0308-5961(02)00092-7

Fallis, D. (2007). Information ethics for the twenty –first century library professionals. *Library Hi Tech*, *25*(1), 23–36. doi:10.1108/07378830710735830

Farelo, M., & Morris, C. (2006). *The working group on e-government in the developing world: Roadmap for e-government in the developing world, 10 questions e-government leaders should ask themselves.* Retrieved December 24, 2008, from http://researchspace.csir.co.za/dspace/bitstream/10204/1060/1/Morris_2006_D.pdf

Farrier, J. (2010, January 11). "Wet Computer" Will Mimic A Biological Brain. Retrieved on July 7, 2010 from http://www.neatorama.com/2010/01/11/wet-computer-will-mimic-a-biological-brain/

Federal Republic of Nigeria. (2001) Nigerian National Policy for Information Technology (IT). Retrieved April 14, 2009 from www.nita.gov.ng/document/nigeriaitpolicy.pdf

Fellenstein, C., & Wood, R. (2000). *Exploring e-commerce, global e-business, and societies, 30.* Upper Saddle River, NJ: Prentice Hall.

Ferraro, M., & Casey, E. (2005). *Investigating Child Exploitation and Pornography.* San Diego, CA: Elsevier.

Fillip, B., & Foote, D. (2007). *Making the connection: scaling telecenters for development.* Washington, DC: Information Technology Applications Center of the Academy for Education Development. Retrived July 10, 2008, from http://connection.aed.org/pages/MakingConnections.pdf

Finocchiaro, G. (2002). Digital Signature and Electronic Signatures: The Italian Regulatory Framework After The d.lgs 10/2002. *Electronic Communication Law Review*, (9), 127.

Foresight Nanotech Institute. (2008). Retrieved on October 10, 2007 from http://www.foresight.org/nano/

Forestier, E., Grace, J., & Kenny, C. (2002). Can information and communication technologies be pro-poor? *Telecommunications Policy*, *26*(11), 623–646. doi:10.1016/S0308-5961(02)00061-7

Forrester, E. C. (2003). *A Life-Cycle Approach to Technology Transition.* Retrieved on June 15 2008 from http://www.sei.cmu.edu/news-at-sei/features/2003/3q03/feature-4-3q03.htm

Foster, I. (2005). *Internet computing and the emerging grid.* Retrieved on October 9, 2007 from http://www.nature.com/nature/webmatters/grid/grid.html

Foster, K., Heppenstal, R., Lazarz, C., & Broug, E. (2008). *Emerald Academy 2008 Authorship in Africa.* Retrieved March 20, 2009 from http://info.emeraldinsight.com/pdf/report.pdf/

Froehlich, T. J. (1992). Ethical consideration of information professionals. *Annual Review of Information Science & Technology*, *27*, 291–32.

Frost & Sullivan. (2007). Cited in *ZAMNET gains global recognition as market leader.* Retrieved on December 8, 2007 from http://www.zamnet.zm/newsys/news/viewnews.cgi?category=30&id=1197109256

Fulford, H., & Doherty, N. F. (2003). The application of information security policies in large UK-based organizations: An exploratory investigation. *Information Management & Computer Security*, *11*(2/3), 106–114. doi:10.1108/09685220310480381

Functional requirements for authority data: A conceptual model. (2009). G. Patton, (Ed.). IFLA Working Group on Functional Requirements and Numbering of Authority Records (FRANAR). Final Report, December 2008. Approved by the Standing Committees of the IFLA Cataloguing Section and IFLA Classification and Indexing Section, March 2009. München, Germany: Saur.

Functional requirements for bibliographic records: Final report. (1998). IFLA Study Group on the Functional Requirements for Bibliographic Records. München, Germany: Saur. Retrieved March 25, 2009, from http://www.ifla.org/VII/s13/frbr/frbr.pdf

Gagliardone, I. (2005). Virtual enclaves or global networks? The Role of information and communication technologies in development cooperation. *PsychNology Journal*, *3*(3), 228–242.

Garnham, N. (1999). Amartya Sen's capabilities approach to the evaluation of welfare: its application to communications . In Calabrese, A., & Burgelman, J.-C. (Eds.), *Communication, citizenship and social policy: rethinking the limits of the welfare state* (pp. 281–302). Oxford, UK: Rowan and Littlefield.

Gasper, D. (1997). Sen's capability approach and Nussbaum's capabilities ethics. *Journal of International Development*, *9*(2). doi:10.1002/(SICI)1099-1328(199703)9:2<281::AID-JID438>3.0.CO;2-K

Gasper, D. (2002). Is Sen's capability approach an adequate basis for considering human development? *Review of Political Economy*, *14*(4), 435–461. doi:10.1080/0953825022000009898

Gavaghan, T. (2010). *DNA computing.* Retrieved on July 6, 2010 from http://EzineArticles.com/?expert=Thomas_Gavaghan

Gaved, M., & Anderson, B. (2006). *The impact of local ICT initiatives on social capital and quality of life.* Chimera Working Paper 2006-6, University of Essex, Colchester, UK. Retrieved October 30, 2006, from http://www.essex.ac.uk/chimera/content/pubs/wps/CWP-2006-06-Local-ICT-Social-Capital.pdf

Gbaje, E. S. (2007). Provision of Online Information Services in Nigerian Academic Libraries. *Nigerian Libraries*, *40*, 1–14.

Gelbstein, E., & Kurbalija, J. (2005). *Internet Governance: Issues, Actors and Divide*. Malta: Diplo Foundation.

Genderit.org. (n.d). *Why gender in ICT policy?* Retrieved November 25, 2009 from http://www.genderit.org/en/beginners/whygender.htm

Geness, S. (2004). E-government, the South African experience. In *Proceedings of workshop of the regional consultation on national e-govt readiness in Gaborone, Botswana*. From 14-16 April 2004. Retrieved July 16, 2004 from http://www.comnet-it.org/news/CESPAM-Botswana.pdf

Genoud, P., & Pauletto, G. (2004). *The e-society repository: Transforming e-government strategy into action.* Paper Presented at the 4th European Conference on E-government, Dublin Castle, Ireland, 17-18 June 2004.

Gerhan, D., & Mutula, S. M. (2005). Bandwidth Bottlenecks at the University of Botswana: Complications for Library, Campus, and National Development. *Library Hi Tech*, *23*(1), 102–117. doi:10.1108/07378830510586748

Gigler, B.-S. (2004 September). *Including the excluded – can ICTs empower poor communities? Towards an alternative evaluation framework based on the capability approach.* Paper for the 4th International Conference on the capability approach, University of Pavia, Pavia. Retrieved May 26, 2006, from http://www.its.caltech.edu/~e105/readings/ICT-poor.pdf

Gillwald, A., & Stork, C. (2008). *Towards the African e-Index: ICT access and usage in 16 African Countries.* Johannesburg, South Africa: The Link Centre. Retrieved September 03, 2008 from http://www.researchictafrica.net/images/upload/Cairo.pdf

Gorman, M. (1978). The Anglo-American cataloguing rules, second edition. *Library Resources & Technical Services*, *3*(22), 209–225.

Gorman, M. (1998). The future of cataloguing and cataloguers. *International Cataloguing and Bibliographic Control*, *27*(4), 68–71.

Gorman, M. ([1979] 1982). Authority control in the prospective catalog. In M. W. Ghikas (Ed.), *Authority control: The key to tomorrow's catalog: Proceedings of the 1979 Library and Information Technology Institutes* (pp. 166-180). Phoenix, AZ: Oryx Press.

Gorniack, K. (1996). The computer revolution and the problem of global ethics. In Bynum & Rogerson (Eds.), *Global information ethics*, (pp.177-190). USA: Opragon press.

Government of Canada. (2006). *On-line forms and services.* Retrieved May 13, 2007, from http://canada.gc.ca/form/e-services_e.html

Government of Mozambique. *Information and Communication Technology Policy Implementation Strategy: Towards the Global Information Society.* Mozambique, June. 2002. http://www.idrc.ca/en/ev-90304-201-1-DO_TOPIC.htm, Retrieved on the 7th of July 2009

Government of Singapore. (2004). *E-citizen: Your gateway to all government services.* Retrieved May 14, 2007, from http://www.ecitizen.gov.sg/

Grace, J., Kenny, C., & Qiang, C. Z. (2004). *Information and Communication Technologies and Broad-Based Development: A Partial Review of the Evidence*. Washington, DC: The World Bank.

Granqvist, M. (2005). Looking critically at ICT4Dev: the case of Lincos. *Journal of Community Informatics, 2*(1), 21–34.

Greene, S. (2006). *Security Policies and Procedures: Principles and Practices*. Upper Saddle River, NJ: Prentice Hall.

Grossenbacher, K. (2000). *Implementing the District Health System in KwaZulu-Natal: A Systemic Inquiry into the Dynamics at the Policy/Practice Interface*. Unpublished masters dissertation, University of Natal, Durban, South Africa.

Grunfeld, H. (2007). *Framework for evaluating contributions of ICT to capabilities, empowerment and sustainability in disadvantaged communities*. Paper presented at the CPRSouth2 (Communication Policy Research) Conference, 'Empowering rural communities through ICT policy and research,' Chennai, India, 15-17 December.

Guba, E. G., & Lincoln, Y. S. (1981). *Effective Evaluation*. San Francisco: Jossey-Bass.

Guidelines for authority and reference entries (1984). Recommended by the Working Group on an International Authority System, approved by the Standing Committees of the IFLA Section on Cataloguing and the IFLA Section on Information Technology. London: IFLA International Programme for UBC.

Guidelines for authority records and references (2001). *Revised by the Working Group on GARE (Rev., 2nd ed.). München, Germany: Saur.*

Guislain, P., Qiang, C., Lanvin, B., Minges, M., & Swanson, E. (2006). Overview . In *Information and Communications for Development: Global Trends and Policies*. Washington, DC: World Bank.

Gull, C. D. (1969). The impact of electronics upon cataloguing rules. *International Conference on Cataloguing Principles, Paris, 9th-18th October, 1961 (1969). Report* (pp. 281-290). London: Clive Bingley.

Gurstein, M. (2003). Effective use: a community informatics strategy beyond the digital divide. *First Monday, 8*(12).

Gurstein, M. (1999). Flexible networking, information and communications technology and local economic development, *First Monday 4* (2). Retrieved March 15, 2003, from http://www.firstmonday.org/issues/issue4_2/gurstein/index.html.

Gurstein, M. (2003, December). Effective use: a community informatics strategy beyond the digital divide, *First Monday, 8*(12). Retrieved June 18, 2004, from http://firstmonday.org/issues/issue8_12/gurstein/index.html.

Gurstein. M. (2007). *What is Community Informatics (and Why Does It Matter)?* Retrieved February 2008, from http://eprints.rclis.org/archive/00012372/01/What is Community Informatics reading.pdf.

Gustaniene, A. (2005). *Gender focused ICT policy making*. Retrieved 2nd November 2009. From: http://www.genderit.org/en/indexshtml?w=a&x=91489

Hafkin, N. (2002). *Gender issues in ICT policy in developing countries: An overview*. Retrieved 9th January 2009, from http://www.un.org/womenwatch/daw/egm/ ict2002/reports/paper--NHafkin.pdf

Hagiya, M. (2000). Theory and Construction of Molecular Computers. Retrieved on July 6, 2010 from http://hagi.is.s.u-tokyo.ac.jp/MCP/.

Hansche, S. (2001). Designing a security awareness program: Part 1. *Information System Security, 9*(6), 14–22.

Harper, T. (2003). *What is Nanotechnology?* 2003 Nanotechnology 14. doi:10.1088/0957-4484/14/1/001. Retrieved on October 10, 2007 from http://www.iop.org/EJ/abstract/0957-4484/14/1/001

Harris, R., & Rajora, R. (2006). *Empowering the poor. Information and communications technology for governance and poverty reduction: a study of rural development projects in India*. UNDP Asia-Pacific Development Information Programme.

Harris, R. W. (2001). *A place of hope, connecting people and organisations for rural development, through multipurpose community telecentres (MCTs) in selected Philippine Barangays; a learning evaluation.* Report for International Development Research Centre (IDRC), Ottawa. Retrieved October 5, 2006, from http://www.idrc.ca/IMAGES/ICT4D/PanAsia/HarrisPhilippineReport.pdf

Hart-Teeter. (2003). The new e-Government equation: Ease, engagement, privacy and protection. *A Report prepared for the Council for Excellence in Government.* Retrieved February 10, 2009 from http://www.cio.gov//egovpoll2003.pdf

Hawker, S. (Ed.). (2002). *Oxford dictionary thesaurus: wordpower guide.* New York: Oxford University Press.

Heeks, R. (2002). *E-government in Africa: Promise and practice.* Manchester, UK: Institute for Development Policy and Management University of Manchester.

Heeks, R. (2006). Theorizing ICT4D research. *Information Technologies and International Development, 3*(3), 1–4. doi:10.1162/itid.2007.3.3.1

Heeks, R. (1999). *Information and communication technologies, poverty and development.* Retrieved 9th January 2009, from http://idpm.man.ac.ik/ idpm/diwpf5.htm

Heslop, H., Davis, S., & Wilson, A. (2002). *An Approach to the preservation of digital records* (pp. 1–10). Australia: Commonwealth of Australia and National Archives of Australia.

Heuck, J. (2005). *Comparative Report on Child Pornography on the Internet in Germany and New Zealand.* Auckland/Freiburg. ECPAT Deutschland e. V and ECPAT New Zealand.

Hjern, B., & Hull, Ch. (1982). Implementation Research as Empirical Constitutionalism. *European Journal of Political Research, 10*(June), 105–115. doi:10.1111/j.1475-6765.1982.tb00011.x

Hone, K., & Eloff, J. H. P. (2002). Information security policy - What do international information security standards say? *Computers & Security, 21*(5), 402–409. doi:10.1016/S0167-4048(02)00504-7

Hoppe, H. U., Joiner, R., Milrad, M., & Sharples, M. (2003). Guest editorial: Wireless and mobile technology in education. *Journal of Computer Assisted Learning, 19*(3), 255–259. doi:10.1046/j.0266-4909.2003.00027.x

Howard, I. (2008). *Unbounded possibilities: observations on sustaining rural information and communication technology (ICT) in Africa.* Retrieved on October 29, 2008, from http://www.apc.org/en/system/files/SustainingRuralICTs_0.pdf

Hsu, J. (2010, January 11). Wet Computer Literally Simulates Brain Cells. Retrieved on July 7, 2010 from http://www.popsci.com/technology/article/2010-01/wet-computer-literally-simultes-brain-cells

http://wougnet.Org/ICTpolicy/ug/ugictpolicy.html. (n.d.). Retrieved on the 20th of June, 2009.

Hudson, H. E. (2006). *From rural village to global village: telecommunications for development in the information age.* Mahwah, NJ: Lawrence Erlbaum Associates.

Hughs, S. (2003). Community Multimedia Centres: Creating Digital Opportunities for all . In Girard, B. (Ed.), *The One to Watch – Radio, New ICTs and Interactivity.* Bonn, Germany: The Friedrich Ebert Stiftung.

Human Development Report Office. (2008). *Human development indices.* Retrieved February 11, 2009, from http://hdr.undp.org/en/media/HDI_2008_EN_Tables.pdf

Human Development Report. (2001). *Making New Technologies Work for Human Development.* New York: Oxford University Press.

ICT Policy for Lesotho. Final 4. (March 2005). Retrieved April 10 2009 from www.Lesotho.gov.is/documents/lesotho-ICT-Policy-Final.p

ICT sub sector in Bangladesh. (n.d.). Retrieved on the 9th of November, 2000 from www.usaid.gov/bd/files/afe-review

Idowu, B., Ogunbodede, E., & Idowu, B. (2003). Information and communication in Nigeria. The health sector experience. *Journal of Information Technology Impact, 3*(2), 69–76.

Ifidon, S. E. (2002). Policy Issues in the Funding of Nigerian University Libraries . In Lawal, O. O. (Ed.), *Modern Librarianship in Nigeria: A Festschrift to mark the Retirement of Chief Nduntuei Otu Ita, University Librarian, University of Calabar, 1977-1997* (pp. 49–54). Calabar, Nigeria: University of Calabar Press.

Ifidon, S.E. (2006, March). *Planning without information: the bane of national development.* Inaugural lecture delivered at Ambrose Alli University, Ekpoma, Edo State, Nigeria.

IFLA Cataloguing Section FRBR Working Group. (n.d.). Retrieved March 25, 2009, from http://www.ifla.org/VII/s13/wgfrbr/

IFLA Meetings of Experts on an International Cataloguing Code. (2009). Retrieved March 25, 2009, from http://www.ifla.org/VII/s13/icc/

Implementation team on global Policy participation. (2002).Global policymaking for information And communications technologies: Enabling meaningful participation by Developing-nation stakeholders. Retrieved October 1, 2009 from http://www.markle.org/downloadable_assets/roadmap_report.pdf

Indian Computer Emergency Response Team. (2006). *Information Security Policy for Protection of Critical Information Infrastructure.* New Delhi, India: Department of Information Technology, Government of India.

Indian Petrochemicals Corporation Ltd. *v.* Union of India, (Decided on: Sept.19, 2006). (2006). MANU/GJ/8490.

Indira, C., & Stone, P. (2005). *International Trade Law.* London: Routledge Cavendish.

Information Development News. (2006)...*Information Development*, *22*(1), 7–21. doi:10.1177/0266666906062685

Information Security Environment in India. *NASSCOM Analysis.* (n.d). Retrieved March 18, 2009, from http://www.nasscom.org/download/ Indian_Security_Envomt_05_06_Factsheet_Final1.pdf

Information Technology Act. (2000). As amended by the Act 10 of 2009.

Information Technology Security Procedure Rules. *(2004).*

Ingjerd, S., & Grøtte, I. P. (in press). [Rural Infrastructural development as a Stages model: a roadmap to new rural infrastructure.]. *Strand.*

Initiative. (2008). Retrieved March 18, 2009, from http://www.nasscom.in/Nasscom/templates/NormalPage.aspx?id=5954

Intellon. (2007). *What is powerline communications?* Retrieved on March 5, 2009 from http://www.intellon.com/technology/powerlinecommunications.php

International Centre for Missing & Exploited Children. (2008). *Child Pornography: Model Legislation & Global Review.* Retrieved on February 15, 2009, from http://www.icmec.org/en_X1/pdf/ModelLegislationFINAL.pdf

International Conference on Cataloguing Principles, Paris, 9th-18th October, 1961. (1963). *Report.* London: International Federation of Library Associations. Reprinted in: International Conference on Cataloguing Principles, Paris, 9th-18th October, 1961 (1969). *Report.* London: Clive Bingley.

International Conference on National Bibliographies, Paris, 1977 (1978). *Final report.* Paris: UNESCO.

International Council on Archives (ICA). (1997 February). guide for managing electronic records from an archival perspective. *ICA Committee on Electronic Records.* Retrieved October 10, 2006 from http://www.ica.org/biblio/cer/guide_eng.html

International Council on Archives (ICA). (April 2005). Electronic records: a workbook for Archivists. Committee on Current Records in an Electronic Environment. *ICA Study 16.*

International Institute for Communication and Development (IICD). (2008). *Introducing Zambia.* Retrieved on September 21, 2009 from http://www.iicd.org/countries/zambia.

International Meeting of Cataloguing Experts, Copenhagen, 1969: Report (1969). *IFLA Annual.* Also published in: Report of the International Meeting of Cataloguing Experts, Copenhagen, 1969 (1970). *Libri, 1*(20), 105-132.

International Records Management Trust (IRMT). (2004). *The e-records readiness tool.* Retrieved October 15, 2006, from www.irmt.org

International standard archival authority record for corporate bodies, persons and families: ISAAR(CPF). (1996). Ottawa, Canada: Secretariat of the ICA Commission on Descriptive Standards.

International Telecommunication Union Task Force on Gender Issues. (2001). *Gender-aware guidelines for policy-making and regulatory agencies.* Retrieved December 23, 2009 from http://www.itu.int/ITU-D/gender/projects/FinalGendAwrnGuidelns.pdf

International Telecommunication Union. (2009). *Special initiatives: Gender.* Retrieved December 23, 2009 from http://www.itu.int/ITU-D/sis/Gender/

International Telecommunications Union. (2005). *What's the state of ICT access around the world?* Retrieved.on February 22, 2009 from http://www.itu.int/wsis/tunis/newsroom/stats/

Internet World statistics. (2007). *Internet Usage Statistics: World Internet Users and Population Stats.* Retrieved September 12, 2007, from http://www.internetworldstats.com/stats.htm

Interoperable Delivery of European E-government Services to Public Administration. *Businesses and Citizens-IDABC.* (2005). Retrieved July 19, 2005, from http://europa.eu.int/idabc/en/chapter/383

ISO/IEC 27000. (2007). *The ISO/IEC 27K Toolkit – Implementation guidance and metrics.* Retrieved January 20, 2009 from http://www.iso27001security.com/html/k_toolkit.html

IT Governance Institute. (2005). Aligning COBIT, ITIL and ISO 17799 for business benefit. *A management briefing from ITGI and OGC.* Retrieved February 4, 2009, from http://www.nysforum.org/documents//itil-6-6-06/.pdf

ITU. (2006). Measuring ICT for Social and Economic Development. *World Telecommunication/Ict Development Report.* Retrieved January 2007, from http://www.itu.int/publications.

ITWeb. (2009). *Meraka boosts e-skills.* Retrieved July 2009, from http://www.itweb.co.za/sections/business/2009/0903241046.asp?S=IT%20in%20Government&A=ITG&O=google, 24 March 2009

Jacobs, S. J., & Herselman, M. E. (2006). Information Access for Development: A Case Study at a Rural Community Centre in South Africa. *Issues in Informing Science and Information Technology, 3,* 295–306.

Jacobson, T., & Servaes, J. (Eds.). (1999). *Theoretical Approaches to Participatory Communication.* Geneva, New Jersey: International Association for Media Communication Research.

Jahankani, H., Antonijevic, B., & Walcott, T. H. (2008). Tools protecting stakeholders against hackers and crackers: An insight review. *International Journal of Electronic Security and Digital Forensics, 1*(4), 423–442. doi:10.1504/IJESDF.2008.021459

Jain, R., & Raghuram, G. (2005). *Study on accelerated provisions of rural telecommunication services (ARTS).* Ahmedabad, India: Indian Institute of Management. Retrieved December 22, 2006, from http://www.iimahd.ernet.in/ctps/pdf/Final%20Report%20Edited.pdf

James, J. (2006). The Internet and poverty in developing countries: welfare economics versus a functionings-based approach. *Futures, 38*(3), 337–349. doi:10.1016/j.futures.2005.07.005

James, T. (Ed.). (2001). *An Information Policy Handbook for southern African, A Knowledge Base for Decision-Makers.* International Development Research Centre.

Jensen, M. (2002). *Information and Communication Technology (ICTs) in Africa – a status report.* UN ICT Task Force "Bridging the Digital Divide in the 21st Century" Presented to the Third Task Force Meeting United Nations Headquarters 30 September – 1 Oct 2002. Retrieved on January 27, 2009 from http://www.unicttaskforce.org/thirdmeeting/documents/jensen%20v6.htm

Jhunjhunwala, A., Ramachandran, A., & Bandyopadhyay, A. (2004). n-Logue: the story of a rural service provider in India. *Journal of Community Informatics, 1*(1), 30–38.

Jidaw Systems Limited. (2007). *The Nigerian National Information Technology Policy.* Retrieved on August 22, 2009 from http://www.jidaw.com/itsolutions/ict4dreview2007.html

Jordan, R. (2002). *Free speech and the constitution.* Retrieved May 11, 2009 from http://www.aph.gov.au/LIBRARY/Pubs/RN/2001-02/02rn42.htm

Jueneman., R R. & Robertson, Jr., R. J. (1998). Biometrics and Digital Signatures in Electronic Commerce 38 *Jurimetrics* 427.

Juraske, I. (2007). *Some success tips for deploying ICTs in Africa*. Retrieved on January 10, 2008 from http://communications-online.blogspot.com/2007/06/some-success-tips-for-deploying-icts-in.html

Kallman, E. A., & Grillo, J. P. (1996). *Ethical decision making and information technology: an introduction with cases*. New York: McGrawHill.

Kamath, A. (2005). *Law Relating to Computers, Internet and E-Commerce: A Guide to Cyberlaws, & the Information Technology Act, 2000 with Rules and Notifications.* Delhi, India: Universal Law Publishing.

Kemoni, H. N., Wamukoya, J., & Kpilang'at, J. (2003). Obstacles to utilization of information held archival institutions: a review of literature. *Records Management Journal, 13*(1), 40–41. doi:10.1108/09565690310465722

Khamis, K. (1999). Making the transition: from backlog accumulation to a new order. Paper presented at XV Bi-annual Eastern and Southern African Regional Branch of the International Council on Archives (ESARBICA) General Conference, Zanzibar, 26-30 July.

Khor, M. (1995). *Globalisation and the Need for coordinated Southern policy Response, Cooperation South.* New York: UNDP.

KNET (Keewaywin Local Government and Economic Day). (2001). Retrieved on November 15, 2008 from http://smart.knet.ca/keewaywin/governance.html

Knight, W. (2007). IBM creates world's most powerful computer. *New scientist.com news service*, June 2007. Retrieved on January 15, 2008 from http://technology.newscientist.com/article/dn12145-ibm-creates-worlds-most-powerful-computer.html

Kosovic, L. (2006). *Gender and ICT policy in Bosnia and Herzegovina: Rethinking ICT development through gender*. Retrieved November 2, 2009, from: http://www.gednerit.org/en/index.shtml?w=a&x=95059

Krishna, A. (2002). *Active Social Capital, Tracing the Roots of Development and Democracy.* New York: Columbia University Press.

Kukulska-Hulme, A. (2008). Mobile Usability in Educational Contexts: What have we learnt? *International Review of Research in Open and Distance Learning, 8*(2), 1–15.

Kukulska-Hulme, A., Sharples, M., Milrad, M., Arnedillo-Sánchez, I., & Vavoula, G. (2009). Innovation in mobile learning: A European perspective. *International Journal of Mobile and Blended Learning, 1*(1), 13–35.

Kumar, R., & Best, M. (2006). Social impact and diffusion of telecenter use: a study from the sustainable access in rural India project. *Journal of Community Informatics, 2*(3).

Kuttan, A., & Peer, L. (2003). *From digital divide to digital opportunity*. Lanham, MD: Scarecrow.

Lanning, K. (1992). *Child Molesters: A Behavioural Analysis*. Washington, DC: National Centre for Missing and Exploited Children.

Laudon, K., & Laudon, C. (2006). *Management Information Systems: managing the digital firm,* (9th ed.). New Delhi, India: Pearson Education.

Lawal, O. O. (2004). *Libraries as Tools for Educational Development*. A Paper presented at NLA 42nd National Conference and AGM at Solton International Hotel & Resort, Akure.

Lee, T. M. (2005). The impact of perceptions of interactivity on customer trust and transaction intentions in mobile commerce. *Journal of Electronic Commerce Research, 6*(3), 165–180.

Lenhart, A., Horrigan, J., Rainie, L., Allen, K., Boyce, A., Madden, M., & O'Grady, E. (2003). *The ever–shifting internet population: A new look at internet access and the digital Divide*. Retrieved May 11, 2003, from http://www.Pewinternet.org/

Lesame, Z. (2005). Bridging the digital divide in South Africa. In N. C. Lesame (Ed), *New Media. Technology and policy in developing countries* (pp. 17-29). Hatfield, Pretoria, South Africa: Van Schaik Publishers.

Lindquist, D., Denning, T., Kelly, M., Malani, R., Griswold, W. G., & Simon, B. (2007). Exploring the potential of mobile phones for active learning in the classroom. *ACM SIGCSE Bulletin, 39*(1), 384–388. doi:10.1145/1227504.1227445

Litho, P. (2005). *ICTs, Empowerment and Women in Rural Uganda: A Scott Perspective.* A paper presented at the "To think is to experiment" SSMA, Centre for Narrative Research, UEL, 22nd April 2005.

Liu, T. C. (2007). Teaching in a wireless learning environment: A case study. *Journal of Educational Technology & Society, 10*(1), 107–123.

Liu, T. C., Wang, H., Liang, T., Chan, T., Ko, W., & Yang, J. (2003). Wireless and mobile technologies to enhance teaching and learning. *Journal of Computer Assisted Learning, 19*(3), 371–382. doi:10.1046/j.0266-4909.2003.00038.x

Livingston, P. (2004). Laptops Unleashed: A Middle School Experience. *Learning and Leading with Technology, 31*(7), 12-15. Retrieved June 15, 2009 from www.usq.edu.au/material/edu5472

Loftus, R. A. (2008). Disconnecting Child Pornography on the Internet: Barriers and Policy Considerations. *Forum on Public Policy Online*, Spring. Retrieved on 15 June 2009 from http://www.forumonpublicpolicy.com/archivespring08/loftus.pdf

Longe, O. B., & Longe, F. A. (2005). The Nigerian Web Content: Combating the Pornographic Malaise Using Content Filters. *Journal of Information Technology Impact, 5*(2), 59–64.

Longe, O. B., & Chiemeke, S. C. (2006). The Design and Implementation of An E-Mail Encryptor for Combating Internet Spam. In *Proceedings of the 1st International Conference of the International Institute of Mathematics and Computer Sciences,* (pp. 1 – 7). Ota, Nigeria: Covenant University.

Longe, O. B., & Chiemeke, S. C. (2008). *Cybercrime and Criminality in Nigeria –What Roles are Internet Access Points in Playing?* Retrieved May 21st, 2009. From: http://www.eurojournals.com/ejes_6_4_12.pdf

Longe, O.B. (2006). Web Journalism In Nigeria: New Paradigms, New Challenges. *Journal of Society and Social Policy.*

Loygren, S. (2003). Computer Made from DNA and Enzymes. Retrievd on 6 July 2010 from http://news.nationalgeographic.com/news/2003/02/0224_030224_DNAcomputer_2.html

Lubetzky, S. (1979). The traditional ideals of cataloging and the new revision. In Freedman, M. J., & Malinconico, S. M. (Eds.), *The nature and future of the catalog: proceedings of the ALA's Information Science and Automation Division's 1975 and 1977 Institute on Cataloging* (pp. 153–169). Phoenix, AZ: Oryx Press.

Lummen, M., & Ruiter, E. (2008). *Product report: Edaikazhinadu project.* Dissertation, Saxion University of Applied Sciences, Hospitality Business School, Education: Tourism Management, Retrieved on November 11, 2008 from: http://dms01.saxion.nl/C12574DB0035EDCF/Al l+documents/52C462DA81331CA4C12574DB006303 12/$File/Afstudeerscriptie%20E.%20Ruiter%20en%20 M.%20Lummen.pdf

Lupe Cristán, A. (2008). Virtual International Authority File: Update 2008. *SCATNews: Newsletter of the Standing Committee of the IFLA Cataloguing Section, 30,* 2–3.

Lyman, M., & Potter, G. (1998). *Organized Crime.* Upper Saddle River, NJ: Prentice Hall.

M.K. Razdan v. The State and Shri Indukant Dixit, Crl. Rev.P. No. 861-62/2005, (Decided on March.03, 2008), Delhi.

Madon, S. (2004). Evaluating the development impact of e-governance initiatives: an exploratory framework. *The Electronic Journal on Information Systems in Developing Countries, 20*(5), 1–13.

Mahrer, H., & Krimmer, R. (2005). Towards the enhancement of e-democracy: Identifying the notion of the middleman paradox. *Journal of Information Systems, 15*(1), 27–42. doi:10.1111/j.1365-2575.2005.00184.x

Malama, F. (2007, April 17). Let's take advantage of VoIP. *The Post.*

Malhotra, Y. (2002). Why knowledge management systems fail? Enablers and constraints of knowledge management in human enterprises. In Holsapple, C. W. (Ed.), *Handbook on knowledge management 1: Knowledge matters* (pp. 577–599). Heidelberg, Germany: Springer-Verlag.

Malinconico, S. M., & Rizzolo, J. A. (1973). The New York Public Library automated book catalog subsystem. *Journal of Library Automation, 1*(6), 3–36.

Malinconico, S. M. (1975).The role of a machine based authority file in an automated bibliographic system. In *Automation in Libraries: Papers presented at the CACUL Workshop on Library Automation, Winnipeg, June 22-23, 1974.* Ottawa, Canada: Canadian Library Association. Cited according to: Carpenter, M. & Svenonius, E. (Eds). (1985). Foundations in cataloging: A sourcebook (pp. 211-233). Littleton, Co.: Libraries Unlimited.

Manch, K., & Wilson, D. (2003). *Objectionable material on the Internet: Developments in enforcement.* Retrieved February 18, 2009, from www.netsafe.org.nz/Doc_Library/netsafepapers_manchwilson_objectionable.pdf.

Mandatory data elements for internationally shared resource authority records: Report of the IFLA UBCIM Working Group on Minimal Level Authority Records and ISADN. *(1998). Frunkfurt/Main, Germany: IFLA Universal Bibliographic Control and International MARC Programme. Retrieved March 25, 2009, from* http://www.ifla.org/VI/3/p1996-2/mlar.htm

Mansell, R. (2006). Ambiguous connections: entitlements and responsibilities of global networking. *Journal of International Development, 18*(6), 901–913. doi:10.1002/jid.1310

Manual, U. N. I. M. A. R. C. *Authorities Format.* (2001). (2nd rev. & enlarged ed.). München, Germany: Saur.

Marcelle, G. (1998). *Strategies for including a gender perspective in African information and communications technologies (ICTs).* Retrieved 9th January 2009, from http://www.un-instraw.org/des/martines.doc

Marcelle, G. (2000). *Gender and information revolution in Africa.* Retrieved November 2 2009, from http://www.genderit.org/en/index.shtml?w=98x=91489

Marcelle, G. (2002). *Gender equality and ICT policy.* Retrieved January 9 2009, from http://www.worldbankorg/gender/digitaldivide/worldbankspresentation.ppt

Marcelle, G. N. (2001). *Getting Gender into African ICT Policy: A Strategic View.* Retrieved July 17th 2009 from http://www.idrc.ca/en/ev-9409-201-1-DO_TOPIC.htIm

Masambu, P. (2007). *A Review of the Postal and Tele-communicatios Sector: June 2006-June 2007.* Kampala, Uganda: Communications Commission.

Masango, D. (2005). *Partnership to bring IT to MPCCs countrywide.* Compiled by the Government Communication and Information System, 3 Oct. Retrieved September 22, 2008, from http://www.buanews.gov.za/news/05/05100316451001.

Masango, D. (2007). *ICT Key for Economic Growth, development.* Compiled by the Government Communication and Information System, 23 Oct. Retrieved September 22, 2008, from http://www.buanews.gov.za/news/07/07102316151004.

Mason, R. O. (1986). Four ethical issues of the information age. *Management Information Systems Quarterly, 10*(1), 5–12. doi:10.2307/248873

Massey, A. P., Ramesh, V., & Khatri, V. (2006). Design, development and assessment of mobile applications: The case for Problem-based Learning. *IEEE Transactions on Education, 49*(2), 183–192. doi:10.1109/TE.2006.875700

Mathiason, J. R., & Kuhlman, C. R. (n.d.). *An International Communication Policy: The Internet, international regulation & new policy structures.* Retrieved November 4, 2009 from http://www.un.org/esa/socdev/enable/access2000/ITSpaper.html

Mauritius, Ministry of Information Technology and Telecommunications. (2007). *National ICT policy 2007-11.* Retrieved on September 21, 2009 from http://www.gov.mu/portal/goc/telecomit/file/ICT%20Policy%202007-2011.pdf

Mazikana, P. C. (1999). Editorial. *ESARBICA Journal, 18,* 4–6.

Mcconatha, D., Praul, M., & Lynch, M.J. (2008). Mobile learning in higher education: An empirical assessment of a new educational tool. *The Turkish Online Journal of Educational Technology (TOJET), 7*(3).

Mcconnell International. (2000). *Cybercrime and Punishment?* Archaic Laws Threaten Global Information.

McNamara, K. (Ed.). (2008). Enhancing the rural livelihoods of the poor: recommendations on the use of ICT in enhancing the livelihoods of the rural poor. *Infodev.* Retrieved on April 1, 2009, from http://www.infodev.org/en/Publication.510.html

McQuail, D. (2000). *McQuail's Mass Communication Theory*. London: Sage Publications.

Meera, S. N., Jhamtani, A., & Rao, D. U. M. (2004). *Information and communication technology in agricultural development: a comparative analysis of three projects from India*. Agriculture Research and Extension Network, ODI Paper No.135. London: Overseas Development Institute. Retrieved August 1, 2006, from http://www.odi.org.uk/agren/papers/agrenpaper_135.pdf

Melkote, S. R., & Steeves, H. L. (2001). *Communication for Development in the Third World: Theory and Practice for Empowerment* (4th ed.). London: Sage.

Metcalf, D. (2005). *mLearning: Mobile learning and performance in the palm of your hand*. Amherst, MA: HRD Press.

Michael, D. (2002), *Globalization Of The World Economy: Potential Benefits And The Cost And A Net Assessment*. Governing Stability Across the Mediterranean Sea: A Transatlantic Perspective, Columbia International Affairs Online. Retrieved on Dec.3,2008 from http://www.ciaonet.org/coursepack/cp09/cp09c.pdf

Microsoft Corporation. (2009). *What is a model*. Retrieved September 19, 2009, from http://msdn.microsoft.com/en-us/library/dd129503 (VS.85, printer).aspx

Mingat, A., Tan, J., & Sosale, S. (2003). *Tools for Education Policy*. Washington, DC: World Bank.

Minges, M. (2001). *The Internet in an African LDC: Uganda Case study*. International Telecommunication Union Study.

Ministry of Corporate Affairs. (n.d.). *MCA21 Stakeholder Handbook*. New Delhi, India: Tata Consultancy. Retrieved on July. 12, 2009, from, http://www.mca.gov.in/MinistryWebsite/dca/help/ProcessHandbook.pdf

Mitra, R. (2000). Emerging state-level ICT development strategies . In Bhatnagar, S., & Schware, R. (Eds.), *Information and communication technology in development* (pp. 195–205). New Delhi: Sage Publications.

Mitrovic, Z. (2006). *ICT and small business development: the importance of government support and community involvement*. Unpublished working paper.

Mnjama, N. (2005). *Archival landscape in Eastern and Southern Africa*.

Moll, P. (1983). Should the Third World have Information Technology? *IFLA Journal*, *9*(4), 296–308. doi:10.1177/034003528300900406

Money Laundering (Prohibition) Act, of the Federal Republic of Nigeria. *(2004)*.

Montero, F., Córcoles, J. E., & Calero, C. (2006). *Handhelds and mobile phones to manage students and resources in classroom: A new handicap to the teacher?* (pp.876-880).

Moore, F. (2000, October). Digital data's future: you ain't seen nothin' yet. *Computer Technology Review*.

Morris, J., Haines, H., & Shallcrass, J. (1989). *Report of the Ministerial Inquiry into Pornography*. Wellington, New Zealand: New Zealand Government.

Mosteshar, S. (1996). *Some legal aspects of internet regulation*. Retrieved November 15, 2009 form http://www.mosteshar.com/sig.html

Muchanga, A. M. (2004). *Statement by Albert M. Muchanga SADC Deputy Executive Secretary Delivered at the Opening of The SADC Regional Seminar on Website Policy Development Held at President Hotel Gaborone*. Retrieved on September 23, 2009 from http://www.sadc.int/archives/read/news/56.

Murtomaa, E., & Greig, E. (Eds.). (1994). Problems and prospects of linking various single-language and/or multi-language name authority files: Notes. *International Cataloguing and Bibliographic Control*, *3*(23), 55- 58.

Musa, P. F. (2006). Making a case for modifying the technology acceptance model to account for limited accessibility in developing countries. *Information Technology for Development*, *12*(3), 213–224. doi:10.1002/itdj.20043

Mutula, S. M. (2009). *Digital Economies, SMES and E-readiness*. New York: Business Science Reference.

Mutula, S. M., & Wamukoya, J. (2009). Public sector information management in east and southern Africa: implications for FOI, democracy and integrity in government. *International Journal of Information Management*, *29*(5), 333–341. doi:10.1016/j.ijinfomgt.2009.04.004

Mutula, S. M., & Wamukoya, J. M. (2007). *Web Information management: a cross-disciplinary textbook.* Oxford, UK: Chandos Publishing.

Mutula, S. M. (2008). Digital Divide in Africa: Its Causes, Amelioration and Strategies . In Aina, L. O., Mutula, S. M., & Tiamiyu, M. A. (Eds.), *Information and Knowledge Management in the Digital Age: Concepts, Technologies and African Perspectives* (pp. 205–228). Ibadan, Nigeria: Third World Information Services Limited.

Mutula, S. M & Ocholla, D. (2009). *Trust, Attitudes and Behaviours in E-Government.* Paper Presented at the African Information Ethics and e-Government Workshop, Held from 23-27 February 2009 at Mount Resort Magaliesburg, South Africa

Mutume, G. (2005). *Africa takes on the digital divide: new information technologies change the lives of those in reach.* Retrieved on November 16, 2007 from http/www.africarecovery.org

Mwesige, P. (2004). Cyber Elites: A Survey of Internet Cafes in Uganda. *Telecommunications: Development in Africa, 21*(1), 83–101.

Naismith, L., et al. (2005). *Report 11: Literature review in mobile technologies and learning. Future lab series.* Bristol, U.K.

NASSCOM. (2008). *NASSCOM 2007-08 Annual Report.* New Delhi: NASSCOM.

National Archive of Australia. (2001). *ISO 15489-1 Information and Documentation-Records Management, Part 1: General, 1, 3.* Amsterdam: International Standards Organization. Retrieved July 10, 2008 from www.naa.gov.au

National Archives of Australian. (2002). A new approach to recordkeeping. *Commonwealth of Australia.* Retrieved May 25, 2007 from www.naa.gov.au

National Council on Disability. (2007). *Over the horizon: potential impact of emerging trends in information and communication technology on disability policy and practice.* Retrieved on May 22, 2007 from http://www.disabilityinfo.gov/

National Skills Registry for IT/ITES Professional. (2007). Retrieved March 18, 2009, from http://www.nasscom.in/Nasscom/templates/NormalPage.aspx?id=51441

National Skills Registry for Knowledge Professionals. (n.d). Retrieved March 18, 2009, from http://www.nasscom.in/upload/29264/NSR_Initiative_Overview.pdf

Navas-Sabater, N., Dymond, A., & Juntunen, N. (2002). *Telecommunications and Information Services for the Poor: Toward a Strategy for Universal Access* (World Bank Discussion Paper No. 432). Washington, DC: The World Bank.

Neumayer, E. (2006). Self-interest, foreign need and good governance: Are bilateral investment treaty programmes similar to aid allocation? *Foreign Policy Analysis, 2*(3), 245–268. doi:10.1111/j.1743-8594.2006.00029.x

New Vision Supplement. (2006, 26th February). Retrieved from Http://newvision.co.ug/D/8/12/613650

Ngulube, P. (2007). The nature and accessibility of e-government in sub Saharan Africa. *International Review of Information Ethics, 7*(09/2007). Retrieved November 25, 2007, from http://www.i-r-i-e.net/inhalt/007/16-ngulube.pdf

Nicol, C. (2003). *ICT Policy: A Beginners Handbook.* Johannesburg, South Africa: Association for Progressive Communication.

Nielsen, L., & Heffernan, C. (2006). New tools to connect people and places: the impact of ICTs on learning among resource poor farmers in Bolivia. *Journal of International Development, 18*(6), 889–900. doi:10.1002/jid.1321

Nigerian National Policy for Information Technology (IT). (2001). Retrieved March 18, 2009 from http://nitda.gov.ng/document/nigeriaitpolicy.pdf/

Niles, S., & Hanson, S. (2003). A new era of accessibility. *URISA Journal, 15,* I.

Njuguna, E. (2006). *ICT policy in developing countries; Understanding the Bottlenecks.* Retrieved March 11. 2008. from http://www.ptc.org/events/ptc06/program/public/Proceedings/Emmanuel%20Njuguna_pa-per_w143.pdf

Norris, C., & Soloway, E. (2004). Envisioning the handheld-centric classroom. *Journal of Educational Computing Research, 30*(4), 281–294. doi:10.2190/MBPJ-L35D-C4K6-AQB8

Norris, C., Sullivan, T., Poirot, T., & Soloway, E. (2003). No Access, No Use, No Impact: Snapshot of Surveys of Educational Technology in K-12. *Journal of Research on Technology in Education, 36*(1), 15-27. Retrieved June 15, 2009 from http://search.epnet.com/direct.asp; http://ez/noxy.us.edu

Nortel., Shashi Kiran., Lareau P., & Lloyd., S. (2002). *PKI Basics - A Technical Perspective.* Retrieved on Nov.19,2008.from http://www.oasis-pki.org/pdfs/PKI_Basics A_technical_perspective.pdf

Nussbaum, M. (2000). *Women and Human Development: The Capabilities Approach.* Cambridge, UK: Cambridge University Press.

Nussbaum, M. (2006). Capabilities as fundamental entitlement . In Kaufman, A. (Ed.), *Capabilities equality: basic issues and problems* (pp. 44–70). New York: Routledge.

Nyasato, R., & Kathuri, B. (2007). High Phone Charges Hamper Region's Growth, Says W Bank. *The Standard.* Retrieved April 10, 2007, from http://www.eastandard.net/hm_news/news.php?articleid=1143967136

O'Brien, J. A. (2006). *Introduction to information systems: essentials for the e-business enterprises,* (9th ed.). New York: McGraw-Hill.

O'Connell, R. (2001). *The Structural and Social Organisation of Paedophile Activity in Cyberspace: Implications for Investigative Strategies.* Accessed October 21, 2002, from http://www.uclan.ac.uk/facs/science/gcrf/crime1.htm

O'Donnell, I., & Milner, C. (2007). *Child Pornography Crime, Computers and Society.* Portland, OR: Willan Publishing.

O'Neil, D. (2002). Assessing community informatics: a review of methodological approaches for evaluating community networks and community technology centres. *Internet research: electronic networking applications and policy, 12*(1), 76-102.

Obileye, O. (2001). *Combating counterfeit drugs: the way forward.* Paper presented at the public hearing of the health and social services committee, Federal House of Representative, National Assembly, Abuja, Nigeria.

Oddy, P. (1986). Name authority files. *Catalogue & Index, 82*(Autumn), 1 & 3-4.

OECD. (2002). Guidelines for the security of information systems and networks: Towards a culture of security. *OECD Information Privacy and Security.* Retrieved February 10, 2009 from http://www.oecd.org/dataoecd//22/15582260.pdf

Office of Film and Literature Classification. (2008). *Report of the Office of Film and Literature Classification for the year ended 30 June 2008.* Wellington, New Zealand: Office of Film and Literature Classification.

Office of Film and Literature Classification [Australia]. (2006). *Enforcement Package.* Retrieved 16 June 2009 from http://www.oflc.gov.au/www/cob/rwpattach.nsf/VAP/(084A3429FD57AC0744737F8EA-134BACB)~896.pdf/$file/896.pdf

Office of Film and Literature Classification. (2007). *Publication of Objectionable Computer Files.* Retrived February 18, 2009, from http://www.censorship.govt.nz/thelaw-occasional-objectionable-computerfiles.html

Ofir, Z. M. *(2003).* Information and Communication Technologies for Development (ACACIA) The Case of Uganda: Final Evaluation Report prepared for IDRC Evaluation Unit.

Ogbo, O. (2003). *Strengthening National ICT Policies in Africa: Governance, Equity and Institutional Issues. Acacia II.* IDRC.

Ojebode, F. I. (2007). Library Funding and Book Collection Development: A Case Study of Oyo State College of Education and Federal College of Education (SP), Oyo. *Gateway Library Journal, 10*(2), 119–126.

Okado, G. (2007). *Formulation of a National ICT Policy: African Technology Policy Studies Network.* Retrieved on the 7th of July 2009 from www.atpsnet.org./pubs/brief/Technopolicy%20Brief%.pdf

Okot-Uma, R. W. (2002). The challenge of the digital divide . In Murelli, E. (Ed.), *Breaking the digital divide: implications for developing countries* (pp. ix–xi). London: Commonwealth Secretariat.

Omoba, O.R. & Omoba, F. A. (2009). Copyright law: influence on the use of information resources in Nigeria. *Library philosophy and practice.*

O'Neil, D. (2002). Assessing community informatics: a review of methodological approaches for evaluating community networks and community technology centres. *Internet research: electronic networking applications and policy 12* (1), 76-102.

OpenNet Initiative. (2007) *Global Internet Filtering*. Retrieved May 8, 2009 from http://map.opennet.net/filtering-pol.html

Oregon State University. (2003). *Archives & records management handbook*. Oregon: OSU Library. Retrieved April 2, 2007 from http://osulibrary.oregonstate.edu/archives/handbook/chapter2/electronic.html

Oriola, T. (2005). Advance Fee Fraud on the Internet: Nigeria's Regulatory Response. *Computer Law & Security Review, 21*(3).

Orlikowski, W. J., & Baroudi, J. J. (2002). Studying information technology in organizations: research approaches and assumptions. (2002. In M.D Myers. & D. Avison, (Eds.), *Qualitative research in information systems,* (pp. 51-77). London: Sage Publications.

Otuteye., E. (2008), *A Systematic Approach to E-Business Security: Ninth Australian World Wide Web Conference*. Retrieved on August. 21, 2009, from http://ausweb.scu.edu.au/aw03/papers/otuteye/

Overå, R. (2006). Networks, distance and trust: telecommunications development and changing trading practices in Ghana. *World Development, 34*(7), 1301–1315. doi:10.1016/j.worlddev.2005.11.015

Overview. (2006). Retrieved Mach 18, 2009, from http://www.nasscom.in/Nasscom/templates/NormalPage.aspx?id=5957.

Oyedemi, T., & Lesame, Z. (2005). South Africa: an information Society? In N. C. Lesame (Ed) *New Media. Technology and policy in developing countries* (pp. 75-97). Hatfield, Pretoria: Van Schaik Publishers.

Oyomno, G. Z. (2003). *Towards a framework for assessing the maturity of government capabilities for 'e-government'*. Retrieved July 20, 2005, from http://link.wits.ac.za/journal/j0401-oyomno-e-govt.pdf

Pais, A. (2006). *Bridging the digital divide by bringing connectivity to underserved areas of the world*. Retrieved October 2, 2009 from http://www.itu.int/osg/spu/youngminds/2006/essays/essay-adrian-pais.pdf

Palfrey, J., & Rogoyski, R. (2006). The move to the Middle: The enduring threat of harmful speech to the end-to-end principle. *Washington University Journal of Law and Policy, 21*, 31–65.

Palmer, J. (2010, January 11). Chemical computer that mimics neurons to be created. *BBC News*. Retrieved on July 6, 2010 from http://news.bbc.co.uk/2/hi/8452196.stm

PanAfrL10n. (2009). *Egypt*. Retrieved on September 21, 2009 from http://www.panafril10n.org/pmwiki.php/PanAfrLoc/Egypt

Panagariya, A. (2000). E-Commerce, WTO and Developing Countries. *World Economy, 23*(8), 959–978. doi:10.1111/1467-9701.00313

Papadopoulous, M., Kafeza, I., & Lessig, L. (2006). *Legal considerations for regulating content on the Internet*. Retrieved from http://www.marinos.com.gr/bbpdf/pdfs/msg52.pdf

Paper delivered at Financial Lecture Series in Lagos, August 10, 2006.

Parfitt, T. (2004). The ambiguity of participation: a qualified defence of participatory development. *Third World Quarterly, 25*(3), 537–555. doi:10.1080/0143659042000191429

Parker, D. B. (1998). *Fighting Computer Crime: A New Framework for Protecting Information*. Chichester, UK: Wiley Computer Publishing.

Parliamentary Office of Science & Technology. (2001). *Regulating Internet content*. Retrieved May 8, 2009 from http://www.parliament.uk/post/pn159.pdf

Patton, G. (2004). FRANAR: A conceptual model for authority data. In Taylor, A.G. & Tillett, B. B. (Eds), *Authority control in organizing and accessing information: Definition and international experience* (pp. 91-104). Binghamton, NY: Haworth Information Press. DOI 10.1300/J104v38n03_09

PEAP. (2000). *Poverty Eradication Action Plan*.

Pearson, J. (2004). Current policy priorities in information and communication technologies in education . In Trinidad, S., & Pearson, J. (Eds.), *Using information and communication technologies in education.* Singapore: Pearson Prentice Hall.

Perkins, M. (2009). Prime Cuts. *Index on Censorship, 38*(1), 129–139. doi:10.1080/03064220802712258

Pinder, P. (2006). Preparing information security for legal and regulatory compliance (Sarbanes-Oxley and Basel II). *Information Security Technical Report, 11*(1), 32–38. doi:10.1016/j.istr.2005.12.003

Plassard, M.-F., & McLean Brooking, D. (Eds.). (1993). *UNIMARC/CCF: proceedings of the Workshop held in Florence, 5-7 June 1991.* München, Germany: Saur.

Plou, D. S. (2005). *Gender and ICT policies: how do we start this discussion?* Retrieved November 2, 2009. from http://www.gender.org/en/index.shtml?w=a&x=90862

Policy, I. C. T. (2003) *The National Information and Communication Policy.* Retrieved from http://ucc.co.ug/nationalIctPolicyFramework.doc

Portable Mass Storage Device With Virtual Machine Activation, U. S. Patent No. WO/2008/021682, PCT/US2007/074399 (2008, Nov. 26). fig.6 Retrieved on Oct. 29, 2008, from http://www.wipo.int/pctdb/en/wo.jsp?IA=US2007074399andwo=2008021682andDISPLAY=CLAIMS

Professional Horizon. (2006). *Important Messages For e-Filing Under MCA21, The Chartered Accountant* (p. 1806). ICAI.

Public Works and Government Services Canada. (2004). *Government online history.* Retrieved July 19, 2005, from http://www.communication.gc.ca/gol_ged/gol_history.html

Quayle, E., Holland, G., Lineham, C., & Taylor, M. (2000). The Internet and offending behaviour: a case study. *Journal of Sexual Aggression, 6*(1/2).

Radloff, J. (2005). *What's gender got to do with IT?* Retrieved November 30, 2009 from http://www.apcwomen.org/node/273

Rajesh Saini *v.* State of Himachal Pradesh, (2008). Cr. LJ 3712

Ramilo, G., & Villanueva, P. (2001). *Issues, policies and outcome: Are ICT policies addressing gender equality?* Retrieved December 27, 2009 from http://www.genderit.org/upload/ad6d215b74e2a8613f0cf5416c9f3865/UNESCAP_Gender_and_ICT_Policy_Research_Report_Feb02.doc

Ramirez, R. (2001). A model for rural and remote information and communication technologies: A Canadian exploration. *Telecommunications Policy, 25*(5), 315–330. doi:10.1016/S0308-5961(01)00007-6

Ramirez, R. (2003). Bridging disciplines: the natural resource management kaleidoscope for understanding ICTs. *The Journal of Development Communication, 1*(14), 51–64.

Ramirez, R. (2007). Appreciating the contribution of broadband ICT with rural and remote communities: stepping stones toward an alternative paradigm. *The Information Society, 23*(2), 85–94. doi:10.1080/01972240701224044

Ramirez, R., & Richardson, D. (2005). Measuring the impact of telecommunication services on rural and remote communities. *Telecommunications Policy, 29*(4), 297–319. doi:10.1016/j.telpol.2004.05.015

Ramuarine, D., & Wilson, J. (n.d.). *Developing National ICT Strategies for Small Island States: Case Study – Trinidad and Tobago.* Retrieved on the 11th November 2009 from www.fastforward.it

Rao, S. S. (2004). Role of ICTs in India's rural community information systems. *Info, 6*(4), 261–269. doi:10.1108/14636690410555663

Rao, J. S. V. (2004). *Law of Cyber-crimes and Information Technology Law.* New Delhi: Lexis-Nexis.

Reddy, R., Arunachalam, V. S., Tongia, R., Subrahmanian, E., & Balakrishnan, N. (2004). *Sustainable ICT for emerging economies: mythology and reality of the digital divide problem – a discussion note.* Retrieved on December 30, 2008 from http://www.rr.cs.cmu.edu/ITSD.doc

Reding, V. (2009). *Internet of the future: Europe must be a key player.* Retrieved May 12, 2009 from http://ec.europa.eu/commission_barroso/reding/docs/speeches/2009/brussels-20090202.pdf

Reed, C. (2007). Taking a side on Technology Neutrality . *SCRIPTed, 4*(3), 263. doi:10.2966/scrip.040307.263

Regulatory Norms in Indian cyber-space. (2006). Retrieved March 18, 2009, from http://www.nasscom.in/Nasscom/templates/NormalPage.aspx?id=6161

Relyea, H. C., & Hogue, H. B. (2003). A brief history of the emergency of digital government in the United States. In Pavlichev & Garson (Eds.), *Digital Government,* (pp. 16-33). Hershey, PA: IGI Global.

Report of Standing Committee on Information Technology. *Raja Sabha Parliamentary Bulletin.* (n.d.). Retrieved on 22 Oct 2008 http://164.100.47.5/Bullitensessions/sessionno/214/221008.pdf

Report of the International Meeting of Cataloguing Experts, Copenhagen, 1969 (1970). *Libri, 1*(20), 105-132.

Republic of Ghana. *National ICT policy and Plan Development committee: Ministerial ICT policy statements.* (n.d.). Retrieved on the 18th of August 2009 from http://www.ict.gov.gh/html/ministerial%20ict%20policy%20statements.htm

Reserve Bank of India. (2002). *Information Systems Security Guidelines for the Banking and Financial Sector.* Retrieved on Feb 22, 2009, from http://www.prsindia.org/docs/bills/1168510210/bill93_2007112393_Press_Release_for_IT_Act_Amendment.pdf

Ribadu, N. (2006). *Money laundry in Emerging economies: Nigeria as a case study.* A

Robeyns, I. (2001). *Understanding Sen's capability approach.* Cambridge, UK: Wolfson College. Retrieved on August 20, 2006, from http://www.ingridrobeyns.nl/Downloads/Under_sen.pdf.

Roco, M. C. (2003). Nanotechnology: convergence with modern biology and medicine. *Current Opinion in Biotechnology.* Volume 14, Issue 3, June 2003. Retrieved on July 7, 2010 from http://www.sciencedirect.com/science/journal/09581669 346

Rogerson, S. & Bynum, T.W. (1995, June 9). Cyberspace: the ethical frontier. *London times.*

Rogerson, S. (1998). The ethics of information and communication Technologies: ICT in business. *IMIS journal, 8*(2), 1-2.

Roschelle, J., & Pea, R. (2002). A walk on the wild side: How wireless handhelds may change computer-support collaborative learning. *International Journal of Cognitive Technology, 1*(1), 145–168. doi:10.1075/ijct.1.1.09ros

Rotberg, R. (2007). The Ibrahim index on African governance: how we achieved our rankings. *African Business, November* (336), 20.

Rowan, M. (2003*). Lesson Learned from Mozambique's ICT Policy Process.* Retrieved on the 7th of July 2009 from www.idrc.ca/en/ev-46223-201-1-DO_TOPIC.html

Russell, I. (2008, Nov.27). Sale of digital signature double in 18 months. *Business Standard.* Retrieved on 27 Jan 2009.Business-standard http://www.businessstandard.com/india/storypage.php?autono=297405.

Sabatier, P., & Mazmanian, D. (1979). The Conditions of Effective Implementation. *Policy Analysis, 5*(Fall), 481–504.

Sabatier, P., & Mazmanian, D. (1980). The Implementation of Public Policy: A framework of Analysis. *Policy Studies Journal: the Journal of the Policy Studies Organization, 8,* 538–560. doi:10.1111/j.1541-0072.1980.tb01266.x

SADC E-readiness Task Force. (2002). *SADC e-readiness review and strategy.* Johannesburg, South Africa: SADC, 1060.

Sadowsky, G., Dempsey, J., Greenberg, A., Mack, B., & Schwartz, A. (2003). *Information Technology Security Handbook.* Washington, DC: The World Bank.

SAE International Emerging Technologies Advisory Board. (2009). Retrieved October 2, 2009 from http://www.sae.org/about/board/committees/etab.htm

Saint-Germain, R. (2005). Information security management best practice based on ISO/IEC 17799. *Information Management Journal, 3*(4), 60–66.

Saipunidzam, M., Mohammad, N. I., Mohamad, I. A. M. F., & Shakirah, M. T. (2008). Open source implementation of M-learning for primary school in Malaysia. In *Proceedings Of World Academy Of Science, Engineering And Technology* (pp.752-756). 34.

Sathish Babu, B., & Venkataram, P. (2009). A dynamic authentication scheme for mobile transactions. *International Journal of Network Security, 8*(1), 59–74.

Sawe, D.J.A. (2005). *Regional e-governance programme: Progress from Tanzania.* 2nd EAC Regional Consultative Workshop held in Nairobi, Grand Regency Hotel from 28-29 June 2005.

Sawhney, H. (1996). Information superhighway: metaphors as midwives. *Media Culture & Society, 18,* 291–314. doi:10.1177/016344396018002007

Schilderman, T. (2002). *Strengthening the Knowledge and Information Systems of the Urban Poor.* London: DFID.

Schischka, J., Dalziel, P., & Saunders, C. (2008). Applying Sen's capability approach to poverty alleviation programs: two case Studies. *Journal of Human Development, 9*(2), 229–246. doi:10.1080/14649880802078777

Schlienger, T., & Teufel, S. (2005). Tool supported management of information security culture: Application in a private bank. In Sasaki, R., Okamoto, E. & Yoshiura, H. (Eds.) *IFIP International Federation for Information Processing. Volume 181, Security and Privacy in the Age of Ubiquitous Computing,* (pp. 65-77). Boston: Springer

Schware, R. (Ed.). (2005). *E-Development: From Excitement to Effectiveness.* Washington, DC: World Bank Group.

Sciadas, G. (Ed.). (2005). *The digital guide to digital opportunities: measuring infostates for development.* Montréal: Claude-Yves Charron.

Seebaluck, R. (2006). *Improving Your Business Environment Through Public and Private Partnerships: Strategies that Work.* Retrieved on September 22, 2009 from http://idisc.infodev.org/proxy/Document.48.aspx

Sehlabaka, C. (2004). The Lesotho government. In *Proceedings of workshop of the regional consultation on national e-govt readiness in Gaborone, Botswana,* 14-16 April 2004. Retrieved July 16, 2005, from http://www.comnet-it.org/news/CESPAM-Botswana.pdf.

Selaimen, G. (2006) *Gender issue at all levels – From policy formulation to implementation.* Retrieved November 2, 2009. From: http://www.genderit.org/ en/index.shtml?w=a&x=93423.

Sen, A. (1985a). Well-being, agency and freedom: the Dewey Lectures: 1984. *The Journal of Philosophy, 82,* 169–221. doi:10.2307/2026184

Sen, A. (1985b). *Commodities and capabilities.* Amsterdam: North-Holland.

Sen, A. (2000). A decade of human development. *Journal of Human Development, 1*(1), 17–23. doi:10.1080/14649880050008746

Sen, A. (2001). *Development as freedom.* London: Oxford University Press.

Sen, A. (2005). Human rights and capabilities. *Journal of Human Development, 6*(2), 151–161. doi:10.1080/14649880500120491

Sesan, G. (2004). *Africa, ICT policy and the millennium development goals.* Retrieved on August 22, 2009 from http://wsispapers.choike.org/africa_ict_mdg.pdf

Sey, A., & Fellows, M. (2009). *Literature review on the impact of public access to information and communication technologies.* CIS Working Paper No. 6. Seattle: University of Washington Center for Information & Society. Retrieved on May 9, 2009, from http://cis.washington.edu/depository/publications/CIS-WorkingPaperNo6.pdf

Shalhoub, Z. K. (2006). Trust, privacy, and security in electronic business: The case of the GCC countries. *Information Management & Computer Security, 14*(3), 270–283. doi:10.1108/09685220610670413

Shao, B., Ma, G., & Meng, X. (2005). The influence to online consumer trust: an empirical research on B2C ecommerce in China. In *Proceedings of the fifth International Conference on Computer and Information Technology,* (pp. 961-965). Washington, DC: IEEE Computer Society.

Sharma, B. R. (2005). *Bank Frauds Prevention and Detection, 61.* New Delhi: Universal Law Publication.

Shelley, L. (1998). Crime and corruption in the digital age. *Journal of International Affairs, 51*(2).

Sheridan, W., & Riley, T. B. (2006). *Comparing e-government and e-governance.* Retrieved December 12, 2006, from http://www.electronicgov.net/pubs/research_papers/SheridanRileyComparEgov.d

Sherratt, D., Rogerson, S., & Fair-weather, N. B. (2005). The challenge of raising ethical awareness: a case based aiding system for use by computing and ICT students. *Science and Engineering Ethics, 11*(2), 299–31. doi:10.1007/s11948-005-0047-7

Shih, H. P. (2004). An empirical study on predicting user acceptance of e-shopping on the web. *Information & Management, 41*(93), 351–368. doi:10.1016/S0378-7206(03)00079-X

Sierra, K. (2006) Foreword. In The World Bank, (Ed.), *Information and Communications for Development: Global Trends and Policies*. Washington, DC: The World Bank.

Singh, D., & Zaitun, A. B. (2006). Mobile learning in wireless classrooms. [MOJIT]. *Malaysian Online Journal of Instructional Technology, 3*(2), 26–42.

Siponen, M. T., & Oinas-Kukkonen, H. (2007). A review of information security issues and respective research contributions. *ACM SIGMIS Database, 38*(1), 60–80. doi:10.1145/1216218.1216224

Small, B. (2007). Sustainable development and technology: Genetic engineering, social sustainability and empirical ethics. *International Journal of Sustainable Development, 10*(4), 402–435. doi:10.1504/IJSD.2007.017912

Smith, A. D., & Rupp, W. T. (2002). Issues in cybersecurity: understanding the potential risks associated with hackers/crackers. *Information Management & Computer Security, 10*(4), 178–183. doi:10.1108/09685220210436976

Snyman, M., & Snyman, R. (2003). Getting information to disadvantaged rural communities: the centre approach. *South African Journal of Library and Information Science, 69*(2), 95–107.

Sofyan, D. (2007). ICT policy in Indonesia . In Obi, T. (Ed.), *E-governance: A global perspective on a new paradigm* (pp. 48–50). Amsterdam: IOS Press.

Song, B. K. (2001). *Content regulation on the Internet*. Retrieved May 1, 2009 http://iml.jou.ufl.edu/projects/fall01/song/page5.html

Souter, D., Scott, N., Garforth, C., Jain, R., Mascarenhas, O., & McKemey, K. (2005). *The economic impact of telecommunications on rural livelihoods and poverty reduction: a study of rural communities in India (Gujarat), Mozambique and Tanzania*. Commonwealth Telecommunications Organisation for UK Department for International Development. Report of DFID KaR Project 8347. London.

South Africa (2005). South Africa Millennium Goals. *Country Report 2005*.

Sribhadung, P. (2006). Mobile devices in elearning. In *Proceedings of the Third International Conference on eLearning for Knowledge-Based Society* (pp.35.1-35.4), Bangkok, Thailand.

Ssewanyana, K. J. (2007). ICT Access and Poverty in Uganda. *International Journal of Computing and ICT Research, 1*(2), 10 - 19. Retrieved from http:www.ijcir.org/volume1-number2/article 2.pdf

State of Punjab and Ors. *v.* Amritsar Beverages Ltd. and Ors., AIR 2006 SC 2820. (2006). (Paragraph 7 & 17, pp. 3488-89).

Statement of international cataloguing principles. (2009, February). Retrieved March 25, 2009, from http://www.ifla.org/VII/s13/icp/

Statement of Principles adopted at the International Conference on Cataloguing Principles Paris, October, 1961. (1971). Annotated ed. with commentary and examples by Eva Verona assisted by Franz Georg Kaltwasser, P.R. Lewis, Roger Pierrot. London: IFLA Committee on Cataloguing.

Stewart, F. (2005). Groups and capabilities. *Journal of Human Development, 6*(2), 185–204. doi:10.1080/14649880500120517

Stewart, F., & Deneulin, S. (2002). Amartya Sen's contribution to development thinking. *Studies in Comparative International Development, 37*(2), 61–70. doi:10.1007/BF02686262

Stovekraft Private Limited, rep. by its Managing Director Rajendra Gandhi v The Joint Director, Directorate of Revenue Intelligence and Ors. *(2007). 214 ELT 179 (Kar)*

Subramanian, M., & Saxena, A. (2008). *Gender mainstreaming of information and communication technologies (ICT) policies and programs. A case study of Ghattisgarth State of India*, Project Report Submitted by WSIS-2 Awarder.

Sullivan, C. (2005). *Internet Traders of Child Pornography: Profiling Research*. Retrieved February 24, 2006, from http://www.dia.govt.nz/pubforms.nsf/URL/Profilingupdate2.pdf/$file/Profilingupdate2.pdf

Sundaram Brake Linings Ltd *v.* Kotak Mahindra Bank Ltd., A. No. 8078 of 2007 in C.S. No. 1072 of 2007, High court of Madras. (2008). MANU/TN/0938.

Sung, M., Gips, J., Eagle, N., Madan, A., Caneel, R., & DeVaul, R. (2005). Mobile-IT education (MIT.EDU): M-learning applications for classroom settings. Blackwell Publishing Limited. *Journal of Computer Assisted Learning, 21,* 229–237. doi:10.1111/j.1365-2729.2005.00130.x

Svenonius, E. (2000). *The intellectual foundation of information organization.* Cambridge, MA: The MIT Press.

Swindells, C., & Henderson, K. (1998) Legal regulation of Electronic Commerce, 3. *The Journal of Information, Law and Technology, 3.* Retrieved on Oct. 29, 2008 from http://www2.warwick.ac.uk/fac/soc/law/elj/jilt/1998_3/swindells/#a3

Sylvester, L. (2001). *The Importance of Victimology in Criminal Profiling.* Retrieved May 21st, 2009, from: http://isuisse.ifrance.com/ emmaf/base/ impvic.html

Sytnyk, O. (2005). *Control of internet content: Purposes, methods, legislative bases.* Retrieved November 4, 2009 from http://www.personal.ceu.hu/students/04/Oleksandra_Sytnyk/testfile.pdf

Talyarkhan, S., Grimshaw, D. J., & Lowe, L. (2005). *Connecting the first mile: investigating best practices for ICTs and information sharing for development.* Rugby, UK: ITDG Publishing.

Tanzania, United Republic of, Ministry of Communications and Transport (2002). *National ICT Policy of Tanzania.* Retrieved on August 22, 2009 from http://www.ethinktanktz.org/esecretariat/DocArchive/zerothorder.pdf.

Tavani, H. T. (2007). Regulating cyberspace: Concepts and controversies. *Library Hi Tech, 25*(1), 37–46. doi:10.1108/07378830710735849

Taylor, A. G. (1984). Authority files in online catalogs: an investigation of their value. *Cataloging & Classification Quarterly, 3*(4), 1–17. doi:10.1300/J104v04n03_01

Taylor, A. G. (1989). Research and theoretical considerations in authority control. *Cataloging & Classification Quarterly, 3*(9), 29–56. doi:10.1300/J104v09n03_03

Taylor, M., Quayle, E., & Holland, G. (2001). Child pornography, the Internet and offending. *ISUMA, 2*(2). Retrieved May 27, 2002, from http://www.isuma.net/v02n02/taylor/taylor_e.shtml

Taylor, W. (2004). Community Informatics: The basis for an emerging social contract for governance, research and teaching. *Building & Bridging Community Networks: Knowledge, Innovation & Diversity through Communication.* Brighton, UK, 31 March and 1-2 April.

Taylor, W. J., Erwin, G. J., & Wesso, H. (2006). *New public policies for the emerging information society in South Africa - a strategic view. Governments and Communities in Partnership conference,* Centre for Public Policy, University of Melbourne, 25-27 September. Retrieved September 2007, from http://www.public-policy.unimelb.edu.au/conference 06/Taylor W.pdf.

The Communication Initiative Network. (2008). *Mozambique ICT4D National Policy.* Retrieved on September 22, 2009 from http://www.comminit.com/en/node/148306.

The Communication Initiative Network. (2009). *Ethiopia ICT4D National Policy* [Draft]. Retrieved on September 22, 2009 from http://www.comminit.com/en/node/148306.

The National ICT policy in Bolivia. (n.d.). Retrieved on the 10th of November, 2009 from http://www.ucd.org/articles/the-national-ict-policy-in-bolivia/?

The World Bank. (2002). *Telecommunications and information services for the poor: Towards a strategy for universal access.* Washington, DC: The World Bank.

Thomas, J. J., & Parayil, G. (2008). Bridging the social and digital divides in Andhra Pradesh and Kerala: a capabilities approach. *Development and Change, 49*(3), 409–435. doi:10.1111/j.1467-7660.2008.00486.x

Thomas, D., & Loader, B. D. (2000). Introduction - Cybercrime: Law Enforcement, Security and Surveillance in the Information Age. In *Cybercrime: Law Enforcement, Security and Surveillance in the Information Age.* New York: Taylor & Francis Group.

Thomas, J. (2006). *Cybercrime: A revolution in terrorism and criminal behaviour creates change in the criminal justice system*. Retrieved November 1, 2009 from http://www.associatedcontent.com/article/44605/cybercrime_a_revolution_in_terrorism_pg12_pg12.html?cat=37

Thomas, P. (2007). Telecom musings: public service issues in India. *info, 9*(2/3), 97-107.

Tillett, B. B. (2004). Authority Control: State of the Art and New Perspectives. In A.G. Taylor & B. B. Tillett, (Eds.), *Authority control in organizing and accessing information: Definition and international experience* (pp. 23-57). Binghamton, NY: Haworth Information Press. DOI 10.1300/J104v38n03_04

Tillett, B. B. (2008). *A Review of the Feasibility of an International Standard Authority Data Number (ISADN)*. Prepared for the IFLA Working Group on Functional Requirements and Numbering of Authority Records, edited by Glenn E. Patton, 1 July 2008. Approved by the Standing Committee of the IFLA Cataloguing Section, 15 September 2008. Retrieved March 25, 2009, from http://www.ifla.org/VII/d4/franar-numbering-paper.pdf

Torero, M., & von Braun, J. (Eds.). (2006). *Information and communication technologies for development and poverty reduction: the potential of telecommunications*. Baltimore: Johns Hopkins University Press.

Transparency International. (2008). *2008 Corruption Perceptions Index*. Retrieved February 5, 2009, from http://www.transparency.org/news_room/in_focus/2008/cpi2008/cpi_2008_table

Tulsian, P. C. (2006). *Business Law*. New Delhi, India: Tata McGraw-Hill.

Twist, J. (2005). *UN debut for $100 laptop for poor*. Retrieved on November 15, 2007 from http://news.bbc.co.uk/2/hi/technology/4445060.stm

UCC. (2006). *Uganda Communication Commission*. Retrieved from www.ucc.co.ug

Uhegbu, A. N. (2007). *The Information User: Issues and Themes*. Okigwe, Nigeria: Whytem Publishers.

UN – United Nations. (2008). *e-Government Survey 2008: From e-government to connected governance*. New York: United Nations.

UNDP – United Nations Development Programme. (2003). *Human Development Report 2003: Millennium Development Goals: a compact among nations to end human poverty*. New York: Oxford University Press.

UNDP. (2001). *Role of UNDP in information and communication technology For development*. Retrieved October 1, 2009 from http://www.undp.org/execbrd/pdf/DP2001CRP8.PD

UNDP. (2003). *Transforming the Mainstream: Gender Mainstreaming Past, Present and future*. Retrieved Dec. 29, 2009 from www.undp.org/gender/policy/htm

UNESCAP. (2009*). Considerations for ICT policy formulation in developing countries*. Retrieved October 1, 2009 from tat/gc/box-ch8.asp http://www.unescap.org/S

UNICEF. (2008). *The State of the World's Children 2009*. New York: United Nations Children's Fund.

UNIMARC/Authorities. *Universal format for authorities*. (1991). Recommended by the IFLA Steering Group on a UNIMARC Format for Authorities, approved by the Standing Committees of the IFLA Section on Cataloguing and Information Technology. München: Saur.

United Nation. (2002). *56th Sess. General Assembly Resolution. A/RES/56/80*. Model Law on Electronic Signatures of the United Nations Commission on International Trade Law.

United Nations Conference on Trade and Development. (2000). *A Positive Agenda for Developing Countries: Issues for Future Trade Negotiations*, (pp. 474). New York: UN Publication.

United Nations Department pf Economic and Social Affairs (2006). *E-government readiness assessment methodology*. Retrieved December 12, 2006, from http://www.unpan.org/dpepa-kmb-eg-egovranda-ready.asp

United Nations Economic Commission for Africa. (2008). Burkina Faso: NICI Policy. Retrieved on August 25, 2009 from http://www.uneca.org/aisi/nici/country_profiles/Burkina%20Faso/burkinapol.htm

United Nations Economic Commission for Africa. (2008). *Rwanda National ICT Policy.* Retrieved on August 17, 2009 from http://www.comminit.com/en/node/148446.

United Nations. (2002). *UNCITRAL Model Law on Electronic Signatures with Guide to Enactment.* New York: United Nations Publication. Retrieved on Jan.02, 2009, from UNICITRAL website http://www.uncitral.org/pdf/english/texts/electcom/ml-elecsig-e.pdf

United Nations. (2005). *Gender equality and empowerment of women through ICT.* Retrieved December 23, 2009 from http://www.un.org/womenwatch/daw/public/w2000-09.05-ict-e.pdf

United Nations. (2008). *UN e-government survey 2008: From e-government to connected government.* Retrieved January 28, 2008, from http://unpan1.un.org/intradoc/groups/public/documents/UN/UNPAN028607.pdf

United States v Williams. No. 06–694. (2008). Retrieved 26 June 2009 from http://www.supremecourtus.gov/opinions/07pdf/06-694.pdf

UnitedNations (UN). (2003). *Tools for Development - Using Information and Communications Technology to Achieve the Millennium Development Goals.* Working Paper, United Nations ICT Task Force, August.

URIM. (1997). *EURIM briefing no 19: The regulation of content on the Internet.* http://www.eurim.org/briefings/brief19.htm

Van Der Stoep, L. (2009). Police target web predators. *Sunday Star Times,* 22 February 2009. Retrieved March 3, 2009, from http://www.stuff.co.nz/stuff/thepress/4856032a26834.html

van Dijk, J., & Hacker, K. (2003). The digital divide as a complex and dynamic phenomenon. *The Information Society, 19*(4), 315–326. doi:10.1080/01972240309487

Vavoula, G. N., & Sharples, M. (2002). KleOS: A personal, mobile, knowledge and learning organisation system. In M Milrad, HU Hoppe & Kinshuk (Eds), *IEEE International Workshop on Wireless and Mobile Technologies in Education* (pp.152–156). Los Alamitos, CA: IEEE Computer Society.

Ved Ram and Sons Pvt. Ltd *v.* Director, (2) AWC 2053 (UP), (2008).

Verona, E. (1983). *Pravilnik i priručnik za izradbu abecednih kataloga. 2. dio: Katalozni opis.* Zagreb, Croatia: Hrvatsko bibliotekarsko društvo.

VIAF. *Virtual International Authority File.* (n.d.). Retrieved June 16, 2009, from http://www.oclc.org/research/projects/viaf/

Vogel, D., Kennedy, D. M., Kuan, K., Kwok, R., & La, J. (2007). Do mobile device applications affect learning? In *Proceedings of the 40TH Annual Hawaii Conference on Systems Sciences,* 1-4.

Von Solms, B. (2006). Information security - The fourth wave. *Computers & Security, 25*(3), 165–168. doi:10.1016/j.cose.2006.03.004

Vrbanc, T. (2007). Choice of author headings for finite internet resources. In M. Willer & A. Barbarić (Eds.), *Međunarodni skup u čast 100-te godišnjice rođenja Eve Verona, Zagreb, 17.-18. studenoga 2005. = International Conference in Honour of the 100th Anniversary of Eva Verona's Birth, Zagreb, November 17-18, 2005* (pp. 459-475). Zagreb, Croatia: Hrvatsko knjižničarsko društvo = Croatian Library Association.

Wagner, E. D. (2005). Enabling mobile learning. *EDUCAUSE Review, 40*(3), 40–53.

Walker, D. W. (2002). *Emerging distributed computing technologies.* Retrieved on August 10, 2007 from http://www.cs.cf.ac.uk/User/David.W.Walker

Walsham, G., & Sahay, S. (2006). Research on information systems in developing countries: current landscape and future prospects. *Information Technology for Development, 12*(1), 7–24. doi:10.1002/itdj.20020

Wang, Z. L. (2008). *What is Nanotechnology?* Retrieved on February 22, 2008 from http://www.nanoscience.gatech.edu/zlwang/research/nano.html

Warren, M. (2007). The digital vicious cycle: links between social disadvantage and digital exclusion in rural areas. *Telecommunications Policy, 31*(6-7), 374–388. doi:10.1016/j.telpol.2007.04.001

Warschauer, M. (2003). Demystifying the digital divide. *Scientific American, 289*(2), 42–48. doi:10.1038/scientificamerican0803-42

Waters, D. (2007). Africa waiting for net revolution. *BBC*. Retrieved on March 10, 2009 from http://news.bbc.co.uk/1/hi/technology/7063682.stm

Watson, C., & Shuker, R. (1998). *In the Public Good? Censorship in New Zealand*. Palmerston North, New Zealand: Dunmore Press.

Weama, T. M. (2005). *Brief history of the development of an ICT Policy in Kenya*. Retrieved from www.scridb.com/.../ict-policy-final-expert-report-pupua-new-guinea

Webopedia. (2008) Retrieved on January 15, 2008 from http://inews.webopedia.com/TERM/d/digital_divide.html

Wei, J., & Lin, B. (2008). Development of a value increasing model for mobile learning. In *Proceedings of the 39th Annual Meeting of the Decision Sciences Institute* (pp. 5251-5256).

Wescott, C. G., Pizarro, M., & Schiavo-Campo, S. (2001). *The role of information and communication technology, in improving public administration*. Retrieved July 21, 2005, from http://www.adb.org/documents/manuals/serve_and_preserve/default.asp

Whatis.com. (2009). *Framework*. Retrieved September 18, 2009, from http://whatis.techtarget.com/definition/0,sid9_gci1103696,00.html

White, B. (2008). *Business management skills-trust building tips for managers*. Retrieved December 24, 2008, from http://ezinearticles.com/?Business-Management-Skills--Trust-Building-Tips-for-Managers&id=679140

Whitman, M. E., & Mattord, H. J. (2009). *Principles of Information Security* (3rd ed.). Boston: Thomson Course Technology.

Whyte, G., & Bytheway, A. (1996). Factors affecting information systems' success. *International Journal of Information Management, 7*(1), 74–93.

Whyte, A. & Macintosh, A. (2002). Analysis and evaluation of e-Consultations. *e-Service Journal, 2*(1), 9-34.

Wiander, T. Savola, R. Karppinen, K. & Rapeli, M. (2006). Holistic information security management in multi-organization environment. In *Proceedings of the International Symposium of Industrial Electronics*, (pp. 2942-2947). Montreal, Canada: IEEE.

Wikibooks. (2008). *Legal and Regulatory Issues in the Information Economy/Censorship or Content Regulation*. Retrieved May 8, 2009 from http://en.wikibooks.org/wiki/Legal_and_Regulatory_Issues_in_the_Information_Economy/Censorship_or_Content_Regulation

Wikipedia. (2009). *Cyberlaw*. Retrieved November 15, 2009 from http://en.wikipedia.org/wiki/Cyberlaw

Wikipedia. (2009). Very-large-scale integration. Retrieved on July 28, 2009 from http://en.wikipedia.org/wiki/Very-large-scale_integration.

Willer, M. (1994). UNIMARC/Authorities: A new tool towards standardization. In M.-F. Plassard & M. Holdt, (Eds.), *UNIMARC and CDS/ISIS: Proceedings of the Workshops Held in Budapest, 21-22 June 1993 and Barcelona, 26 August 1993*, (pp. 19-36). München, Germany: Saur. The paper was also presented at the following seminars organized by IFLA UBCIMP Office: *UBC/UNIMARC: A seminar on Universal bibliographic Control and UNIMARC*, Lietuvos Nacionalnie Martyno Mazvydo Biblioteka, Vilnius, 2-4 June 1994, Lithuania; *UNIMARC Workshop "Crimea '95"*, Jevpatoria, Ukraine, 12-16 June 1995; *IFLA International Seminar Authority Files: Their Creation and Use in Cataloguing*, St. Petersburg, October 3-7, 1995. For reports see *International Cataloguing and Bibliographic Control*.

Willer, M. (1999). Formats and cataloguing rules: developments for cataloguing electronic resources. In *Program, 1*(33), 41-55.

Willer, M. (2004). UNIMARC format for authority records: Its scope and issues for authority control. In A.G. Taylor, & B. B. Tillett, (Eds.), *Authority control in organizing and accessing information: Definition and international experience*, (pp. 153-184). Binghamton, NY: Haworth Information Press. DOI 10.1300/J104v38n03_14

Willer, M. (2007). IFLA UBCIM Working Group on FRANAR recommendations for potential changes in the UNIMARC Authorities format. In Plassard, M.-F. *UNIMARC & friends: Charting the new landscape of library standards: Proceedings of the International Conference held in Lisbon, 20-21 March 2006* (pp. 61-68). München, Germany: Saur.

Willer, M. (2009, in press). Third edition of *UNIMARC Manual: Authorities Format*: Implementing concepts from the FRAD model and IME ICC Statement of International Cataloguing Principles. In *International Cataloguing and Bibliographic Control, 4*(38).

Williamson, O. E. (1981). Calculativeness, trust and economic organisation. *The Journal of Law & Economics, 26*, 453–486.

Wilson, C. (2000). Holding Management Accountable: A New Policy for Protection Against Computer Crime. In *National Aerospace and Electronics Conference . Proceedings of the IEEE, 2000*, 272–281.

Wilson, D. (2002). Censorship in New Zealand: the policy challenges of new technology. *Social Policy Journal of New Zealand, 19*, 1–13.

Wilson, D. (2007). Responding to the Challenges: Recent Developments in Censorship Policy in New Zealand. *Social Policy Journal of New Zealand, 30*, 65–78.

Wilson, D. (2008). Censorship, new technology and libraries. *The Electronic Library, 26*(5). doi:10.1108/02640470810910710

Wilson, D., & Andrews, C. (2004). *Internet Traders of Child Pornography and Other Censorship Offenders in New Zealand*. Retrieved February 24, 2009, from http://www.dia.govt.nz/pubforms.nsf/URL/profilingupdate.pdf/$file/profilingupdate.pdf

Winn, J. K. (1998). Open Systems, Free Markets, and Regulation of Internet Commerce. *Tulane Law Review, 1177*, 72.

Wolak, J., Finkelhor, D., & Mitchell, K. (2005). The Varieties of Child Pornography Production . In Quayle, E., & Taylor, M. (Eds.), *Viewing Child Pornography on the Internet. Lyme Regis*. Dorset, UK: Russell House Publishing.

World Bank. (2008). *The Little Data Book on Information and Communication Technology*. Washington, DC: The World Bank.

World Bank. (2006). *2006 information and communications for development: Global trends and policies*. Retrieved October 1, 2009 from http://www-wds.worldbank.org/external/default/WDSContentServer/WDSP/IB/2006/04/20/000012009_20060420105118/Rendered/PDF/359240PAPER0In101OFFICIAL0USE0ONLY1.pdf

World Bank. (February, 2005). Fostering trust and transparency through information systems. *PREM notes, 97*.

World Information Society Report. (2006). Digital opportunity index 2005. Retrieved February 13, 2007, from http://www.itu.int/osg/spu/publications/worldinformationsociety/2006/World.pdf

WSIS (World Summit on the Information Society). (2008). WSIS follow up Report 2008. Advanced unedited draft (for comment). Note by the Secretariat. Retrieved on August 12, 2008, from http://www.unctad.org/en/docs/none20081_en.pdf

Xiong, J. A. (2006). *Current status and needs of Chinese e-government users*. Carbondale, IL: Southern Illinois University.

Zeithaml, V. A., Parasuraman, A., & Berry, L. L. (1999). *Delivering quality service: Balancing customer perceptions and expectations*. New York: The Free Press.

Zhao, X., & Okamoto, T. (2008). A personalized mobile mathematics tutoring system for primary education. *The Journal of the Research Center for Educational Technology, 4*(1), 61–67.

Zheng, Y., & Walsham, G. (2008). Inequality of what? Social exclusion in the e-society as capability deprivation. *Information Technology & People, 21*(3), 222–243. doi:10.1108/09593840810896000

Zhu, H. (2002). *The interplay of web aggregation and regulations. (Tech.Rep.)*. Cambridge, MA: MIT Sloan School of Management.

Zulu, S. F. C. (1994). Africa's survival plan for meeting the challenges of information technology in the 1990s and beyond. *Libri, 44*(1), 77–94. doi:10.1515/libr.1994.44.1.77

Zulu, S. F. C. (2008). Intellectual Property Rights in the digital Age . In Aina, L. O., Mutula, S. M., & Tiamiyu, M. A. (Eds.), *Information and Knowledge Management in the Digital Age: Concepts, Technologies and African Perspectives* (pp. 335–354). Ibadan, Nigeria: Third World Information Services Limited.

Zurita, G., & Nussbaum, M. (2004). A constructivist mobile learning environment supported by a wireless hand-held network. *Journal of Computer Assisted Learning*, *20*(4), 235–243. doi:10.1111/j.1365-2729.2004.00089.x

About the Contributors

Esharenana E. Adomi holds BEd, MEd. MLS and PhD degrees. He attended University of Ibadan, Ibadan and Delta State University, Abraka both in Nigeria. He was secretary of Nigerian Library Association(NLA) 2000 – 2004 and currently the chairman, NLA, Delta State Chapter. He was Acting Head, Department of Library and Information Science, delta State University, Nigeria. January 2008 – February 2009. He received the 2004 Award for Excellence of the Most Outstanding Paper published in *The Electronic Library,* 2003 volume with an article entitled: " A survey of cybercafés in Delta State, Nigeria" co-authored with two other colleagues. He is a member of Editorial Advisory Board, *The Electronic Library*, previously contributing Editor, *Library Hi-Tech News* and currently the editor of Delta Library Journal. He has published over 45 articles in reputable national and international journals, chapters in books and two textbooks. He is also the editor of *Security Software for Cybercafes* and *Handbook of Research on Information Communication Technology: Trends, Issues and Advancements* – published 2008 and 2010 respectively by IGI Global, Hershey, PA. His interests lie in ICT policies, community informatics, information/internet security, Internet/web technology and services, and application of ICTs in different settings.

***

Edwin I. Achugbue is a lecturer in the Department of Library and Information Science, Delta State University Abraka, Nigeria. He holds a Diploma in Library science, B.Sc (Ed) in Library Science, M.Sc in Library and Information Science. He is currently a PhD student in the Department of Library and Information Science, Delta State University, Abraka, Nigeria.

C. E. Akporido is the Librarian of the College of Health Sciences, Delta State University, Abraka, Nigeria. She holds an MLS from University of Ibadan, Nigeria. Her research interests are in information needs, resources and seeking behaviour

C. Annamalai is an Indian working as an Information & Communication Technology (ICT) Specialist in SEAMEO RECSAM, Penang, Malaysia. He has published articles on ICT in the national and international journals. He is a peer reviewer of International Journal on Education using ICT (IJEDICT) and Journal of Information Systems Education (JISE). He is having a total of twenty years of experience in Information Communication Technology teaching at the University level as well as worked as Projects Manager, Systems Manager in the software industry. He is currently pursuing a PhD in Technology Management focusing on Enterprise Resource Planning (ERP) systems.

Okon E. Ani is academic librarian, currently, the Head of Systems Development, University of Calabar Library, Calabar, Nigeria. He holds, B.Sc.(Hons) and M.Sc. degrees in Physics from University of Calabar, Calabar, Nigeria, and M.Inf.Sc. (Masters in Information Science) from University of Ibadan, Ibadan, Nigeria. He has published widely in both local and international journals. He has research interest in Information and Communication Technology and its applications in libraries, IT use, and digital/virtual libraries.

Udo Averweg is employed as an Information Technology (IT) Research Analyst at eThekwini Municipality, Durban, South Africa. He entered the IT industry during 1979 and holds a Masters Technology degree in Information Technology (cum laude), a second Masters degree in Science from the University of Natal and a third Masters degree inCommerce from the University of KwaZulu-Natal, South Africa. As an IT practitioner, he isa professional member of the Computer Society of South Africa. He has published more than100 research outputs: some of them have been delivered at local conferences, some haveappeared as chapters in textbook and some research findings have been presented atinternational conferences on all five continents. Udo has recently been appointed as anHonorary Research Fellow at the University of KwaZulu-Natal. South Africa. DuringJanuary 2000 Udo climbed to the summit of Africa's highest peak, Mount Kilimanjaro(5,895 metres), in Tanzania.

Sandra Babatope Ihuoma holds a Bachelors degree in Library Science from Delta state University, Abraka Nigeria and Masters Degree in Library and Information Science from Delta State University, AbrakaNigeria. She is anAssociate member ofChartered Institute ofLibrary and Information Professionals (CILIP) United Kingdom. She has worked in special libraries such as the Nigerian Navy Engineering College (NNEC) Library and Nigerian Stored Products and Research Institute (NSPRI).Her interest is in library automation, cyber security, library consortium, database management system, application development, surveillance system technology, networking, knowledge management, information and communication technology (ICT) and its application to the library and modern society.

CarolAzungi Dralega is a Senior Researcher with Western Norway Research Institute. She obtained herPhDand Master ofPhilosophy degrees from the department ofMedia and Communication, University of Oslo in Norway and a B.A. Arts (Hons.) in Mass Communication from Makerere University, Uganda. She has previously worked as a journalist, sub-editor, lecturer and research consultant.

Margaret B. Edem is currently Senior Librarian and Head of Resource Development, University of Calabar Library, Calabar, Nigeria.

Nelson Edewor is a practicing librarian with Delta State Polytechnic, Ozoro, Delta State, Nigeria. He had his university education at Delta State University, Abraka, Nigeria, where he earned a B.Ls in library science and an M.Sc in library and Information Science. His research interest covers library and Information science policies; knowlege management; database administration; information needs; virtual learning; corporate governance; strategic information management and information handling and seeking behaviour. Nelson Edewor has published various scholarly articles in reputable national and international journals.He had served in various capacities, ranging from; e-library manager; reader services librarian, knowledge manager and and reference services librarian. He is currently the Technical services Librarian at the Delta State Polytechnic Library, Ozoro, Delta State, Nigeria. He is a Consultant to the Local Government on Library and Information Sevices.He is a member of the Nigerian Library Association as well as Associate of the Institute of Strategic Management, Nigeria. Edewor was the Immediate past assistant secretary of the Nigerian library Association,Delta State Chapter,Nigeria

Geoff Erwin is currently a Director of The Information Society Institute (TISI), based in Cape Town, South Africa. He has worked for and with government, Universities and corporates in Australia, UK, Europe, United States of America and Africa. He has been in many roles related to ICT and socio-economic development, including programmer, systems analyst, system integrator, project manager, software developer, regional manager for an international computer services company, Dean of a University Faculty and Director of a Research Centre. He has written several ICT textbooks, many refereed journal and conference papers and book chapters. His current research interests are the social appropriation of ICT for local benefit and the development of e-skills for an inclusive Information Society. He has never climbed Mount Kilimanjaro.

Swapneshwar Goutam is final year law scholar at Hidayatullah National Law University, India, expected BA LLB. Hons. 2010. He holds D.C.A., Diploma in Computer Applications from Rashtriya Computer Saksharta Mission, (Visakhapatnam, A.P) He is a member of the International Human Rights Defense Organization, Norway. He won fifth position in an international essay competition conducted on 'Human rights', by International Human Rights Defence organization, Norway july/2008. He participated in international conferences organized by WTO, WHO International Social Security Association, Kosha and other repute law related forums.He has contributed to round table forum on, How to prevent and identify the needs and solutions? , by European Affairs Delegate, Accor Services and also being consulted to review manuscripts for reputed, "The African Journal of Business Management". He is an author of sixteen articles on the legal issues published in national and internationally repute peer reviewed journals and also authored a chapter entitled Unregistered Trademark and Practices, to be published in an ensuing book entitled 'Burden of proof'. Mr. Goutam received a letter of appreciation for real contribution to the important topic of constitution and cyber law from eminent Jurist MR. K K Venugopal, Padma Bhushan, Senior Council, Supreme Court of India.

Helena Grunfeld has over 30 years experience in the ICT sector in Australia, working for the incumbent telecommunications operator Telstra, and participating in the establishment of a new carrier, Uecomm. She has also completed numerous consulting assignments for various organisations, including co-authoring a book on number portability with Ovum and working on project related to costing and pricing through the International Telecommunication Union for the regulatory authority, ARCOM in Timor Leste. Helena has an undergraduate degree from the School for Social Work and Public Administration in Lund, Sweden and completed a Master of Business Administration at the Royal Melbourne Institute of Technology in 1989. She is now a research scholar and PhD candidate at the Centre for Strategic Economic Studies at Victoria University, Melbourne, Australia. The conceptual framework presented in this chapter forms the core of her thesis.

Sriram Guddireddigari is currently pursuing a PhD around the development of a framework for understanding the nexus between information and communications technologies and the social networks of Indian migrants in France. This follows a Masters in Electrical Engineering specialising in wireless and mobile communications from Columbia University and a Bachelors in Electrical Engineering from the University of Melbourne. Professionally, he has been involved with Infoxchange Australia as projects officer where he developed the sustainability of initiated social enterprises, assisted in the design and development of a mesh wireless community network, and organised a community technology conference. Before that, he was involved with Engineers Without Borders Australia in coordinating their international Information for Development projects and in spearheading a computer training program assisting refugees. He has also worked with Midas Communication Technologies first in their business development division and then as product manager on an end user Internet access device

Basil Enemute Iwhiwhu is a lecturer in the Department of Library and Information Science, Delta State University, Abraka, Nigeria. He holds a B. Sc Ed honours (Chemistry) and a Master degree in Library, Archival and Information Studies from the University of Ibadan, Ibadan, Nigeria. He also holds a Diploma in Computer Science. His area of research interests are Archives and Records Management, Information Management, Knowledge Management, ICT Application to Information Resources and Services. He is a member of the Nigerian Library Association (NLA) and the current Public Relations Officer of the Delta State Chapter of this professional body.

Vijay Kumar has a Diploma Computer Engineering from the Institute of Jeevana Jhothi, Chennai, and a Diploma of Advanced Systems Hardware and Networking from the Institute of SISI, Pondicherry. He is currently working with The East West Foundation of India as a systems administrator, conducting computer classes and maintaining systems hardware, networking and is also monitoring the educational software for children in the community. Vijay is currently pursuing a Bachelor of Computer Science at Madras University through distance learning.

Benita Marian is an Honorary Consultant for the East West Foundation of India (TEWFI) and has been actively involved in coordinating the activities in the community through the Post Tsunami Health Care and Research Project from 2005-2009. She is also an Honorary Advisory Committee Member for the Uluru Children's Home. Her involvement in the current research is because of her interest in the development of marginalised communities and in the role of ICT education in the empowerment of the rural community. Benita Marian has a doctorate in Social Work and is at the faculty in Social Work in Stella Maris College at Chennai, India. She is a Member, Board of Studies in Social Work at the University of Madras. She is an external examiner for Pre – Doctoral and Doctoral theses. She also coordinates a project 'Towards a More Connected World - Piloting an Innovative Indo-US Professional Exchange Programme', which makes use of Web based technology to connect students in the USA and India for interactiive exchanges as part of a course.

Stephen M. Mutula is an associative professor in the Department of Library and Information Studies, University of Botswana where he serves as the head of department. He holds a PhD in information science (University of Johannesburg, South Africa), Master's degree in information science (University of Wales, UK), Postgraduate diploma in computer science (University of Nairobi, Kenya) and Bachelors degree in Education-Mathematics and Chemistry (Kenyatta University, Kenya). He has published extensively in international refereed journals and books. He is a first co-author of a book titled: *Web information management: A cross disciplinary textbook* published by Chandos Publishing, London, 2007. He is also the author of 'Digital Economies: SMEs and E-readiness published in 2009 by IGI. He is an honorary research fellow of the University of Zululand, South. He has won several international excellence awards for his extinguished research work from various academic societies and such as the Emerald Literati Club (UK).

Alex Ozoemelem Obuh holds a Bachelors degree in Computer Science, Masters degree in Library and Information Science plus professional certification in database management system. He attended University of Benin, Benin City, Nigeria and Delta State University, Abraka, Delta State, Nigeria. He has published book chapters and several articles in both national and international journals. He has participated as lead lecturer in seminars and workshops. He has worked in several IT firms and is currently a Programmer with the Department of Library and Information Science, Delta State University, Nigeria. His interest is in database management system, application development, system security, web technology, networking, management information systems and artificial intelligence.

John Peter has as Master of Social Work from Madurai, Institute of Social Sciences. He is currently working with The East West Foundation of India as a social worker, organising community health and development activities and is also monitoring the education of children in the community attending the local schools. John Peter is currently pursuing a Master of Business Administration at Madurai Kamarajar University through distance learning.

Tracy E. Rhima is a librarian in Delta State University Library, Abraka, Nigeria. She holds B.A (HONS) in French and an MSc (LIS) from Delta State University Abraka, Nigeria. She commenced her professional career as graduate assistant in 2005 with the Delta State University Library, Abraka where she works in the Collection Development Division. She is a member of the Nigerian Library Association (NLA). She has published some articles in the area of Library and Information science.

Ramamurthy Subramanian is an Assistant Professor of Law at the Hidayatullah National Law University, India. He studied international criminal law at the Universities of Leipzig (Germany), Vienna (Austria) and the King's College, London (UK). He previously was with the Terrorism Prevention Branch, Division for Treaty Affairs, United Nations Office on Drugs and Crime (UNODC), Vienna and the School of Technology, Law and Development, West Bengal National University of Juridical Sciences, India. He has special research interests in issues of science and technology law.

Ng Khar Thoe is a Specialist in Research and Development (R&D) Division, SEAMEO RECSAM, Penang, Malaysia. She has the academic/professional qualifications of B.Sc.Ed. (Hons.) (USM), Dip. In H.E.N. and M.Ed. from UK. She has many years of experience in science teaching via ICT integration with involvement in international programmes incorporating R&D activities as resource person. She has contributed to more than 50 research-based publications on Science Education via ICT integration. Some papers were presented in conference/workshop/seminar, published in the conference proceedings, training manuals, book chapters and as articles for national/international refereed journals. She has completed the first doctoral thesis entitled *"Evaluative Studies on Problem-Solving Curriculum with Development and Validation of Tools for Cultural Scaffolding in the Management of Project/Problem/ Programme-based Learning"* from USA. Currently she is completing her second doctoral thesis to study a PBL programme that enhance secondary students' Higher Order Thinking and Motivation under open and distance learning mode.

Sitalakshmi Venkatraman is a Senior Lecturer in the Graduate School of Information Technology and Mathematical Sciences at University of Ballarat. She has 22 years of work experience - in industry, developing turnkey IT projects, and in academics, teaching and researching in a variety of IT topics. Currently, she teaches E-Commerce, IT Project Management and Advanced Software Engineering for the graduate programs. Her current research focus is on Malware and Digital Forensics and she supervises PhD students within the university's Internet Commerce Security Laboratory. She has published several technical research papers in internationally well-known refereed journals. She has peer-reviewed research articles for journals and conferences and has been serving the editorial board of IJCSE. Sitalakshmi Venkatraman is the corresponding author for the chapter 'A Framework for ICT Security Policy Management' and can be contacted at: s.venkatraman@ballarat.edu.au

Wahyudi Wahyudi is an Indonesian working as a Research and Development Specialist in SEAMEO RECSAM, Penang, Malaysia. He has published research based articles on Science and Mathematics Education in the national and international journals. He is having many years of experience in science teaching and research. He has his PhD in Science Education from Curtin University, Australia and Master of Education from State University of New York, Buffalo. His areas of interest include classroom action research, science education, learning environment and curriculum evaluation. He is one of the editors of the Journal of Science and Mathematics Education in South East Asia, the official journal of SEAMEO RECSAM, Penang, Malaysia.

Mirna Willer is Associate Professor at University of Zadar, Zadar, Croatia since 2007. She worked from 1980 to 2007 as systems librarian, standards officer and senior researcher at the National and University Library in Zagreb, Croatia responsible for implementing UNIMARC bibliographic and authority formats on Library's LIS software, and for incorporating national cataloguing rules into the formats. Among other international body memberships, she was a standing member of the IFLA Permanent UNIMARC Committee since its establishment in 1991 until 2005 (chair of Committee from 1997 to 2005), since then she is its consultant and honorary member. Currently, she is a member of the IFLA Working Group on FRANAR (Functional Requirements and Numbering of Authority Records), the Working Group responsible for the conceptual model FRAD.

David Wilson is a senior officer of the New Zealand House of Representatives. Previously, he has worked in policy roles in the New Zealand's censorship agencies - the Office of Film and Literature Classification and the Department of Internal Affairs. David was an adviser to the parliamentary committee that amended New Zealand's censorship laws in 2004. He has also advised the Independent State of Samoa on modernising its censorship office. David holds Masters degrees in history and management. He has presented conference papers on media regulation and is the author of published papers on that subject. David lives in New Zealand's capital city, Wellington.

Saul F. C. Zulu is a Lecturer in the Department of Library and Information Studies, at the University of Botswana. He holds a Bachelor of Arts degree in Library Studies with Sociology and Political Science from the University of Zambia, Masters degree in Librarianship (IT Applications) obtained from the University of Wales at Aberystwyth, a Masters degree in Archives and Records Management from the University of Denver, Colorado, U.S.A and Masters degree in Librarianship and Information Management from the University of Denver, Colorado, U.S.A. Mr. Zulu has previously worked at the University of Zambia, where he served in various capacities, including as Head of Department of Library and Information Studies at the University of Zambia and Acting University Librarian, University of Zambia Library

Index